I0897129

Companions in Geography

East and West

CULTURE, DIPLOMACY AND INTERACTIONS

Edited by

Chuxiong George Wei (*University of Macau*)

VOLUME 1

The titles published in this series are listed at *brill.com/ewcd*

Companions in Geography

East-West Collaboration in the Mapping of Qing China (c.1685–1735)

By

Mario Cams

BRILL

LEIDEN | BOSTON

Cover illustration: Detail from the copperplate version of the *Overview Maps of Imperial Territories*, 1717 or 1719, Royal Library of Belgium LP VB 11.283 E (2), fol. 10., Reproduced with permission of the Royal Library of Belgium.

The Library of Congress Cataloging-in-Publication Data is available online at http://catalog.loc.gov
LC record available at http://lccn.loc.gov/2017011277

Brill has made all reasonable efforts to trace all rights holders to any copyrighted material used in this work. In cases where these efforts have not been successful the publisher welcomes communications from copyright holders, so that the appropriate acknowledgements can be made in future editions, and to settle other permission matters.

Typeface for the Latin, Greek, and Cyrillic scripts: "Brill". See and download: brill.com/brill-typeface.

ISSN 2467-9704
ISBN 978-90-04-34535-5 (hardback)
ISBN 978-90-04-34536-2 (e-book)

Copyright 2017 by Koninklijke Brill NV, Leiden, The Netherlands.
Koninklijke Brill NV incorporates the imprints Brill, Brill Hes & De Graaf, Brill Nijhoff, Brill Rodopi and Hotei Publishing.
All rights reserved. No part of this publication may be reproduced, translated, stored in a retrieval system, or transmitted in any form or by any means, electronic, mechanical, photocopying, recording or otherwise, without prior written permission from the publisher.
Authorization to photocopy items for internal or personal use is granted by Koninklijke Brill NV provided that the appropriate fees are paid directly to The Copyright Clearance Center, 222 Rosewood Drive, Suite 910, Danvers, MA 01923, USA. Fees are subject to change.

This book is printed on acid-free paper and produced in a sustainable manner.

For my dearest companions:
my husband, Huang Xingrui,
and my parents, Anny Nijs and Jos Cams

Pour l’enfant, amoureux de cartes et d’estampes,
L’univers est égal à son vaste appétit.
Ah! que le monde est grand à la clarté des lampes!
Aux yeux du souvenir que le monde est petit!

CHARLES BAUDELAIRE (1821–1867)

Contents

Acknowledgements

This book would never have materialized without the help and support of several *compagnons de route*. In the first place, I cannot be more grateful to my advisor, Nicolas Standaert, who many years ago awakened my enthusiasm for early Sino-European contacts through his inspired way of teaching. Throughout my time as a Ph.D. student, he gave me the space to develop my own ideas, all the while making sure I got all the support needed to make full use of that space. Together with his own companions in history, Noël Golvers and Ad Dudink, he has built the "Leuven history group," known throughout the field for producing dizzying quantities of sterling scholarship on the history of the Jesuit missions to China. I am in no small measure indebted to their work and support. Another leading scholar and friend who has shaped my work is Catherine Jami. Her insight and sincere interest in my project motivated me to go further and dig deeper. Other scholars I wish to thank are Ronnie Po-chia Hsia at Pennsylvania State University, Michael Szonyi and Mark Elliott at Harvard University, Sun Xiaochun at the Chinese Academy of Sciences, and my new colleagues at the University of Macau, with special mention for George Chuxiong Wei, the editor of this series, and Rui Martins, who generously provided extra funding in support of this publication. This brings me to those who have more directly impacted the final text of this book. Matthew Edney, Matthew Mosca, Fresco Sam-Sin, and John Cirilli helped to substantially improve the manuscript in its final stages.

Perhaps even more important throughout these years were my companions in the truest sense of the word. The University of Leuven's sinology department, where I spent nearly a decade of my life, provided an intellectually challenging and inspiring environment for a student and an aspiring teacher. The Ph.D. students it nurtured, in the first place Chen Yanrong, who has become a close friend and scholarly ally, but also Lee Ting-mien, Annick Gijsbers, and Nie Tao, took on the roles of my staunchest critics and my most important comrades in sharing the frustrations, hardships and hopes of a doctoral candidate. Special thanks must also go to Carine Defoort, who gave me the opportunity to teach classes in Classical Chinese, continuously challenging my skills and forcing me to refocus my sometimes wavering attention on my teaching responsibilities. Beyond the University of Leuven, I am grateful to have befriended Wu Huiyi, Fresco Sam-Sin, Wu Hsinfang, and Gregory Afinogenov, among others.

Last, but most important, are my companions in everyday life. In the first place my husband Huang Xingrui, who by his presence and care always encouraged me to continue and whose companionship has been a major source of inspiration for this book; my loving parents, Anny Nijs and Jos Cams, without whose unconditional love and support I could not have pursued my scholarly interests in the first place; and my grandmother Mia Das, whose candy-laden table and warm heart have been a source of stability for as long as I can remember.

To all of them, and to the many others I failed to mention: thank you!

Mario
Macao, January 2017

List of Illustrations

Figures

Tables

INTRODUCTION

Towards a New Cartography of Cross-Cultural Circulation

> Nos messieurs les Tartares sont tous fort bonnes gens; mais ils sont paresseux et laissent à leurs valets le soin de tirer la corde; ils ne songent qu'à se munir contre les maux divers qui accompagnent les voyageurs en cette vie. Le P. Cardoso est le plus aimable Portugais que j'aye encore vu: nous vivons en une union de cœur et d'affection très grande. Il me charge de vous faire ses complimens, comme je vous prie de faire les miens aux P. de Mailla et Henderer vos chers Compagnons de Geographie, sans oublier messieurs vos Tartares.[1]
>
> Our Tartar gentlemen are all very good people; but they are sluggish and leave the pulling of the rope to their servants; they are only concerned with arming themselves against the various difficulties that accompany travelers in this life. Father Cardoso is the kindest Portuguese I have met thus far: we live in a union of heart and great affection. He asks me to give you his regards, as I hope you will give mine to Father de Mailla and Henderer, your dear companions in geography, without forgetting your Tartar gentlemen.

These words close a letter written by French Jesuit Pierre Vincent de Tartre (1669–1724) and sent to Jean-Baptiste Régis (1663–1738) on June 12th, 1712. In their capacities as envoys appointed by the Kangxi 康熙 emperor (r. 1661–1722), both men were members of two different teams of mapmakers surveying the vast territories of the Great Qing.[2] At the time of the letter's writing,

1 Archives de la Province de France (hereafter APF), *doss. pers.*, De Tartre. The letter appears in part in Brucker (1890–1).

2 The Kangxi or Elhe Taifin emperor, was the fourth emperor of the Great Qing, a state founded in 1636 but considered as a Chinese dynasty beginning in the year 1644, when the Manchu Qing armies crossed the Great Wall and conquered the former Ming capital of Beijing.

© KONINKLIJKE BRILL NV, LEIDEN, 2017 | DOI 10.1163/9789004345362_002

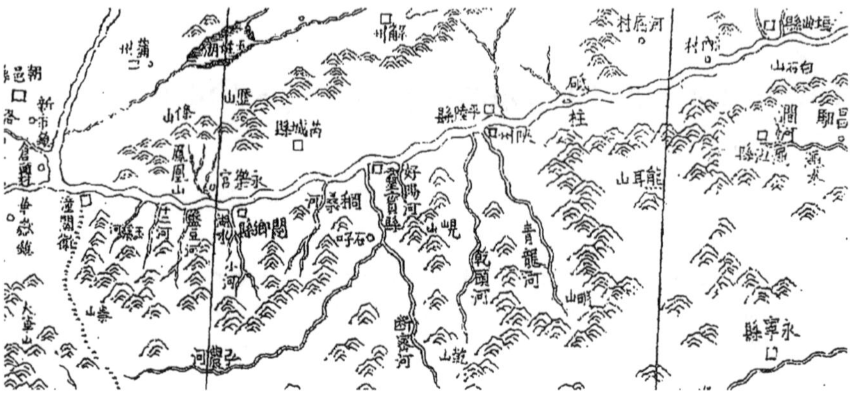

FIGURE 1 *Detail of the lands surrounding the Yellow River where it demarks the border between Shanxi and Henan provinces.*
TAKEN FROM WANG & LIU (2007), VOL. 1.

both teams were independently surveying the border (formed by the Yellow River) between the provinces of Shanxi and Henan near the city of Shanzhou 陝州 (today's Sanmenxia 三門峽). The letter first voices some of the frustrations and hardships de Tartre and his fellow team members experienced while drawing the provincial map of Shanxi, but later it expresses deep sentiments of friendship and companionship, elaborating on the collaboration including both European missionaries and Qing officials of different ranks. With its deep attention to the social dimension of mapmaking, de Tartre's letter propels us directly into the heart of this study, in which I intend to shed light on the unprecedented circulation of cartographic practices and related material goods between Europe and the Qing at the turn of the eighteenth century. The comprehensive mapping of Qing China, initiated at the behest of the Kangxi emperor, both sets the stage and forms the backbone of this study.

The expression "companions in geography," used by de Tartre in his letter to Régis and providing the title for this book, encapsulates in a broader sense the almost haphazard coming-together of diverse individuals in pursuit of variegated personal goals. The modern French word "*compagnon*," cognate and direct ancestor of the modern English "companion," yet often translated into English as "fellow," "friend," or "mate" as well as "partner" or "associate," itself descends from the Latin "*companionem*," a compound of "*con-*" ("with") and "*panis*" ("bread"), meaning literally "bread-fellow" or "messmate," one with whom one shares food. Etymologically, the expression thus evokes the connection between two or more individuals through the act of sharing. This idea of material goods mediating and shaping the connections between individuals is the guiding light behind my reconstruction of this immense Qing mapping project.

My study owes much to recent scholarship concerning both early Sino-European contacts and the history of science. The Jesuits' active involvement in cartography has for decades held center-stage in the scholarship on early Sino-European relations, and the different frameworks employed in studying these exchanges between Europe and Asia have shaped the field profoundly. At the same time, an emphasis in the history of science on knowledge circulation has led to important new insights on the structures and behaviors of social networks, the roles of material goods, and the production of scientific knowledge. In combining features from both fields, this study will focus on the circulation of material goods that relate to cartographic practice, permitting us to trace out basic connections between individuals through common artifacts. Based on this, I propose a careful and gradual reevaluation of past scholarship and suggest ways to move beyond four conceptual dichotomies that pervade the literature: traditional versus scientific practice, global versus local networks, the circulation versus the production of knowledge, and China versus Europe.

Situating the Study

In *The Landscape of History*, John Lewis Gaddis draws a parallel between cartography and history, in that both "provide a way of reversing divisibility, of retrieving unity, of recapturing a sense of the whole, even though it can never be the whole."[3] The cartographer represents with a map a landscape at a particular moment in time, whereas the historian represents some aspect(s) of change and continuity over some span of past time through narrative. Each seeks a representation of space and time (respectively) at once accurate and general and which is instrumental precisely because it is merely a representation. If this comparison is useful, and I believe it is (at least to some extent), the history of cartography occupies a special intersection in history of two varieties of representation of time and of space, superimposed upon and entangled with one another.

Perhaps in small part because of the difficulties inherent to studying complex interplays between representations over time and space, throughout much of the twentieth century the field of historical cartography focused

3 Gaddis (2004), 32.

almost exclusively on such technical issues as surveying methods and projection, enabling scholars to evaluate what they conceived as strictly linear progress, the precision of cartographic depictions increasing steadily over time. In light of cross-cultural contacts, however, in which different cartographic traditions often confronted one another, the object of evaluation typically shifted to the success of transmissions of cartographic knowledge, seen in turn as a one-way phenomenon, with more "scientific" European cartography displacing the more "cultural" or "traditional" cartographic practices that had developed elsewhere in the world.

The Sino-European context in particular has inspired scholars to evaluate the success of such one-way transmissions of cartographic knowledge, crafting weighty statements as to whether one could discern a lasting impact on the practice of cartography in China, as Laura Hostetler has elucidated.[4] Such studies were useful in that they explored sometimes radically differing perspectives even while operating solely within the sphere of intercultural contacts. In his *Methodology in View of Contact between Cultures*, Nicolas Standaert discusses these past frameworks at length, focusing on scholarship relevant to Matteo Ricci's (1552–1610) world map and so illustrating the privileged position of cartography in the study of early Sino-European contacts.[5] In his review of past scholarship, Standaert distinguishes among the three frameworks of transmission, reception, and invention, preoccupied as they are with respectively: the success of the transmission from the perspective of the transmitter; the success of the transmission from the perspective of the receiver; and the ways in which the constructed realities of the transmitter shaped the transmission's success.

Standaert argues that we must move beyond such evaluative frameworks while retaining their useful insights. Thoroughly studying the cultural encounter from multiple perspectives simultaneously, he proposes careful reconstruction of the networks of interaction and communication that allowed the encounter to take place, thereby circumventing the strong tendency to evaluate such an encounter simply in terms of the success of a transmission. The resulting "interaction and communication framework" represents an effort at mitigating the supposed gap between the European "self" and the Asian "other" essential to the three previous frameworks. We engage this gap by shifting the focus of attention from "input," the question of who contributed what, to "output," the question of the encounter's results.

4 Hostetler (2001), 6–25.

5 Standaert (2002).

An important concept in contemplating the outcomes of an intercultural encounter within this framework, to which Standaert and others have frequently referred, is that of the interstitial or the in-between: the dynamic space or opening created by the encounter.[6] This locus, abstract or physical, makes interaction possible and permits mediation of the sensitivities, hopes, and needs of different actors and the formulation and production of creative new answers. In devoting our attention to the in-between of a historical encounter, we confront the ground zero of historical processes, the very field of tensions that is the context for this process. Even so, we can never fully reconstruct what took place in-between. Only the sediment of the interaction, to use an unlucky metaphor, has been passed down to us through history, often consisting of what different actors (re)produced or (re)negotiated in-between. This "sediment" can range from letters, maps, or other material objects to surviving rituals.[7] Because such products of the in-between constitute the traces of interaction, studying their trajectories from origin over the course of negotiation and creative adaptation to reevaluation allows the historian better to approach the actual process of interaction.

There is of course no need to limit oneself to a binary opposition between two cultures or to see each only as transmitter or receiver. It makes little sense to divide actors conceptually into two cultural blocs if one aims to study the results of their interaction within the in-between space created by any given encounter. More often than not, the complexities of the actors involved in any interaction, intercultural or otherwise, preclude clear-cut categorization, saying nothing of self-identifications that actors may or may not themselves have experienced towards one group or another. Ideally, each actor should be considered in his or her own right, minimizing unnecessary pigeonholing whenever possible so as to appreciate the multiplicity and complexity involved in human interactions and historical processes, and allowing us better to approach what took place in-between. In other words, when studying a shifting network of actors and the process of interstitial creativity, I believe we must minimize and ideally eliminate preconceptions as to the nature and structure of that network or its individual actors, especially in the onset of our inquiries.

One particular insight from Bruno Latour's Actor-network-theory (ANT), which I do not use as such in this study, can aid us in exploring this point. Latour writes that "it is crucial that enquirers do not in advance, and *in place* of the actors, define what sorts of building blocks the social world is made

---

6 Fa-ti Fan, for example, uses the term "cultural borderlands," a concept sharing many characteristics with Standaert's. Fan (2007), esp. 215.

7 Standaert (2008).

of."[8] In Latour's eyes, groups are not unitary entities in themselves, but never-silent, constantly changing associations. Latour thus warns against starting off with a list of fixed groups to study, encouraging scholars instead to begin with the traces left behind by constant group (re)creation. "Social links have to be traced *by the circulation of* different vehicles that cannot be substituted by one another," rather than thinking in terms of predefined groups.[9] By following these connections as closely as possible, by letting them guide us free of fixed ideas as to what the social should look like and by paying attention to the smallest of details, one can, according to Latour, slowly reassemble the social without neglecting its complexities and contingencies.

This idea is one of the methodological insights Catherine Jami has introduced into the field, in exploring the circulation of mathematical knowledge between and within Europe and the Qing in her book *The Emperor's New Mathematics*.[10] As Jami points out, not only did the Kangxi emperor identify himself on multiple occasions as a Manchu as opposed to a Han Chinese, but the Manchu language itself also played a specific role in the circulation of information, both oral and written. This calls into question the position of Manchu identity and language within the binary opposition between China and Europe that permeates the field, a crucial insight inspired partly by the so-called New Qing History School of the early 2000s.[11] More crucially, Jami places the individual at the center of her inquiries and reveals how circulating information between individuals—and, in this case, across cultures—produces knowledge. In other words, the production of knowledge is the process rather than the result of circulation. While perhaps specific to the overlap between early Sino-European contacts and the history of science, Jami's work clearly illustrates why the analytical macro framework of two interacting entities—of China on the one hand and of Europe on the other—is hardly supportable when one studies the specifics of the circulation or production of scientific knowledge.

Delineation and Approach

The circulation of cartographic knowledge between East Asia and Europe as discussed in this study arose through a convergence of interests among several

8 Latour (2005), 24.
9 Latour (2005), 41.
10 Jami (2012).
11 Waley-Cohen (2004).

key actors. First, in the latter half of the seventeenth century, the French *Académie royale des sciences*, supported by the French crown, sought ways to expand its knowledge of the world as part of its broader goal of advancing the prestige of the French state. Second, the Jesuits hoped that a deeper involvement in Qing imperial science would lead to greater court protection for their mission in the Chinese provinces, since the Kangxi emperor had signaled that he wished to employ more experts in mathematics and astronomy. Third, the emperor himself actively pursued techniques of empire-building in which geography and cartography played an important role, as evidenced by his personal interest in land surveying. Due to these converging interests, in 1685 a group of French Jesuit missionaries, hoping to advance their mission with the help of their knowledge of mathematics and its many practical applications, journeyed to the East as officially appointed correspondents of the *Académie.* As a result, an extended network of individuals evolved, spanning both ends of the Eurasian continent and allowing, among other things, the circulation of material goods related to cartographic practices.

Beginning in 1708 the emperor gradually commissioned a series of grand scale land surveys, a direct outcome of these converging interests, following a test-of-concept mapping of routes north of the Great Wall and the establishment of a new practice of surveying land. These surveys took nearly a decade to complete and involved teams of dozens of mapmakers traversing Qing territories and drawing regional maps on the basis of precise observations and calculations. Each team consisted of officials from different Qing institutions as well as several European missionaries. The surveys began relatively slowly in the Manchu homelands and the Mongolian steppes beyond the Great Wall that the Manchus had only recently incorporated into the empire, but efforts accelerated once they reached the Chinese provinces, where several teams of surveyors were able to operate simultaneously and where local officials proved indispensable in providing necessary assistance. When the last team of surveyors finally returned to the capital, the editing and integration of regional maps was already well under way. In late 1717, the emperor received the first printed version of the resulting atlas, known in Chinese as *Huangyu quanlan tu* 皇輿全覽圖 or *Overview Maps of Imperial Territories*. In geographic coverage, quantities of personnel involved, resources allocated, and time required to complete the surveys, this collaborative project was without doubt the largest mapping endeavor the world had ever seen.[12]

12 The first full and detailed survey of France was only complete in 1744, while a large-scale Russian mapping project concluded one year later. Hostetler (2001), 73. Neither of these

Once the actual surveys were complete, the "network in cartography" that had formed between Europe and the Qing gradually shifted its focus to transmitting the resultant maps to Europe and gathering cartographic data for areas beyond the scope of the surveys. This led to an intense period of circulation in the two decades or so after 1718 by an extended network of intellectuals and officials operating mainly in Beijing, Paris, and Saint Petersburg. A large corpus of maps, travelogues, and geographic descriptions travelled in every direction along this network, dominated by, but certainly not limited to, maps and data resulting from the land surveys. In East Asia, the court strictly controlled circulation, while in Europe much of the material eventually became concentrated in Paris, first put aside, then organized, edited, and finally published in 1735 in a four-volume "description" of continental East Asia. Forty-one maps were included, all executed by French cartographer Jean-Baptiste Bourguignon d'Anville (1697–1782).[13]

Thus, the period spanning the three stages of build-up, practical execution, and aftermath outlined above lasted about half a century, taking the French Jesuits' departure from France in 1685 as the starting point and the 1735 publication in France of d'Anville's maps as the virtual end. To be sure, this rather abstract and nondescript "network in cartography," as I have somewhat reluctantly dubbed it above, along which objects and knowledge passed, was in constant flux over these fifty years, its participants, structures, and dynamics all accordingly adapting and shifting constantly. Perhaps we would do better, then, to consider it not as a network in the traditional sense but rather as a highly fluid and interactive web of social relations, only a small part of an all-encompassing web of social connections perpetually reinventing itself. In this light, this study's major goal consists of reassembling (to use Latour's terminology) precisely that part of the web of social relations connected through the circulation of material goods related to cartography. In subscribing partly to Latour's perspective on what the "social" looks like, I hope to capture its contingent and complex nature and thus to improve our understanding of the circulation of cartographic knowledge between Europe and East Asia.

My approach centers on studying the movements of material goods linking individuals within this web of social relations. In Latour's Actor-Network-Theory, not only human participants but material objects as well can function as actors or mediators within networks provided they modify states of affairs. If we view material objects as actors in the sense that they can effect change

atlases would surpass the *Overview Maps of Imperial Territories* in any of the aspects mentioned above.

13 du Halde (1735).

and are "participants in the course of action,"[14] they can help us, no less than individual human actors, to gain access to the fleeting connections associated with them. Although I value its insights, this study departs from Actor-Network-Theory, however, in that it does not consider unique material objects alongside other actors. Rather, it tells the story of *sets* or *families* of material goods passing through the hands of individuals, thereby connecting them across territorial and cultural boundaries. In the process, these actors collectively reinterpret and reproduce material goods into new variations, inspiring in turn further adaptations or negotiations.

The trajectory of any such set of material goods can through their circulation constitute concrete proof of a connection between two or more participants, traceable through the bringing-together of scattered data from a diverse variety of source materials. In this study, I will undertake such a microhistory, specifically of the movements of surveying instruments and maps between and within Europe and continental East Asia. Within this circulation of material goods, instruments and maps followed significantly similar trajectories, albeit in opposite directions: (1) the instruments used in the surveys first traveled from Europe to Asia via different routes, and later from the Qing court in Beijing, where the Imperial Workshops (Chinese: *zaobanchu* 造辦處 / Manchu: *Weilere arara ba*) stored and reproduced them, into the provinces. On the other hand, (2) the resulting maps first passed from the provinces to the Qing court, where the Imperial Workshops reproduced and kept them, and later from Asia to Europe along different routes. By considering the varying transmission routes of both instruments and maps, I am able to pursue my inquiries with two clear advantages: first, in my ability to trace material objects from "origin" to "endpoint" in terms of concrete exchanges surrounding the production of the *Overview Maps of Imperial Territories* and, second, to backtrack spatially, so to speak, the trajectories followed by the resulting maps.

This approach reveals largely unstudied patterns and groupings of actors: closely tracing the trajectories of instruments and maps within a well-defined timeframe creates a cartography of material movement enabling a different perspective on the web of social relations underlying their circulation. An image borrowed from Michael Markieta, carefully mapping out global flight paths and sea routes against a black background, can help to illustrate this point. In his effort to "emphasize the significant cross-boundary *movement* of persons and cargo,"[15] Markieta created an intriguing cartography exuding far more

14 Latour (2005), 71.

15 Emphasis added. Quote taken from http://www.spatialanalysis.ca/2012/an-esri-mapbook-submission-that-never-was/#more-1745 (August 29th 2014, since offline).

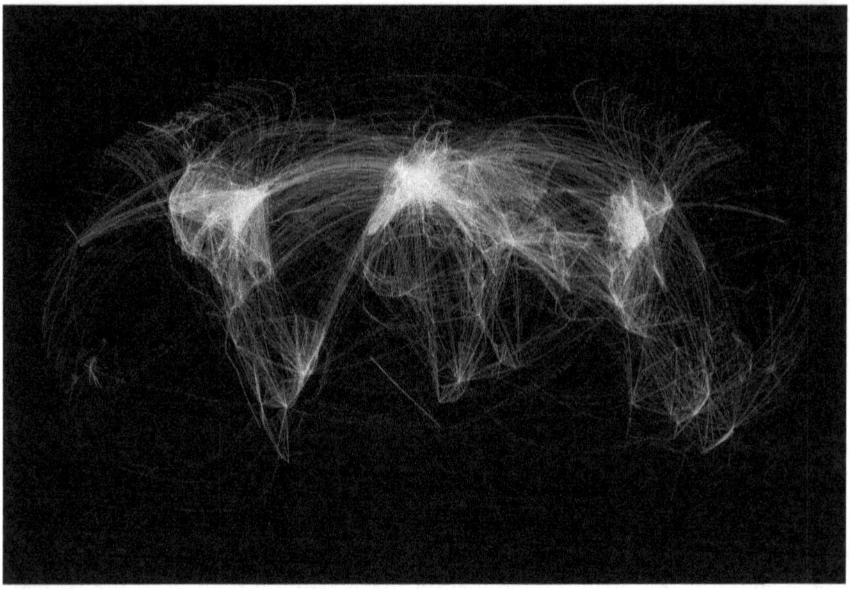

FIGURE 2 Global Air and Sea Routes *by Michael Markieta.*

than a simple network of global transportation routes. Perhaps surprisingly, the shapes and contours of continents and oceans remain easily recognisable even where global transportation paths are less dense. In fact, the macro-geographical shapes are distinguishable precisely because of those "peripheral" transportation routes. At the same time, Markieta's map inherently highlights the primary nodes and clusters of activity in the global transportation network and emphasizes their interconnection, fostering in turn insights relevant to such other topics as economic activity or social mobility, to name only two.

I argue that tracing the trajectories of instruments and maps leads to a very similar sort of (textual) cartography and thereby yields enhanced access to the dynamic web of social relations. First, it leaves blank what instruments and maps do not connect and in so doing elides any preconceptions about the structure and participants of the underlying "network," including the long-standing purported dichotomy between everything and everyone identifiable either as Chinese or as European. Second, it creates greater space within my narrative for what some scholars might identify as more "peripheral" actors and patterns. In other words, without understating the importance of the main contributors and of the major centers of circulation, this type of cartography forces us to examine the bigger picture by simply following the most basic dynamic connections. I therefore devote an equal share of my attention to so-called "peripheral" connections and clusters of actors, which, in consequence, emerge as equally essential, perhaps even indispensable, in supporting the overall circulation of materials related to cartographic practice.

In social network theory, the concept of "weak ties" can help to clarify this point. Weak ties facilitate the mutual exchange of information between different clusters or subgroups while at the same time integrating them. In social networks lacking such weak ties, "new ideas will spread slowly, scientific endeavors will be handicapped, and subgroups that are separated by race, ethnicity, geography, or other characteristics will have difficulty reaching a modus vivendi."[16] Weak links thus act as brokers, effectively restricting or facilitating the flow of new information between distant clusters within their social network and prove key to analyzing the circulation or production of knowledge. In the field of early Sino-European contacts, it is well-established that the Jesuits occupied this position by bridging the geographic and cultural distance between Europe and East Asia. I argue that my approach can uncover more such weak links by closely tracing the trajectories of material goods. For the mapping project under discussion, weak links included individuals who worked in the logistic headquarters of the Imperial Workshops, Qing officials in the farflung teams of mapmakers, scientists in Saint Petersburg, and many others.

Thus, when we trace the most basic connections between individuals through the movements of material goods, we can avoid from the start conceptual macro units of analysis which might otherwise divert our gaze from less conspicuous clusters of actors and the weak links connecting them. My approach therefore calls for a careful, gradual reevaluation of scholarship presupposing a clear-cut cultural divide between Chinese and Europeans. Emphasizing the most basic connections between individuals reminds us of the simple fact that individual human beings executed this Qing mapping project, sometimes and to different degrees backed by powerful institutions but always acting in the context of their highly personal social connections and their own particular backgrounds. Needless to say, these individuals were operating within a profoundly challenging cross-cultural environment, in that their social connections spanned two or more vastly differing regimes of values. Precisely this context, however, experienced through personal relations but mediated in part by material objects, induced the particular mode of creativity that resulted in the events and outcomes described in this study. The cross-cultural context therefore undeniably shaped and defined those individuals through their shared experiences, as well as thereby blurring the boundaries between those elements that were "Chinese" and those that were "European." Indeed,

16 Kadushin (2012), 31.

if this study's main concern is parsing and comprehending the circulation of knowledge as a *result* of cross-cultural contacts, then such conceptual dichotomies and macro categories of analysis tend to become obsolete and, in some cases, may even obstruct efforts at understanding them better.

In sum, I believe that this study, in employing the approach outlined above, offers a fresh perspective, particularly in its focus on material objects as mediating connections among individuals. This entails three major advantages: First, through close study of the connections between individuals through transfers of instruments and maps, I achieve a horizontally enhanced understanding of the overall web of social relations guiding their circulation. This translates into the emergence of new clusters, actors, and patterns, further specifying the roles and positions of all actors involved, including the Jesuit missionaries. Second, in bypassing wherever possible interpretive macro categories in the onset of my inquiries, my approach engages previously overlooked contingencies and complexities inherent to circulating physical objects. Third, this study emphasizes the slow and gradual transformations of instruments and maps over time, through their reproduction, reinterpretation, and reworking via social connections and within different environments. As a result, knowledge circulation and knowledge production merge and grow indistinguishable.

Cartography and the Jesuit Missions to China

This study owes greatly to the scholarly tradition once known as "missionary studies" but now typically termed "Sino-Western cultural relations" or "early Sino-European contacts." These new names reflect a longstanding evolution in the field from deep Eurocentrism towards a more intercultural approach. It is my intent here to provide an overview of what cartography meant for the Catholic missions to China (and vice versa), based on the very detailed scholarship that frequently characterizes the field, and to provide a more general introduction to the exchanges in cartography between Europe and continental East Asia in the decades preceding the mapping project under discussion.

During the seventeenth and eighteenth centuries, cartography in the form of ornate published maps was of interest mostly to elites, the common traveler generally lacking occasion to cultivate the faculty of mental displacement necessary for reading such maps. Such a person, travelling between two major cities, would simply make the journey in stages, town by town, sometimes guided by a city's street names and gate names indicating the direction or the next large town. Only educated men and women (so far as we know) pondered questions pertaining to a larger, cartographic, view of the world, such as its size, or felt the strong pull of the blank spaces on maps of the world. Jesuit education in particular, with its prominent focus on both Euclidean geometry and its practical applications, primed students for the practice of cartography.[17] Unsurprisingly, many Jesuit missionaries of the seventeenth and eighteenth centuries became involved in gathering geographic data and in producing maps, not least because good maps were useful aides in proselytizing and in better streamlining the geographically extended Jesuit missions.[18]

Jesuit missionaries first reached Ming China in the latter half of the sixteenth century. From the start, they participated in cartographic exchanges, albeit in different ways and for different purposes. Jesuits Matteo Ricci and Michele Ruggieri (1543–1607), known as the first missionaries to gain a foothold in the Chinese provinces, were no exceptions. Ricci's maps are well-known and a large body of scholarship has explored both their more general interpretative implications and their specific technicalities.[19] The eight different versions of Ricci's world map, including the *Yudi shanhai quantu* 輿地山海全圖 and the *Kunyu wanguo quantu* 坤輿萬國全圖, began from European atlases but constituted cultural translations, seen by many as entirely new maps. Ricci's adaptations included placing East Asia at the center of the map and gradually adding significant explanatory text in successive editions. For Chinese elites, Ricci's maps constituted rather bizarre representations in sharp contrast with their own understanding of the world, introducing Chinese readers for the first time to the concept of projection and hemispheres, East-Asia depicted as merely one small part of a vast whole, the inclusion of the Americas, and other new elements that originated from European cartographic practice. Later, Jesuits Guilio Aleni (1582–1649), Francesco Sambiasi (1582–1649), and imperial astronomer Ferdinand Verbiest (1623–1688) further edited Ricci's map and produced updated world maps often entitled *Wanguo quantu* 萬國全圖 or *Kunyu quantu* 坤輿全圖.[20]

Ruggieri, on the other hand, initiated a more or less opposite movement of cartographic material. Ruggieri drew maps based on Chinese cartographies, which took the form not of large stand-alone maps or elaborate atlases but often appeared within regular books on geography consisting primarily of

17 At the Collegio Romano, for example, the curriculum for the second year of study included geography and spherical mathematics. In the third year, students studied the theory of the planets. von Collani (1999), 232. After 1599, this focus would continue to mark Jesuit education in the Ratio Studiorum.

18 On the importance of "territorial management" for the Jesuit missions, see Vu Thanh (2009).

19 Standaert (2002).

20 Heirmann et al. (2009); Lin (1994). For a full overview of cartography in the Jesuit missions to China, see Bernard (1935).

FIGURES 3.1 & 3.2 *The 1602 edition of Ricci's* Kunyu wanguo quantu.
LIBRARY OF CONGRESS, GEOGRAPHY AND MAP DIVISION.

text.[21] The maps by Ruggieri include pictograms of towns, villages, and outposts, place names added in Roman script, and depictions of major rivers. The collection includes a general map of Ming China as well as separate maps of individual provinces, with alternative or experimental versions added for several provinces. On some maps, Ruggieri took over the grid system he found in Chinese works and attempted to delineate external borders in line with European cartographic practice. These manuscript maps eventually reached Europe, undoubtedly as a completely new representation of East Asia, much as Ricci's maps would have impacted their audience in Asia. One major difference, however, is that Ricci's cartographic work became relatively widespread in later decades due to its printing in several editions, while Ruggieri's maps vanished into Italian archives until well into the twentieth century.[22]

Nearly half a century passed before Jesuit missionaries took the next major step in terms of cartographic exchanges. Around 1650, Jesuits Michael Boym (1612–1659) and Martino Martini (1614–1661), both exceptionally mobile missionaries, found themselves on opposite sides of the Ming-Qing transitional wars. Boym had traveled far inland to the court of the last Ming emperor, some of whose relatives had become Christians, whereas Martini was able to travel in the areas controlled by the Manchus of the Qing. Both missionaries made it back to Europe and drew maps of East Asia based on Chinese-language compilations, to which they added important data based on their own rudimentary observations and information from local contacts, an important difference compared to Ruggieri's maps. Boym's work consists of eight manuscript maps, whereas Martini published an entire atlas as he passed through Amsterdam on his way to Rome. The ornately decorated *Novus Atlas Sinensis* (1654) was an eye-opener for European readers, including both an improved general map of East Asia, a detailed map for each province, and extensive textual descriptions. These maps clearly indicate borders both administrative and external, and, for certain areas, included improved data based on in-person observations.[23]

Although we must not overestimate the role of cartography in the early Jesuit missions to China, it played an important role in transmitting concrete knowledge of the "other" between the continents. Ricci's maps and their later editions constituted a surprising new representation of the world for audiences in Asia, mediating personal relations with literati and, particularly in their published form, serving as an important source of geographic knowledge of

21 For an overview on recent scholarship of Ruggieri's maps, see Yao (2014).

22 Lo Sardo (1993).

23 Martini (2003).

FIGURE 4 *Map of the capital province of Zhili from Martino Martini's* Novus Atlas Sinensis *(1655).*
LIBRARY OF CONGRESS, GEOGRAPHY AND MAP DIVISION.

the wider world; similarly, new cartographic depictions of the Far East by Martini became very popular in Europe, informing the European reading public's knowledge of the region. Both maps became widespread, in part because they presented knowledge of and by the other in what we might call cultural translations, that is, presented so as to speak to the visual expectations of the target audience. Taken at face value, other missionaries' representations were less accessible to the general public, perhaps one reason they remained unpublished despite also contributing to the circulation of cartographic knowledge between Europe and East Asia. In any case, it is clear that Jesuit missionaries fully controlled this mutual, relatively wide, and to some degree equal exchange, taking the initiative in redrawing and editing existing maps—undoubtedly with help from their local contacts—and in promoting their wider circulation.

In the seventeenth century, when these maps began circulating, Jesuit missionaries in the Chinese provinces had achieved a certain degree of recognition from elites and took active part in local literati culture. As a result of their extensive contacts, Jesuits also procured imperial interest, first for the technical and artistic skills displayed as part of their strategy to gain a foothold in the provinces, including but not limited to cartographic and textual representations of regions largely unknown to educated Chinese, such as those depicted on Ricci's world maps.[24] Particularly after the Qing conquest, a number of Jesuits served as high officials or artisan-experts at the imperial palace, instructing the Kangxi emperor and his sons on mathematics and astronomy, improving the crucial state calendar, building European-style cannon, etc. Throughout his reign, the emperor signaled he wished to employ more experts in mathematics and astronomy. The Jesuits thus understood that missionaries with particular training in the sciences could contribute greatly to their proselytizing mission, hoping that the presence of more trained missionaries would increase Jesuit influence at court and perhaps lead to court protection for their missionary work in the provinces (which local administrators and public distrust frequently hampered). The comprehensive mapping project discussed in this study took shape within this general context of scientific exchange, significantly dependent on Jesuit mobility between Europe and East Asia.

Chapter Overview

Material objects, geography, and chronology all undergird the structure of this book's three chapters, recapitulating the consecutive stages of the Qing mapping project: "build-up," "land surveys," and "aftermath." The narrative opens in the first chapter with the origins and intercontinental journeys of different surveying instruments between France and the Qing and closes in the third chapter with the proliferation and movements of the resultant maps in the opposite direction. The two intersect in the second chapter, in which surveying teams bearing instruments from the capital outward to the hinterland (or the provinces and outlying regions of the Qing) employed the observations and measurements acquired therewith to begin drawing maps on the spot, which they in turn forwarded to the capital. I hope that this book's structure, inspired by chronology, material goods, and geography, serves to remind the reader of its main contribution, namely, exploring circulation over time by studying the movement of material goods.

24 For Jesuit-authored textual material on geography as taught in Europe at the time, an important work to mention is the *Zhifang waiji* 職方外紀, first compiled by Diego de Pantoja (1571–1618) and Sabatino de Ursis (1575–1620).

The first chapter explores the build-up towards the mapping project. It begins in the instrument workshops in Paris of the second half of the seventeenth century and its strong links with the French *Académie royale des sciences*. When, in 1685, French Jesuit missionaries trained and financed by the *Académie* journeyed to Asia for the first time, a number of surveying instruments from the Parisian workshops accompanied them. As I follow specific instruments about the globe, the focus shifts gradually to the Jesuits' politics of gift-giving and to the reception of Parisian instruments at the Qing court, leading to the instruments' emulation and reproduction in the Imperial Workshops at the behest of the emperor himself, transforming and adapting prototypes from Paris for a new environment. Finally, I follow the instruments' use in new frontier regions throughout the late 1680s and 1690s, in service to the interests of the Qing state (and eventually leading to the re-standardization of the Qing's most basic unit of length, the *chi* 尺). Chapter 1 thus illustrates how the Jesuit order, the French *Académie royale des sciences*, and the Kangxi emperor found one another and how the mapping project constitutes one of the major outcomes of this unique convergence.

The second chapter focuses centrally on the land surveys conducted between 1708 and 1718. First, European sources enable a case study of surveying techniques and progress during the Great Wall survey, as well as permitting exploration of surveying in regions outside the Great Wall, including Manchu, Mongolian, Tibetan, and Korean lands. Next, I turn to Qing sources that speak to the logistics involved in the final stages of the land surveys: the surveying of the fifteen provinces. I reconstruct these surveys' progress, timing, the role the emperor and his central administration played, as well as the formation and composition of the different teams of surveyors. Chapter 2 continues by focusing on the backgrounds of the individuals in these teams, discussing the inclusion of various Qing officials and European missionaries, and concludes by exploring a missing link, the Qing Imperial Workshops, which emerge as the crucial logistical force behind the entire project. The whole underlines the unique and collaborative character of the project, with special attention to the diversity of the individuals and institutions involved.

The third and final chapter deals with the interrelatedness and the spread of the different resulting printed maps and atlases, all directly based on data gathered during the land surveys. As a first step, I discuss the different editions of atlases produced at the Qing court, as well as the strongly reduced versions included in textual compilations and later renditions under subsequent emperors. Second, I explore the transmission of Qing cartographic material to Europe, tracking different maps and atlases from Beijing to Paris, which leads me to focus on the European incorporation of the Qing atlas produced by French cartographer Jean-Baptiste Bourguignon d'Anville, revealing the relationship between his collection of maps and the different editions of the original Qing atlas. In a third step, finally, I attend to the involvement of scientists in Saint Petersburg in the circulation of cartographic material between East Asia and Europe. As such, Chapter 3 gauges the impact of the cross-cultural circulation of maps and atlases on mapmakers worldwide, shaped by the connections between individuals and spanning the entire Eurasian continent.

In among these three chapters I insert two intermissions, briefly leaving instruments and maps behind in order to reexamine certain issues from the missionary perspective so as to connect to and reevaluate past scholarship related to the Qing mapping project. The first intermission explores the place of the mapping project in the Jesuit mission, commenting on different missionaries' approaches with regard to their commitment to Qing cartography. The second focuses on the missionary work undertaken during the surveys as European missionaries utilized their status as imperial envoys to aid local Christian communities. Thus, while departing methodologically in subject matter, the two intermissions build on insights newly garnered from the chapters to explore further some of the issues scholars active in the field of missionary studies have raised. In doing so, they contribute to our knowledge of the mapping project while also further contextualizing it.

To recapitulate, finally, this study aims for a deeper understanding of how the circulation of cartographic material and cartographic practice played out between and within Europe and East Asia around the turn of the eighteenth century. It does so by tracing the trajectories of two kinds of material objects, instruments and maps, closely following the processes of their many adaptations and partially reassembling the constantly changing web of social relations that emerges from tracking their peregranations. The result is a new cartography of cross-cultural circulation revealing unstudied patterns and actors while emphasizing their basic connectedness.

CHAPTER 1

Instruments for the Emperor: New Frontiers, New Practices

各省輿圖、載入一統志、事殊重要。或一省有大圖而無小圖。或一省有里數程途、而無疆界。或一省有疆界、而無里數程途。各省皆不一例、何也。[25]

康熙25年6月12日

As for incorporating the maps of each province into the *Descriptive Account of the Unified* [*Great Qing*], this is a matter of great importance. Either a province has an overview map but no regional maps; or, a province has the road distances recorded, but not its borders; or, a province has its borders recorded, but not its road distances. All the provinces are inconsistent [in their representations], why is this?

July 31st, 1686

Towards the end of the seventeenth century, Jesuit missionaries' participation in scientific and technical matters at the Qing court grew more organized and more specialized. Unlike earlier missionaries, the French Jesuits who reached Beijing in 1688 enjoyed both the support of French King Louis XIV (r. 1643–1715) and financing from his *Académie royale des sciences*. To aid in the explicit scientific goals, including systematic data-gathering of all kinds, that had become integral to the French Jesuits' proselytizing mission, the *Académie* and their patrons provided them with state-of-the-art surveying instruments. This chapter follows the trajectories of those instruments from the Parisian workshops that crafted them to the French Jesuits who transported them to Asia to the imperial palace where the emperor demanded his workshops reproduce the Parisian instruments, and thence out into the field.

25 *Kangxi chao shilu, juan* 126. The (*Da Qing*) *Yitongzhi* (大清一統志) constituted the official imperial/dynastic geography.

© KONINKLIJKE BRILL NV, LEIDEN, 2017 | DOI 10.1163/9789004345362_003

As well as exploring the different uses and adaptations of these instruments across cultural and political boundaries, this chapter will also discuss the Kangxi emperor's long-standing interest in cartography. The emperor paid careful attention to the lessons provided by the Jesuit Ferdinand Verbiest, one of his tutors, who instructed the emperor in European-style mathematics, of which practical geometry, the very basis for land surveying, was an important component. The emperor saw in these lessons a tool for the improved, more consistent mapping of his rapidly expanding empire. Thus, immediately after the arrival of the French Jesuit missionaries, the emperor displayed a keen interest in the Parisian instruments and in the methods of practical geometry implicit in their design, eventually ordering his Imperial Workshops to reproduce the Parisian prototypes. At the same time, surveying teams tried out some of the original instruments in the field, surveying the empire's new frontiers and later prompting the emperor to order the precise measurement of one degree of latitude just south of the capital. This eventually led the court to restandardize the Qing's most basic unit of length, the *chi*, profoundly impacting scientific practice at the Qing court and paving the way for the consistent mapping of all Qing territories over the following decades.

This chapter focuses entirely on cartographic *practice*, although of course the practice is far from entirely extricable from its mathematical and astronomical underpinnings, known at the Qing court as "Western learning" or *xixue* 西學. Recent scholarship has devoted tremendous attention to Western learning under the Kangxi emperor, particularly with regard to the European-style astronomy and mathematics taught at court.[26] No study, however, has yet comprehensively examined the practices of cartography or of land surveying during the Kangxi years. I therefore attempt in this chapter precisely such a discussion of the evolution of cartographic practice, separate from more theoretical concerns. I argue that the circulation of surveying instruments paved the way for the emperor to integrate selected new techniques of land surveying, as developed at the Parisian *Académie*, into the existing framework of cartographic practices at the Qing court, in effect establishing a new cartographic practice in East Asia. Surveying instruments thus lead into the exploration of cartographic practice at the Kangxi court, eliding when possible reflections on the instruments' theoretical underpinnings in favor of their practical use in the field. In all, Chapter 1 will show how the Qing mapping project discussed in this study arose directly from converging interests among the Jesuit order, the *Académie royale des sciences*, and the Kangxi emperor and how the circulation

26 Jami (2012). Elsewhere, Jami proposed a distinction between "science as action" and "science as discourse." Jami (2007). This study fully focuses on the former.

of surveying instruments, and thus of the practices they prescribed, shaped the project fundamentally.

1.1 An Instrumental Convergence of Interests

In late seventeenth-century Paris, the instrument workshops of Butterfield, Chapotot, Bion, and others, produced state-of-the-art surveying instruments in a more systematic and specialized manner than in previous decades. Louis XIV's establishment (in 1666) and patronage of his *Académie royale des sciences* had inspired instrument-makers to set up shop throughout the city, catering first to the scientists attached to the *Académie*, but soon also to other elites, rendering Paris unquestionably one of Europe's great centers of instrument manufacture. A number of such instruments travelled along with the French Jesuit mission to China, supported and financed by the *Académie*, passing eventually from the hands of Jesuit missionaries into those of the Qing emperor and his court artisans. This, it turned out, would pave the way for the absorption of observational practices developed at the *Académie* into the framework of Qing cartographic practice.

1.1.1 *The* Académie *and the Instrument Market in Paris*

At the turn of the eighteenth century, instruments useful in land surveying were in great demand because the *Académie*'s primary objectives during its early days included measuring French territories and tracing a meridian through Paris. European land surveys of previous decades and centuries had made use of numerous ingenious instruments but their production had never before reached such levels, either in quantity or in precision, as in Paris of the late seventeenth century. Another factor in favor of the Parisian instrument market was the fact that French elites had begun commissioning instruments executed in silver or gilded brass, as conspicuous consumption, that is, for the "acquisition of goods in order to set oneself apart or to compete with others."[27] Parisian instruments had become material symbols of contemporary French statecraft and scientific enquiry, thus serving purposes beyond the merely instrumental.

What set such instruments apart from earlier generations of larger, mostly stationary specimens was, on one hand, their design as functionally mobile units for fieldwork and, on the other, their precision, frequently featuring refined graduation bars and telescopic sights. Before the invention of the

27 Schechner (2001), 211.

telescope, instrument-makers installed simple alidades, small sighting-holes or pointers, to help in aiming instruments in a given direction, but the advent of the telescope in the early seventeenth century not only increased precision, it also facilitated the production of smaller, more portable, instruments, thereby greatly improving observational mobility.[28] Moreover, whereas previous generations of astronomers and other scientists had often had to design their own instruments, by the late seventeenth century a new generation of highly specialized instrument-makers had developed the technical know-how and infrastructure necessary for producing telescopic sights commercially.[29] This shift towards more specialized instrument-production also enabled artisans to create instruments incorporating multiple functionalities for an enhanced range of applications—as well as easier usage: quadrants or semi-circles often now

FIGURE 5 *Details of a Chapotot mobile quadrant's telescopic sight.*
PHOTO: CLAUDE MENNINGER. RÉGION GRAND EST—ALSACE CHAMPAGNE-ARDENNE LORRAINE—INVENTAIRE GÉNÉRAL.

28 Daumas (1972), 27–56; Golvers & Nicolaidis (2009), 93.

29 The production of telescopes required such specialized skills that some workshops further outsourced the job to others. Turner (2002), 380.

came with compasses built into their plane surfaces, for example, and leveling instruments now sometimes included telescopic sights mounted on top. In short, the Parisian instrument workshops embodied a more systematic and specialized production of a new generation of more precise, more delicate, and more complex instruments (figures 5 & 6).

With the *Académie royale des sciences* and a handful of instrument workshops well-established by the seventeenth century's final decades, a solid market developed in Paris for drawing tools, quadrants, semi-circles, levels, sundials and other such precision instruments. Three particular workshops, especially favored by the academicians, other elites, and travelers at the turn of the eighteenth century, represented the best Paris had to offer in this regard.[30] The first was that of Michael Butterfield, an Englishman who worked in France between 1665 and 1724 making sundials and astronomical rings (then in great demand, such that others pirated them).[31] Elites often commissioned

FIGURE 6 *Details of a Chapotot mobile quadrant's graduation bar, plumb line, and signature.* PHOTO: CLAUDE MENNINGER. RÉGION GRAND EST—ALSACE CHAMPAGNE-ARDENNE LORRAINE—INVENTAIRE GÉNÉRAL.

30 Daumas (1972), 71–83.

31 Daumas (1972), 79; Schechner (2011), 215.

sundials incorporating different precious metals; Butterfield preferred to use silver rather than gilded brass for his best pieces.[32] Louis (active 1670–1700) and Jean (active 1676–1721) Chapotot, father and son, headed another popular instrument workshop, producing a wide variety of instruments, smaller and larger ones represented in equal measure, mostly executed in brass and known for the quality of their structural parts (figure 7). The third renowned instrument-maker was Nicolas Bion (1652–1733), who enjoyed an excellent reputation throughout Europe and even published an instrument construction manual in 1709, entitled *Traité de la construction et des principaux usages des instruments de mathématique*. All of these men received the prestigious title of *Ingénieurs du Roy* from the French court.[33] Further testimony to the fame of these workshops was the fact that Russian Tsar Peter the Great (r. 1682–1725) visited all three workshops in 1717 and bought a great variety of instruments to supply his own Academy of Sciences in Saint Petersburg.[34]

Perhaps the most important instrument these workshops produced, at least in retrospect, was the mobile quadrant (*quart de cercle mobile* or *quart de nonante mobile*). "Of all the instruments of astronomy," Jérôme Lalande (1732–1807) wrote in the mid-eighteenth century, "the portable quadrant is that whose use is the oldest, the most wide-spread, the most indispensable, the most convenient."[35] By that time, the instrument had become a symbol for the sciences as practiced by the *Académie*, embodying the evolution in instrument-production in the latter half of the seventeenth century. The author of the first description of this type of mobile quadrant was Jean Picard (1620–1682), who led the *Académie*'s very first stage in tracing a meridian through Paris in the years 1669–1670, accurately measuring the length of just over one degree of latitude to the south of Paris by triangulating from carefully set baselines. Based on these measurements, Picard became the first to reach a reasonably accurate estimate of the size of the earth, one of the great scientific debates of the day and one of the *Académie*'s *raisons d'être*.[36]

32 Schechner (2011), 212.

33 Conley (2012). Note that these men did not necessarily produce these instruments personally. Turner (2002), 380.

34 Kniajetskaia & Chenakal (1975), 250 & 257. The article includes a list of instruments bought at the Chapotot workshop. The online catalogue of the State Hermitage Museum includes several instruments from all three workshops.

35 "Le quart de cercle mobile est de tous les instruments d'Astronomie, celui dont l'usage est le plus ancien, le plus général, le plus indispensable, le plus commode." Quoted by Turner (2002), 373.

36 Levallois (1988).

FIGURE 7 *Trade Card of Louis Chapotot, Engineer, À la Sphère, 1670–1700. Engraving on paper (21.2 × 20 cm). Waddesdon (National Trust) Bequest of James de Rothschild, 1957. Acc. No. 3686.2.74.186.*

PHOTO: UNIVERSITY OF CENTRAL ENGLAND DIGITAL SERVICES. NATIONAL TRUST, WADDESDON MANOR.

In his treatise on the subject, entitled *Mesure de la terre* (1671), Picard describes the new mobile quadrant he designed for the purpose of tracing the meridian through Paris (figure 8). Major improvements established by Picard include both the precision and the flexibility of the quadrant: the instrument features telescopic sights fitted onto the quadrant to increase observational precision, a footrest for easy set-up or taking-down, and a joint permitting easy tilting of the instrument for taking measurements both horizontally and vertically. As such, Picard's mobile quadrant could easily perform, with increased precision, the two measurements most essential to land surveying: finding the angular heights of heavenly objects such as the pole star, or the sun's elevation at noon, necessary for determining one's latitude; and the angular distance between two points in the landscape for triangulation from a baseline, which Picard himself employed in measuring the terrestrial degree near Paris. By increasing observational precision as well as the quadrant's ease of use in the field, Picard's instrument greatly impacted contemporary techniques in both land surveying and cartographic practice.

The technical requirements described by Picard were only achievable by contracting highly experienced and specialized craftsmen. The workshops of Butterfield, Chapotot, and Bion all appear to have been privileged partners in the construction of Picard-style mobile quadrants. In fact, scholars have argued that Michael Butterfield was actually the first to realize Picard's design. In subsequent decades, however, it was Chapotot the younger who specialized in producing the structural parts for the mobile quadrants commissioned by the *Académie*, manufacturing great numbers of them on demand.[37] Moreover, all three workshops supplied scientific instruments to one of the *Académie*'s most notable scientists, Giovanni Domenico Cassini (1625–1712). At the King's invitation, Cassini settled in Paris in 1669, where he joined the newly founded *Académie* as the Paris Observatory's first director.[38]

Cassini's achievements included improving the technique for observing the satellites of Jupiter. Galileo Galilei (1564–1642) had been the first to time the emersions and immersions of Jupiter's four largest moons (having also discovered them in 1610 after constructing one of the earliest telescopes), but Cassini perfected the method and produced tables enabling precise determination of

---

37 Joseph-Nicolas Delisle, who will feature prominently in Chapter 3, ordered many of these quadrants. Daumas (1972), 73–78, esp. 76–77; Turner (2002), 380.

38 His son Jacques Cassini ("Cassini II," 1677–1756), his grandson César François Cassini ("Cassini III," 1714–1784), and his great-grandson Jean Dominique Cassini ("Cassini IV," 1748–1845) would each succeed him in this post.

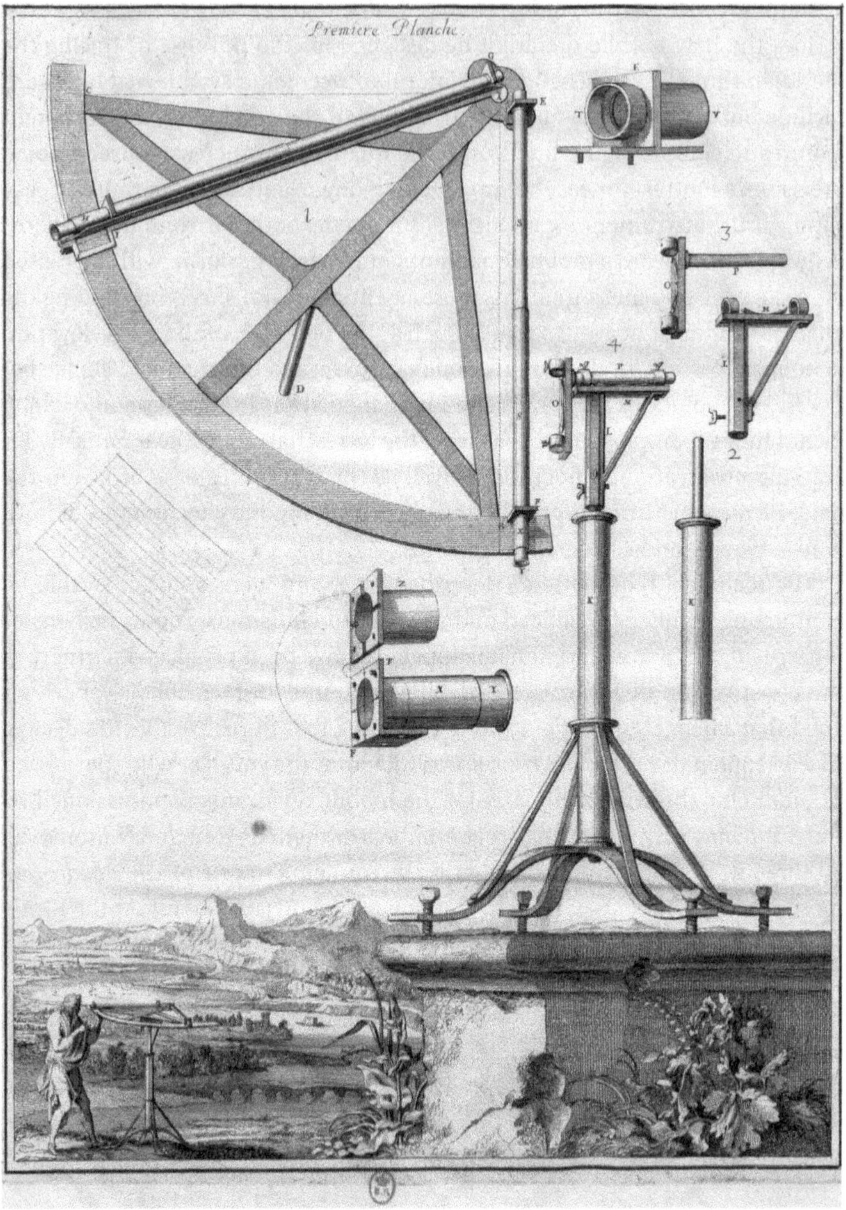

FIGURE 8 *Drawing from Picard's Mesure de la terre (1671) explaining the construction of the mobile quadrant.*
BIBLIOTHÈQUE NATIONALE DE FRANCE.

longitudes *vis-à-vis* Paris. Although this method constituted a significant leap forward in solving the problem of longitude, it required repeated and stationary observations in which small observational mistakes could produce large margins of error. Even so, a revolution in cartographic practice took place when Picard-style trigonometric land surveying, using carefully set-out baselines for triangulation, combined with Cassini's technique of determining longitude based on observing Jupiter's sattelites: from 1679 to 1682, Picard (in collaboration with Philippe de la Hire (1640–1718)) led a project that dramatically corrected the cartographical outline of French territories (figure 9).[39] The first full map of France based on large-scale application of these new practices, effected by laying a skeleton of triangles across France as illustrated on a 1744 map known as *La Description Géometrique de la France* (1744) (figure 10), however, waited until Cassini's grandson was director of the Observatory.

1.1.2 *The King's Mathematicians' Interest in Cartography*

Following the Qing conquest of the Chinese provinces in the mid-seventeenth century, Jesuit missionaries, who had traveled to Asia on Portuguese ships and operated under the auspices of the Portuguese crown, had quickly gained access to the Qing court. Under the Qing's Shunzhi 順治 (r. 1643–1661) emperor, the Jesuits built on experiences gained in the 1620s, when they had introduced calendrical reforms based on the Tychonic astronomical system to the Ming state.[40] The Jesuits thus continued enhancing the calendar in Beijing, serving China's new Manchu rulers through their knowledge of astronomy as they had the Ming, to maintain support for their proselytizing mission in the provinces. Later, in the 1660s, having withstood resistance from within the Qing administration to their involvement in composing the state calendar, missionaries in Beijing further consolidated their access to the Qing court under the Kangxi emperor and began providing the court with a further range of specific skills and techniques relating to astronomy and applied mathematics. As a result, by the 1680s a network based on scientific exchanges had started forming in support of the Jesuit missionaries' ever-more complex and specialized involvement in scientific practice. As Noël Golvers has pointed out, this emerging network connected "Peking, over Macao, to the young European

39 Pelletier (2013), 58–64.

40 Tycho Brahe (1546–1601) was a Danish nobleman and astronomer who had improved the Ptolemaic model of the universe. Unlike the Ptolemaic model, which held that the moon, the sun, and the planets all orbit the earth, Brahe proposed that the planets orbit the sun, while the sun and the moon orbit the earth. The Tychonic model served many Roman Catholic astronomers as a viable alternative to the Copernican.

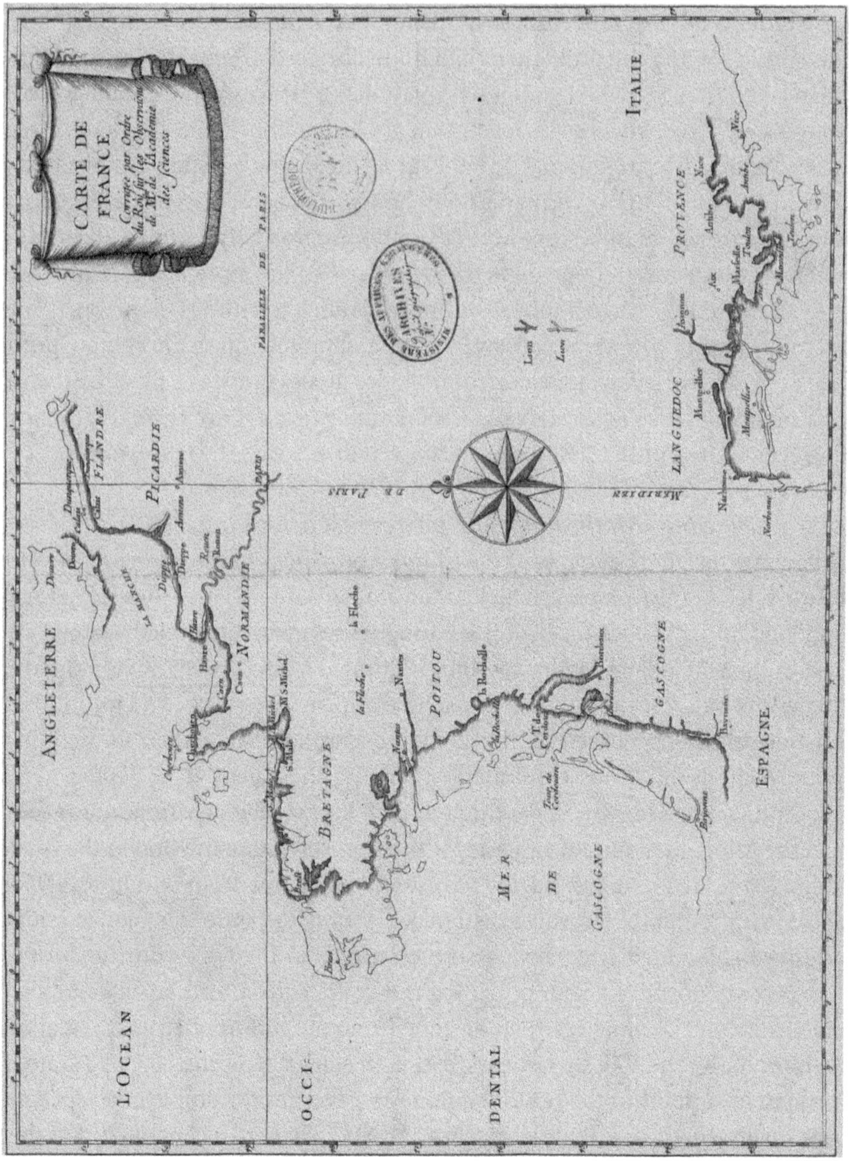

FIGURE 9 Carte de France. Corrigée par ordre du Roy sur les observations de Mss. de l'Académie des Sciences (*c. 1684, printed 1693*).
BIBLIOTHÈQUE NATIONALE DE FRANCE, CARTES ET PLANS, GE DD-2987 (777).

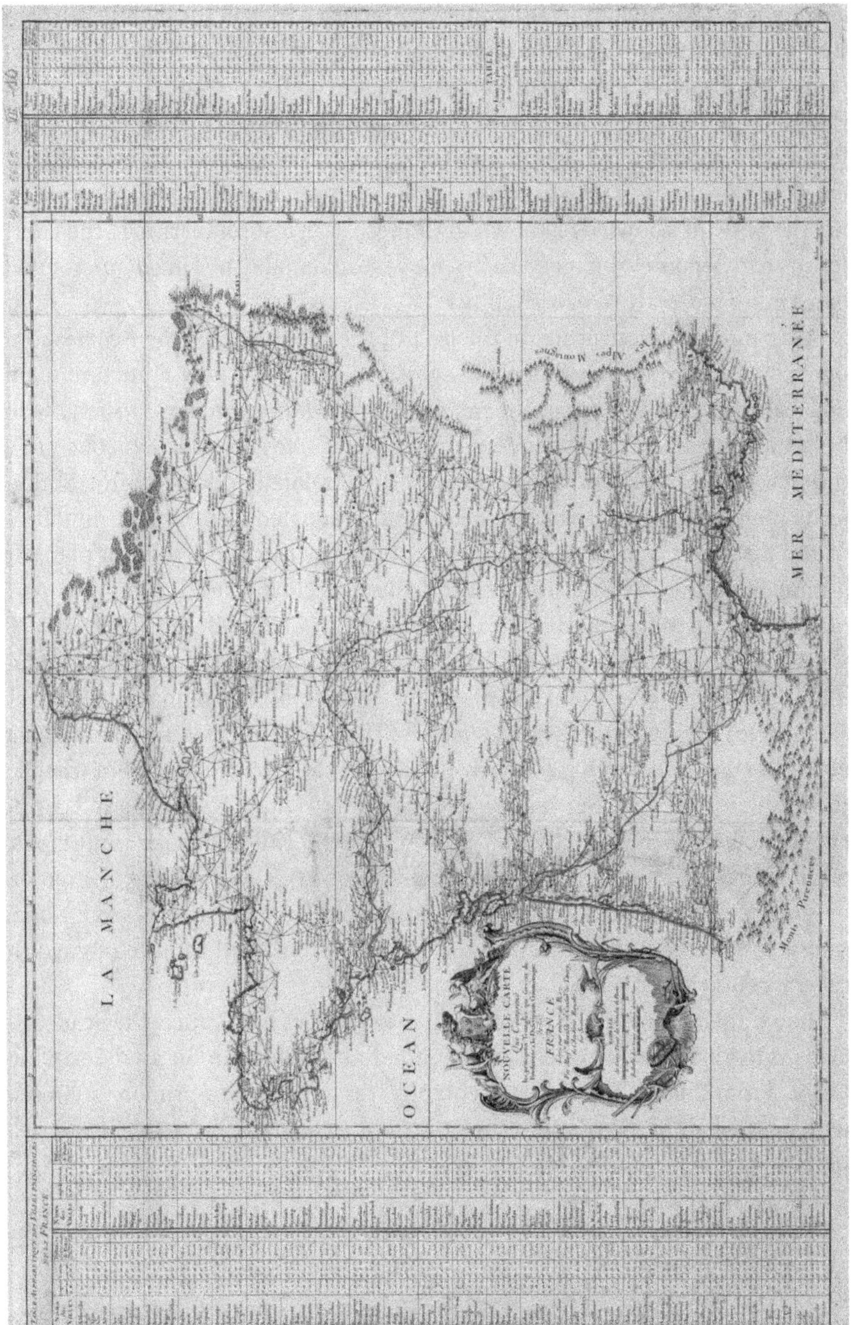

FIGURE 10 Nouvelle carte qui comprend les principaux triangles qui servent de fondement à la Description géométrique de la France. Levée par ordre du Roy par Messrs. Maraldi et Cassini de Thury, de l'Académie royale des Sciences (*1744*). BIBLIOTHÈQUE NATIONALE DE FRANCE, CARTES ET PLANS, GE BB-565 (A7,10).

academic world, [including in] France, a practice which in the eighteenth century became much more common… and complex."[41] This emerging connection to the French scientific community deepened after the return to Europe of Philippe Couplet (1623–1693), sent to France with the full backing of the court Jesuits to plea for more missionaries trained in the sciences. Fully aware of the emperor's keen interest in European-style astronomy and applied mathematics, the body of knowledge known in Chinese as "Western learning," the court Jesuits had become convinced such a move would make the Qing emperor and elites more receptive to Christianity.[42]

After much maneuvering on the part of the missionaries, their correspondents, and their patrons, Louis XIV agreed to sponsor a new French mission and commanded the appointment of several *Mathématiciens du Roy*, who became recognized correspondents of the *Académie* (the most obvious institutional base of support for all French science-related endeavors, in addition to its members generally having been Jesuit-educated). Prominent members of the *Académie* and of the Jesuit order in Paris thus worked out a plan for a mission to the Far East, the first to feature explicit scientific goals as part of its broader religious mission (also the first not under the patronage of the Portuguese crown). In 1685, Jean de Fontaney (1643–1710), Claude Visdelou (1656–1737), Jean-François Gerbillon (1654–1707), and Joachim Bouvet (1656–1730) were among those selected for the mission. They travelled to Asia aboard a ship also carrying a French embassy to the king of Siam, from whence the Jesuits would then continue on to Qing China.[43] In their writings and correspondence, the "King's Mathematicians" did not hesitate to make continuous, often lyrical, comparisons between "their" Louis XIV and the Kangxi emperor, describing the latter as "in the Orient today what Louis the Great is in Europe, that is to say, the greatest prince, the most accomplished to have reigned in several centuries."[44]

Before this privileged group of Jesuit missionaries left France, the academicians provided a list of questions intended as a guideline for studying Qing China. From this list, it appears geography was among their primary interests, with nearly half of the questions relating to the geography of areas in and surrounding Qing territories, for example:

---

41 Golvers (2014), 139.

42 Du & Han (1992), 268–269.

43 In Siam and en route, the Jesuits conducted several observations in line with the standards set by the *Académie*, using Parisian instruments. Landry-Deron (2011) & Hsia (2009) discuss these at length. For the details regarding the selection process of the "King's Mathematicians", see Landry-Deron (2001).

44 de Thomaz de Bossierre (1994), 98 & 102–103; Hsia (2009), 51–72; Jami (2014), 58.

> [2] Whether the reverent fathers, the Jesuits, have not made some considerable observations of the longitudes and latitudes of China...
> [13] Of the size and people of the principle cities such as Beijing, Nanjing, Quantong,[45]
> [14] and of the entire Kingdom...
> [21] Of East and West Tartary...[46]
> [32] Of Korea, if its King is tributary and able...
> [27] [and of the] provinces of China.[47]

Before reaching Qing China, the four Jesuit correspondents of the *Académie* together with fellow French Jesuit Louis Lecomte (1655–1728), had to decide how to organize their scientific explorations. They agreed that each of them should be responsible for one or more fields of inquiry. Jean de Fontaney, leader of the expedition, took charge of "the history of Chinese astronomy and geography, and the daily observations of the sky corresponding to those made at Paris in the Observatory," and so assumed responsibility for inquiries pertaining to geography and cartographic practice.[48] As the only French Jesuit also a full member of the *Académie*, de Fontaney was the best-suited for conducting observations which had to correspond to the *Académie*'s institutional standards. After stopovers at the Cape of Good Hope and in Siam, where they made several observations with their state-of-the-art instruments, they arrived at Ningbo, far from Portuguese-controlled Macao, finally reaching Beijing in February 1688,[49] where they immediately gained an audience with the emperor, who soon decreed that Bouvet and Gerbillon might stay at court, whereas

45 "Quantong" here refers to either Guangzhou (Canton) or Mukden (today's Shenyang), since it was sometimes used to refer to the lands comprising today's Liaoning province.

46 The name "Tartary" then referred to almost the entire area between the Chinese provinces and Muscovy, inhabited mostly by nomadic peoples, corresponding approximately to modern Northeast China (or Manchuria), Mongolia, Xinjiang, and parts of Siberia. By East and West Tartary, the academicians referred to Qing territories north of the Great Wall, respectively inhabited by Manchus and Mongols.

47 "[2] Si les RR.PP Jésuites n'ont pas fait quelques observations considérables des longitudes et des latitudes de la Chine... [13]De la grandeur et du peuple des principales villes comme Péking, Nanking, Quantong, [14] et de tout le Royaume... [21] De la Tartarie orientale et occidentale... [23] De la Corée, si son Roy est tributaire et puissant... [27] [Des] Provinces de la Chine." Pinot (1932), 7–9; Landry-Deron (2002), 150–159. Questions 2, 13, 14, 15, 16, and 20–28 (out of 28) all relate to geography.

48 From a letter by de Fontaney to the academicians in Paris, written on November 8th, 1687. Quoted in Hsia (2009), 112.

49 Hsia (2009), 84–100.

de Fontaney, Visdelou and Lecomte were free either to travel or to stay in the Chinese provinces.[50]

Undeterred, de Fontaney conducted several observations of the satellites of Jupiter during the first years following the French missionaries' arrival in China, in the manner developed by Cassini (making use of a well-established network of missionary bases throughout the Chinese provinces). He thus gathered important observational data from Beijing, Jiangzhou (today's Jiujiang) and other cities, fully commensurate with the standards of the *Académie*, which required stationary and repeated observations with the most precise Paris-made instruments available.[51] By timing the immersions and emersions of Jupiter's satellites and comparing them to Cassini's tables, the French Jesuits could determine longitudes *vis-à-vis* Paris for several of the Qing's most important cities. The new data they shared with Gerbillon and Bouvet in Beijing and sent to the academicians in France. Moreover, from the writings of Louis Lecomte it is clear that de Fontaney and the others also recorded observations en route throughout their travels, both before and following their stay in Beijing.[52] Lecomte mentions the recording of detailed information along routes from Ningbo to Beijing and on to Jiangzhou, his first destination after leaving the capital.[53] Whether or not these observations led to the drawing of a map is unclear.

In Jiangzhou, the three French Jesuits not allowed to stay at court split into teams: Visdelou and Lecomte continued to the Shanxi-Shaanxi region while de Fontaney traveled to Jiangnan, where the emperor would later consult him on astronomical matters during his tour of the region in 1689.[54] In his correspondence, Antoine Gaubil (1689–1759) would later note that Lecomte also made useful observations, among others in Xi'an.[55] Soon after, the French Jesuits decided that Lecomte, procurator of the French mission, would return to France to solicit more support. Presented with an opportunity to travel through the southern provinces with his companion, de Fontaney joined Lecomte on his way to the port city of Guangzhou. While traveling south, both men cooperated

50 Jami (2012), 112–116.

51 Cassini had given the king's mathematicians the necessary tables for determining longitude based on the method of observing Jupiter's satellites. Jami (2014), 60. In 1690, Bouvet and Gerbillon would perform repeated observations in order to determine latitude of Beijing, eventually set at 39°53′N by Gerbillon. Subsequent observations by others differed only slightly. Gaubil (1970), 614–615.

52 Hsia (2009), 58–66 & 110–116; Jami (2014), 59–62.

53 Lecomte (1990), 525.

54 Han Qi (1996); Jami (2012), 131–133.

55 Gaubil (1970), 96.

in drawing a map of the river courses connecting Nanjing to Guangzhou. It took them two to three months, as reported by Lecomte in his published account:

> I also have the course of the rivers that connect Nanjing with Guangzhou. It was a work of two to three months, and a very tiring one when one wants to do things carefully. The map is eighteen *pieds* long [c. 5.75 m], and each minute occupies more than four lines, or one third of a *pouce* [c. 9 mm]. Thus, all the bends and the width of [each] river, the tiniest islands and smallest villages are marked exactly. We always had a compass at hand and now and then on our way we carefully observed the meridian height of a few stars so as to correct our estimates and determine the latitudes of the main cities of the country as accurately as possible.[56]

A reduced copy of this map appears in the collection of cartographic material of Joseph-Nicolas Delisle (1688–1768), now preserved at the French National Archives.[57] Joseph-Nicolas was the younger brother of Guillaume Delisle (1675–1726), *Premier Géographe du Roy* and full member of the *Académie*. When the Russian Tsar Peter the Great offered Joseph-Nicolas the position of head astronomer of the Russian Academy of Sciences in 1725, he accepted, bringing many Paris-made instruments with him to Saint Petersburg, including some produced by the Chapotot workshops.[58] Tsar Peter later charged Delisle, in his official capacity, with the task of preparing an atlas of the fast-expanding Russian Empire, which would come to be known as the *Atlas Russicus* (1745). The fact that a copy of the map drawn by Lecomte and de Fontaney appears in Delisle's collection indicates of the existence of a large-scale exchange of cartographic material among Beijing, Paris and Saint Petersburg in the 1720s, which Chapter 3 will discuss at length.

The map drawn by de Fontaney and Lecomte indicates the latitudes for every town depicted, mostly positioned on or near rivers, as well as the intervening

---

56 "J'ai aussi le cours des rivières qui menent de Namkim jusqu'à Canton. C'est un travail de deux ou trois mois et très fatigant quand on veut faire les choses avec quelque soin. La carte est de dix-huit pieds de long et chaque minute y occupe plus de quatre lignes ou un tiers de pouce ; ainsi tous les détours, la largeur de la rivière, les moindres îles et les plus petits villages y sont exactement marqués. Nous avions toujours la boussole à la main et nous prenions soin d'observer de temps en temps sur la route, les hauteurs méridiennes de quelques étoiles pour corriger notre estime et déterminer au plus juste la latitude des principales villes du pays." Lecomte (1990), 526; Jami (2014), 61–62. One French *pied de Roy* corresponds to c. 32.5 cm. The *pied* equaled twelve *pouces*.

57 Archives nationales (hereafter AN), Cartes et plans, 6JJ4A:13b. For more on Joseph-Nicolas Delisle, see Chapter 3.

58 Daumas (1972), 76–77; Kniajetskaia & Chenakal (1975).

distances, presumably gleaned from local gazetteers or route books. A short commentary written directly on the map states one should calculate these distances, recorded in *li* 里, equating 250 *li* to one degree of latitude. This was indeed the standard of the time, first proposed by Matteo Ricci and later elaborated fully in the astronomical works of court Jesuits Adam Schall von Bell (1592–1666) and Ferdinand Verbiest, each of whom was responsible for calendar-making at the Qing's Directorate of Astronomy (Chinese: *Qintianjian* 欽天監 / Manchu: *Abka be ginggulere yamun*) under the Shunzhi and Kangxi emperors, respectively.[59] Thus, combining information from Chinese geographic descriptions with direct observations of latitudes and longitudes, this map constitutes a major improvement on European knowledge of the locations of towns and cities in the region. While de Fontaney and Lecomte traveled through the southern Chinese provinces, mapping their way for the *Académie*, the two French missionaries who had been allowed to stay at court, Gerbillon and Bouvet, played an important role in Qing imperial cartography, as discussed in this chapter's second and third parts.

1.1.3 *Paris-made Instruments for the French Mission*

European sources contain many references to instruments given to the emperor by the French Jesuits, as well as of instruments the missionaries carried with them or kept in their residencies for personal use. In the decades preceding the arrival of the French mission, missionaries in Beijing had themselves designed and constructed a wide variety of instruments, a common practice among early seventeenth-century astronomers. In the context of the Jesuit missions to Asia, the best example of this practice is undoubtedly Ferdinand Verbiest's efforts with regard to improving the instruments used in calendar-making at the Beijing observatory.[60] Other missionaries, such as Tomás Pereira (1645–1708) and Antoine Thomas (1644–1709), also made substantial contributions to instrument design and production at the Qing court as well as gifting European-made instruments to the emperor.[61] The arrival of more French Jesuits conveying precision instruments produced by the specialized workshops of Paris, in fact, initiated a steadier flow of instruments between Europe and East Asia. Generally speaking, this allowed a new, better-connected generation of missionaries to rely more on European-made precision instruments and less on their own skills in instrument design.

---

59 Chu (1999), esp. 397. For 250 *li* for one degree of latitude as proposed by Ricci, see Huang & Gong (2004), 21–22.

60 Halsberghe (1992); Zhang (2003); Golvers & Nicolaidis (2009).

61 Jami (2012), 153; Golvers (2014), 141.

When the king's mathematicians left France, they took with them some of the latest instruments, a record of the most important of which appears in *Voyage de Siam* (1686), a book recounting the French embassy to Siam of which they were part during their journey's first stages:

> We commissioned... two quadrants, one with a radius of eighteen *pouces* [c. 49 cm], the other of 26 [c. 70 cm]; three large seconds pendulums; an instrument for finding, at once the right ascension & the declination of stars; [and] an equinoctial square, marking the hours to the minute [sic], & featuring a large compass at the bottom for finding the declination of the needle at all times of day. All of these instruments would serve for astronomical observations. We also took two semi-circles divided very precisely at six minutes intervals for geometrical operations, one of which only featured alidades, & the other telescopes. The Duke of Maine... gave us a third, which was much larger, divided three minutes intervals, and which he had made for his private use.[62]

These instruments, serving both as necessary apparatus for the Jesuits' own scientific enquiries and as diplomatic gifts to the emperor, were bound to trace varying trajectories after arriving at the Qing court. Some instruments the Jesuits needed for observing the satellites of Jupiter, which de Fontaney, Visdelou and Lecomte took along when they left Beijing for Jiangzhou. Later, when Lecomte returned to Europe from Guangzhou, they decided that a number of the instruments they carried should return to Beijing in further support of the mission and of Bouvet and Gerbillon's activities at court. This shipment of instruments from Guangzhou to Beijing on August 14th, 1691 included "an astronomical ring, indicating at once the hour, the minute, the height of the sun and the declination of the pole... a proportional compass, two ordinary

62 "on fit faire... deux Quarts de nonante, l'un de dix huit pouces de rayon, l'autre de vingt-six; trois grandes Pendules à secondes, un instrument pour trouver en même-têms l'Ascension droite & la déclinaison des Estoiles, un Quadran Equinoctial qui marquoit les heures jusques aux minutes, & qui portoit au bas une grande Boussole pour trouver à toutes les heures du jour la déclinaison de l'Ayman. Tous ces instrumens devoient servir aux Observations Astronomiques. On prit encore deux demi-Cercles divisez fort exactement de six en six minutes, pour les Opérations de Géometrie, l'un estoit à pinnules seulement, & l'autre à lunettes. Monsieur le Duc du Maine... nous en donner un troisiéme beaucoup plus grand, & divisé de trois en trois minutes, qu'il avoit fait faire pour son usage particulier." Tachard (1686), 9–10. The Jesuits also carried other "machines" and instruments less relevant to this study. Jami (2012), 102–108 & 151–156. The Duke of Maine was Louis-Auguste de Bourbon (1670–1736), son of Louis XIV.

compasses, a ruler, a ruling-pen, a semi-circle, [and] a sphere," indicating that the king's mathematicians also carried with them many smaller instruments not mentioned in the *Voyage de Siam*.[63] Moreover, before Lecomte set sail from Guangzhou, Gerbillon explicitly asked that he make sure even more instruments be forwarded to the capital.[64]

A third example of instruments reaching the Qing court through the agency of French Jesuit missionaries appears in the diaries of Joachim Bouvet: Bouvet mentions Visdelou and de Fontaney, recalled to court in 1693, giving the emperor a number of watches and instruments. These included:

> A curious large watch in a golden box, a new invention; a quadrant of the kind by Mr. Butterfield, the second one that we have offered to His Majesty, who recognizes its quality and precision… a pendulum clock used for observation of the kind by Mr. Thuret, with the indication of seconds; a [sighting] level, reflecting and with a telescope, a new invention;[65] another small level; a large proportional compass; a [proportional] compass for artillery use, divided according to the different size calibers; a number of other kinds of small refined compasses; [and] a small case for mathematics.[66]

Bouvet emphasizes further that the arrival of newly invented instruments particularly pleased the emperor, giving the example of a pendulum clock made after a prototype by Christiaan Huygens (1629–1695), inventor of the pendulum clock.[67] Meanwhile, they offered somewhat less valuable instruments to two of

63 "un anneau astronomique qui donne en même temps l'heure et la minute, la hauteur du soleil et déclinaison de l'aimant,… un compas de proportion, deux compas ordinaires, un équerre, un demi-cercle, un tire-ligne, une sphere…" de Thomaz de Bossierre (1994), 51.

64 de Thomaz de Bossierre (1994), 100.

65 Isaac Thuret (c. 1630–1706) and Jacques Thuret (1669–1738), father and son, both *Horlogers ordinaire du Roy*.

66 "Une grosse montre à boëte d'or fort curieuse et de nouvelle invention, un quart de Nonante de la façon du Sr Butterfield, c'est le second que nous avons offert à Sa Majesté qui en connoit la bonté et la justesse,… deux pendules à secondes pour les observations de la façon du Sr. Thuret, un niveau à reflexion et à lunette de nouvelle invention. Un autre petit niveau, un grand compas de proportion, un compas d'artillerie divise selon le calibre des pieces de differentes grandeurs, plusieurs autres sortes de petis compas fins. Un petit etuy de mathematique…" von Collani (2005), 99.

67 von Collani (2005), 108.

the emperor's sons, one of them crown prince Yinreng 胤礽 (1674–1725), who later requested a clock like the one they had given to his father. It appears that, when Bouvet returned to Europe later that year as an imperial envoy, different members of the imperial family had made similar requests for watches and instruments closely resembling those described above.[68] Clearly, the emperor was not the only person to whom the Jesuits presented these instruments, as Bouvet suggests that Paris-made instruments became prized possessions for members of the Qing imperial clan at the end of the seventeenth century, much as they had among French elites.[69]

Contemporary European sources thus make it clear that, initially, a large number of Paris-made instruments reached the court through the agency of the king's mathematicians themselves, some of whom later returned to Europe to solicit further elite support for their mission. Bouvet, sent back to Europe in 1693 as imperial envoy and returning only in 1699, appears particularly to have made a substantial contribution to the first transfers of scientific matériel from Paris to Beijing: not only returning with more instruments but also bringing a number of maps from France, constituting some of the most up-to-date cartographies available in Paris at that time. At least two of these maps, one world map and one map of Asia, both by Nicolas de Fer (1646–1720), survive in the First Historical Archives in Beijing. Entitled *Mappe-monde, ou Carte générale de la terre, divisée en deux hémisphères suivant la projection la plus commune* (1694) and *L'Asie ou tous les points principaux sont placez sur les observations de Mrs. de l'Académie royale des sciences* (1696), both maps include substantial textual content in the margins, much of it translated into Chinese and simply pasted over the original text.[70] The maps also carry yellow labels with Chinese-language renditions of place names simply pasted atop the originals. It is particularly noteworthy that de Fer had employed in his map of Asia a projection method developed by Nicolas Sanson (1600–1667), known as the Sanson-Flamsteed projection, which Wang Qianjin has identified as the projection used in the copperplate atlas produced as a result of the Qing mapping project central to this study.[71] It thus appears that imperial interest in these

68 von Collani (2005), 100 & 194.

69 Schechner (2001), 211–212. French elites, including academicians like Cassini and de la Hire as well as royalty such as Louis Auguste (1670–1736), son of Louis XIV, donated some of the instruments the missionaries gave to the emperor. Gaubil (1970), 634.

70 *Aomen lishi ditu jingxuan* (2000), 26–27 & 30–31. Untouched copies of these maps are available for consultation at BNF, Cartes et plans, 06472 B & 00092 B.

71 Wang (1991). Discussed in Chapter 3.

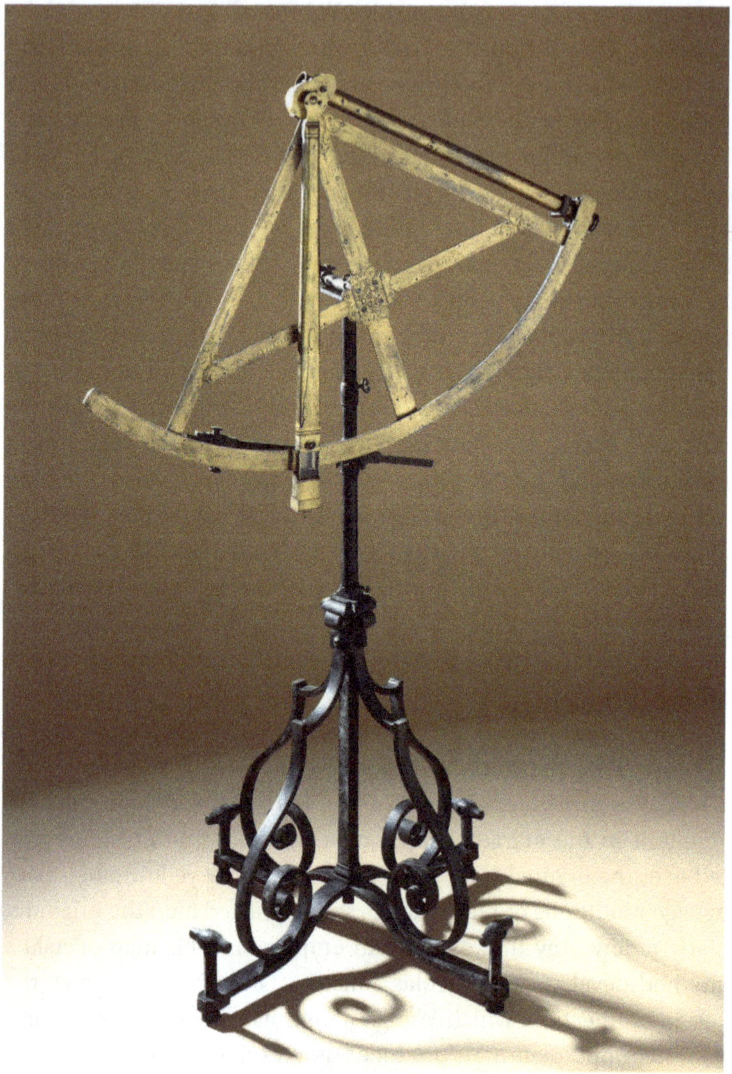

FIGURE 11 *Quadrant by Butterfield (radius c. 58 cm) similar to the one given to the emperor by the king's mathematicians.*
SCIENCE & SOCIETY PICTURE LIBRARY, SCIENCE MUSEUM.

French cartographies of Asia directly informed the Qing mapping project in later decades.

The number of instruments gifted to curry imperial favor escalated rapidly from the 1690s to the 1720s as more French missionaries arrived at court and amid the court Jesuits' ongoing efforts to obtain more curiosities from Europe via all possible channels. In the first years of the eighteenth century, Gerbillon even asked the secular missionaries of the Paris Foreign Missions Society to aid him in acquiring more instruments for that purpose,[72] in return for which he promised to assist them from Beijing should they meet any difficulties in the provinces.[73] As a result of such efforts, a substantial number of the Paris-made instruments, today in the collections of the Palace Museum in Beijing, reached the imperial capital. At the same time, a few other Paris-made pieces, including telescopes, reached the court via Russian embassies to the Qing.[74] Museum catalogues further confirm that the Butterfield, Chapotot, and Bion workshops are particularly well-represented among the extant instruments once kept at the Qing court, including some specifically tailored to cartographic practice (figure 12).[75] Yet, even as many instruments flowed into the court, others remained in the possession of the missionaries themselves for their personal use. In a letter dating back to the 1750s, Jesuit missionary Antoine Gaubil confirms that the French residence in Beijing still harbored two quadrants, one of them produced by Chapotot, both donated to the French missionaries by Louis XIV in the early years of the eighteenth century.[76]

To summarize, the French Jesuit mission was the first backed by a powerful non-religious institution, the *Académie royale des sciences*, and thus the first including explicit scientific goals among its mandates, prominent among them the geographic study of continental East Asia. The mapmaking activities of French Jesuits de Fontaney and Lecomte in the Chinese provinces between 1688 and 1691 confirm this. At the same time, among the key material progeny of this aspect of the French mission were the many Paris-made instruments that reached Beijing in the decades following the French Jesuit missionaries' arrival. These instruments, acquired from the specialized Parisian instrument workshops established to cater to the *Académie*, signaled the gradual displacement of large, stationary Tychonic astronomical instruments by a new generation of mobile precision instruments tailored to practical surveying. These paved the way for the construction of new cartographies, a valuable tool in a time when strong monarchs sought new ways to control their vast territories and peoples, a need felt at both ends of the Eurasian continent.

72 The Paris Foreign Mission Society, known in French as the Missions étrangères de Paris, originated between 1658 and 1663.

73 de Thomaz de Bossierre (1994), 140.

74 Guo (2015), 123–124. He writes of two Paris-made telescopes donated by the Fyodor Golovin embassy in 1686.

75 *Qinggong xiyang yiqi* (2011), 21, 23, 29–32, 70–71, 76, 83, 128–132. Guo (2015).

76 Gaubil (1970), 633. Gaubil further mentions that both instruments survived the earthquake of 1730 without much damage.

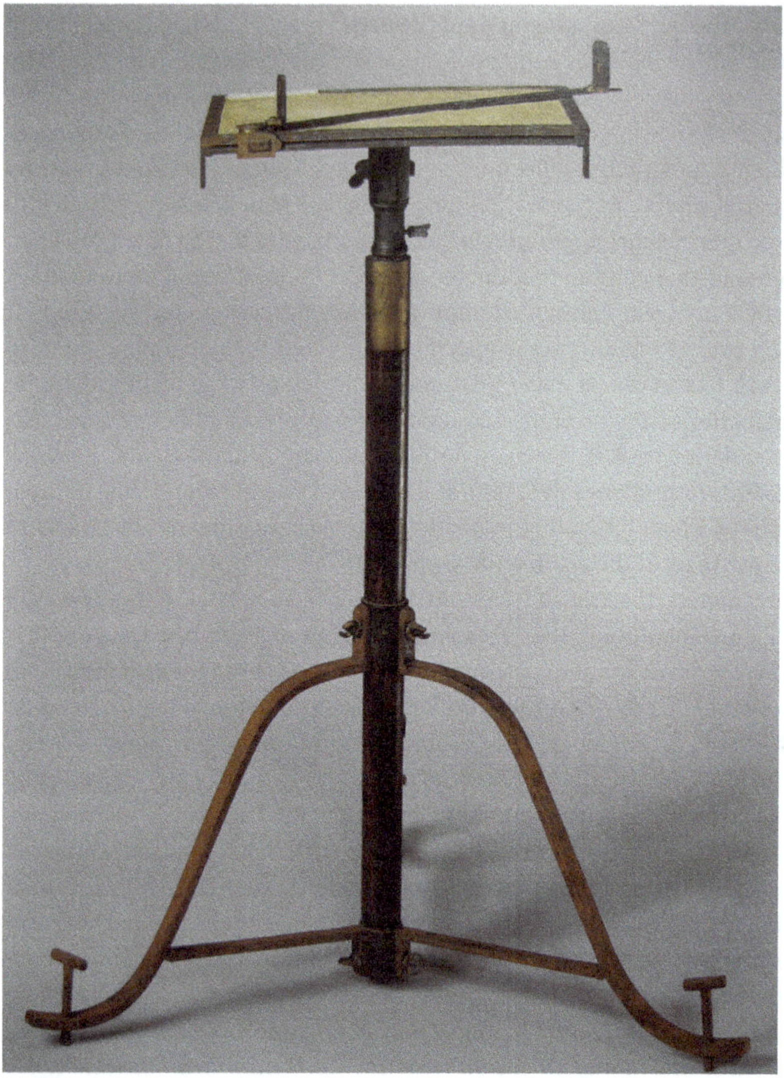

FIGURE 12 *"Plane alidade for drawing" kept in the Qing court collection, originally produced by the Chapotot workshop in Paris. The Qing court's workshops manufactured the footrest.*
PALACE MUSEUM.

1.2 Improving Cartographies: An Emperor's Quest

Once the Qing had consolidated their control over the former Ming territories through the suppression of the Three Feudatories in 1681 and the conquest of western Taiwan in 1683, the Kangxi emperor ordered new maps prepared for a comprehensive gazetteer of all Qing territories. At the same time, he began to take his Jesuit mathematics tutor, Ferdinand Verbiest, along with him on his outings beyond the Great Wall. The emperor asked Verbiest, whom he saw as an imperial expert in Western learning, to create topographical charts and route maps during these imperial outings, the emperor's first experiments with cartographic practice in the field, at least as taught by the Jesuits. Verbiest had been the emperor's astronomer as well as his tutor and had in that capacity designed and supervised the production of new instruments and mechanical pieces at court. Following the death of Verbiest, other missionaries would take up his role at the Qing court, while over the same period Russian expansion down the Amur River formed the general context for ongoing cartographic efforts by the emperor and his missionary experts. Thus, this section discusses Qing cartographic practice in the years straddling the arrival of the French missionaries at court.

After revisiting and exploring how the arrival of the French mission shaped the emperor's study of Western learning as well as his understanding of its practical applications, our spotlight turns to Jean-François Gerbillon, one of the newly arrived French missionaries who took part in the 1689 Qing-Russian border negotiations at Nerchinsk. In this capacity, Gerbillon determined positions on the Siberian frontier and the way thither with the help of Paris-made instruments. Perhaps intrigued by Gerbillon's work in this regard, the Kangxi emperor soon sought to appropriate the Paris-made instruments he had used and the techniques they prescribed by ordering his Imperial Workshops to reproduce and adapt the Parisian prototypes. In a bid to display imperial authority over cartographic practice, the emperor personally evaluated the quality of the adaptations and practiced their different functions and applications. This would eventually lay the foundations for integrating selected techniques of land surveying as practiced at the *Académie*, into the preexisting framework of Qing cartographic practice.

1.2.1 *The Kangxi Emperor's Cartographic Aspirations*

After Qing forces poured through the gates of the Great Wall and took the former Ming capital of Beijing in 1644, the conquest of Ming territories had only just begun. In the years following the fall of Beijng, Qing forces slowly but steadily seized control of the Chinese provinces, a process completed in 1683, more than twenty years into the reign of the Kangxi emperor. The role of cartography in the Qing conquest of the Chinese provinces has not received detailed study, but some evidence indicates map-making in decisive moments: Two years before the Qing conquest of western Taiwan, for example, the court produced a map of the island including extensive labeling in

Manchu.[77] There is also ample evidence of maps of Portuguese-controlled Macao featuring Manchu annotations.[78] In the period of relative stability after the conquest of the Chinese provinces, the emperor turned his attention to consolidating Qing territories. From the 1680s onwards, therefore, increasing cartographic consistency became a prominent part of imperial policy, eventually inspiring the emperor to explore personally various scientific practices that might help him manage his rule and increase his control over the different peoples and vast territories comprising the Qing. In short, cartography soon became a potent tool of statecraft.

Immediately following the conquest of the Chinese provinces, the Kangxi emperor ordered maps for the *Descriptive Account of the Unified Great Qing* (Chinese: *Da Qing yitongzhi* 大清一統志 / Manchu: *Daicing gurun i uherileme ejehe bithe*), a comprehensive gazetteer of all Qing territories completed only in 1746. In 1683, the emperor dispatched a team to map the course of the Yellow River and, in 1684, had the Manchu homelands up to the Amur River surveyed by another team in order to establish new horse relay stations in the face of Russian expansion in the region.[79] By 1685, each of the provinces had submitted a series of maps to the palace towards the production of a first version of the *Descriptive Account*. However, as evidenced by the quote included at the beginning of this chapter, the emperor decried the maps' fundamental inconsistencies. Some provinces only provided general maps, whereas others, such as Guangdong, also submitted prefectural maps. Moreover, only some administrators had recorded road distances on their maps, and had presumably done so using a practice dating back to Han times, with graduated ropes, afterwards accounting for uneven terrain and bends in the road via an ancient method of geometric calculations known in Chinese as *gougu* 句股.[80] The imperial response to these maps, reflecting the accumulated knowledge of inherited cartographic practice, included ordering in 1692 the editing and redrawing of the maps from the first version of the *Descriptive Account* for greater cartographic consistency.[81] At the same time, the emperor pushed for better maps of areas beyond

77 The map is dated 1681 and appeared without reference as part of a blog post: http://jones1635.blog.sohu.com/239890748.html (May 8th, 2015).

78 Roque di Oliveira (2006).

79 Cheng (2009), 55; Ha-si-ba-gen (2010), 94; Cheng (2015), 11.

80 Yee (1994), 110–113; Feng (1998); Han (2016).

81 Liu (2012), 47.

the Chinese provinces, including the Manchu homelands and the Mongolian territories known today as the province of Inner Mongolia.

The emperor's cartographic concerns are apparent already in the two tours of areas north of the Great Wall he undertook in the years 1682 and 1683. Historians see these tours, following decades focused on conquering the Chinese provinces, as an effort to refocus his attention on the Manchu homelands, showcasing his authority over them. On both occasions, he ordered his trusted tutor and one of the Directorate of Astronomy's head astronomers, the Flemish Jesuit Ferdinand Verbiest, to travel along in the imperial entourage.[82] This was the first time a missionary became involved in Qing cartographic field practice. Known first and foremost as the successor to Adam Schall von Bell, Verbiest had tutored the emperor during the 1670s, teaching him, among other subjects, practical geometry and world geography.[83] In his capacity as imperial astronomer, Verbiest had also constructed instruments, principally large stationary specimens for the imperial observatory but also such smaller devices as astrolabes and quadrants, at the emperor's behest, such that the emperor saw Verbiest as an expert in both instrument-making and the geometric calculations involved in their correct use.[84] The emperor thus asked Verbiest to put his teachings into practice by conducting basic observations during these imperial outings. From March 23rd to June 9th, 1682, Verbiest accompanied the emperor on his journey to the other Qing capital of Mukden (Shengjing 盛京 or Fengtian 奉天) and on to Kirin (Jilin 吉林), determining latitudes—using an unidentified small instrument to observe the height of the pole star, probably a sundial or small quadrant—for many of the places the imperial party visited.[85] Later, from July 6th to September 15th, 1683, Verbiest joined the emperor on a hunting trip to the Mulan 木蘭 imperial hunting

82 Verbiest's accounts of the journeys appear edited in du Halde (1735), 4:74–86. See also Golvers (2016).

83 As a result of his tutoring sessions in geography, Verbiest produced an edited version of the Ricci map, showcasing the two hemispheres and entitled *Kunyu quantu* 坤輿全圖. A treatise on world geography accompanied the map. Both exist in several versions. Lin (1994).

84 Halsberghe (1992); Feng (2002), 450–456.

85 Pang (2003); Golvers (2016). Small maps made on the basis of information included in Verbiest's letters to Europe appear in the upper-right corner of two larger maps: the *Carte Nouvelle de la Grande Tartarie* by Nicolas Witsen (Amsterdam, c. 1705) and *L'Asie: Selon les nouvelles observations de Mess.rs de l'Académie des Sciences* by Pieter van der Aa (Amsterdam, s.d.). Mukden, today called Shenyang 沈陽, was the Manchu capital from 1625 until the fall of Beijing in 1644.

grounds, during which he mostly occupied himself calculating the heights of mountains along the route and the distances between them.[86]

In an extant letter, Verbiest confirms that he joined the Kangxi emperor on two tours north of the Great Wall, in what he calls "East and West Tartary," and that the emperor asked him to draw up both a route map and a topographical map.[87] In 1683, the topography along the route appears to have caused difficulties for surveying—Verbiest explains that he often could not see the site where he had conducted observations on the previous day, complicating the calculation of distances based on trigonometry. He asks his correspondent, identified by Noël Golvers as Antoine Thomas, to devise a mathematical formula to address the issue. Thus, the emperor considered these land surveys a rather serious matter, perhaps directly in connection with the compilation of the *Descriptive Account*. It seems the emperor meant to test the practices of Western learning as a practical tool for exploring the land and exerting his authority over it.[88] Later on in the 1680s (according to a 1703 narrative by Antoine Thomas), the emperor asked Verbiest to make new maps of all Qing territories as he had for Tartary, starting with the Chinese provinces and indicating latitudes and longitudes.[89] If Verbiest indeed received this request, which remains doubtful, it would seem that the emperor saw European-style land surveying techniques as a possible solution to the problem of cartographic inconsistency. The sudden death of Verbiest, the emperor's personal expert in practical geometry and instrument operation, in 1688, just weeks before the arrival of the French Jesuits, however, would have necessitated the shelving of any such plans for the time being. While Verbiest is the earliest known missionary to attempt land surveying at the behest of an emperor concerned with improving cartographic practice, it is clear that a keen interest in mapping certain parts of Qing territories existed well before the arrival of the French Jesuits in 1688.

Meanwhile, in the years following the Qing conquest of the Chinese provinces, Russian expansion into Siberia along the Amur River had led to clashes between Qing and Russian troops at the Russian fortification of Albazin. Qing troops had attacked the fort a first time back in 1652, after Albazin's founding,

86 du Halde (1735), 4:74; Bernard (1935), 453; Jami (2014), 63; Golvers (2016). As well as determining latitudes and the heights of mountains, Verbiest also calculated the widths of rivers along the way.

87 Golvers (2000).

88 On the role of imperial touring in the construction of Qing rule, see Chang (2007).

89 Bosmans (1926), 160.

but soon after decided to withdraw. They attacked again in 1658, this time killing the Russian commander and obliterating the fortress. After the conquest of the Chinese provinces, in 1685, Qing troops destroyed the rebuilt Albazin fortress a second time (the Russians then rebuilt the fortress yet again). In 1686, Qing bannermen decided to change strategies and laid siege to the fortress, after which the Russians sent an ambassador to negotiate a treaty.[90] Developments in Qing cartographic practice during the latter half of the 1680s focused closely on this Amur River frontier with Muscovy, which the Qing court saw as a major threat. This helps to explain why the emperor took his chances with a new generation of missionaries who not only worked for him within the walls of the imperial palace but also corresponded with and participated in the emerging European scientific community.

Two of the primary European contributors to Qing cartography were French Jesuit Jean-François Gerbillon and Antoine Thomas, in many ways Verbiest's successor. Thomas was born in Namur (in modern Belgium) and entered the Society of Jesus in Tournai (Doornik). The Society selected him for the Far East in 1678, but he spent all of 1679 teaching mathematics in Coimbra, during which time he composed his *Synopsis Mathematica* (1685), a mathematics textbook including sections on geography and practical geometry. In 1680, he boarded a ship for Goa and two years later, following short stopovers in Malacca and Siam, arrived in Macao, where he conducted a number of astronomical observations.[91] During Thomas' stay in Macao, Verbiest contacted him about the mathematical problems encountered drawing a route map in the emperor's entourage. In the summer of 1685, finally, Verbiest called Thomas to the capital, where Thomas immediately started working under him at the Directorate of Astronomy. When the Kangxi emperor resumed his study of mathematics in the early 1690s following Verbiest's death, Thomas was an obvious choice and tutored the emperor on a range of mathematical subjects. Memorials relating to Thomas' deathbed, when the emperor personally ensured Thomas got the best medicine available at court, confirms the personal connection Thomas forged with the emperor.[92]

90 The Eight Banners constituted the main military-social structure or organization among the Manchus of the Qing. Elliott (2001).

91 Bosmans (1924–5).

92 Puente-Ballesteros (2012). These "memorials" were official documents between high-ranking officials and the imperial throne, which we will consider more deeply in Chapter 2.

Perhaps in spite of working under the patronage of the Portuguese crown, Thomas was part of a new and emerging network of missionaries who enjoyed contacts with the academicians in Paris.[93] At the same time, as discussed above, the other main actor with regard to cartographic practice after Verbiest, Jean-François Gerbillon, was one of the newly arrived French Jesuits and an appointed king's mathematician, a direct correspondent of the *Académie*. With their more structural involvement in the sciences, both men would come to shape the development of imperial cartographic practice under the Kangxi emperor. In the 1690s, taking his Jesuit tutors along on tour and on campaign became the emperor's custom, not only so that they could perform observations and calculations but also as sources of information on topics of all kinds, from the customs and geography of different European countries to meteorological phenomena.[94] For the purpose of this study, however, I will discuss only those instances concerning cartographic practice.

1.2.2 *Qing Statecraft and Cartographic Practice*

The root of Gerbillon's involvement in Qing cartographic practice lies in events at court soon after his arrival. Together with Tomás Pereira, Gerbillon twice joined entourages of Qing envoys to negotiate a peace treaty with Russia.[95] In May 1688, a team of sixty to seventy officials set out for the first time for Selinginsk, a town south of Lake Baikal, with the aim of meeting the Russian ambassadors sent to negotiate a treaty at the border. In July, however, after nearly four months of exhausting travel through the Ordos region and into the Mongolian deserts, an attack by the Zunghar Mongols on the Khalka Mongols (themselves in conflict with the Russians) forced them to turn back; since the 1650s, the Russians had established several forts along the Amur and Selenga rivers, just north of the lands controlled by, respectively, the Qing and the Khalka. With both the Zunghar and the Russians threatening the Qing frontier, historians believe Kangxi wished to prevent the two forging an alliance. The Russians, for their part, were also eager to negotiate a peace treaty and a common border to establish permanent trade, but continuous attacks by the Zunghar on the Khalka, whose lands lay between the Qing and Selinginsk, meant both parties were forced to call off the meeting.

In 1689 the mission set out again, this time for the Russian fortress of Nerchinsk (Nibcu in Manchu), located on the middle reaches of the Amur

93 Golvers (2014), 139.

94 Antonucci (2012), esp. 117.

95 Gerbillon's accounts of both journeys appear edited in du Halde (1735), 4:87–251.

FIGURE 13 *Sketch Map of East Asia c. 1690.*

River (so the Qing envoys could bypass Khalka territories on their way north). As in the 1688 attempt, the envoys took a number of Paris-made instruments along for observational purposes. Gerbillon in his edited diaries mentions that he had three instruments at his disposal in Nerchinsk: two mobile quadrants, the largest 32.5 cm in diameter, and one semi-circle.[96] Negotiations occurred in Latin, shaped by the intermediaries (the two Jesuits on the Qing side and a Polish translator on the Russian), although at some point during the complex and unprecedented negotiations it seems that some negotiators switched to Mongolian in an unsuccessful attempt to circumvent Jesuit mediation. The Jesuits, for their part, showcased their ability to calculate and communicate to the Russians, on behalf of the Qing emperor, the positions of a number of

96 du Halde (1735), 4:172–186.

settlements. At a crucial point during the negotiations the two sides exchanged maps (the precise maps involved remain unidentified), an inportant step towards the final agreement.[97] As several scholars have noted, the Jesuits were personally invested in these negotiations, as the possibility of a permanent overland trade route between Moscow and Beijing raised hopes of a more secure line of communication with Europe, circumventing the treacherous sea voyage between Western Europe and Southern China.[98]

While Tomás Pereira's involvement in the embassy makes ready sense (being one of the emperor's most trusted missionaries), the emperor's selection of Gerbillon is perhaps less straightforward.[99] The French Jesuit had arrived at court within only the last year and could not have acquired sufficient language skills in either Chinese or Manchu. It seems likely that Gerbillon's skills in certain observational techniques must have been key to his membership in the Qing negotiating team. Gerbillon recorded almost daily the latitudes and distances traveled on the way to Nerchinsk. Once there, his repeated observations of the height of the sun at noon and of the pole star at night allowed him to determine, as accurately as possible, Nerchinsk's latitude. At the same time, he gathered data (presumably obtained from his Russian counterparts) about Selinginsk and Albazin, the Russian fortress a few days downstream. Before departing Nerchinsk, the Jesuits dispatched—with the help of the Russians—some of these data directly overland to Paris for publication as early as 1692.[100] Gerbillon may also have produced a map of the area, which I have not been able to trace.

Another map incorporating information gathered during the Nerchinsk negotiations—drawn by Antoine Thomas—reached Rome as early as 1692 and survives today in the Archivum Romanum Societatis Iesu.[101] Covering all of Eurasia and dated 1690, it clearly features the three Russian settlements of

97 Perdue (2010).

98 Sebes (1962). For an in-depth analysis of the search for overland routes, see Roque de Oliveira (2012).

99 In one of his letters, Gerbillon mentions the choice of Pereira over Thomas due to his superior language skills. de Thomaz de Bossierre (1994), 41.

100 de Thomaz de Bossierre (1994), 40. The atlases discussed in Chapter 3 also include all three of these places, although not others Gerbillon might plausibly have located along the route to Nerchinsk.

101 Archivum Romanum Societatis Iesu (hereafter ARSI), 105:7. The map is almost certainly an edited version of a map by Verbiest, itself a rendition of a route map given to him in 1676 by a Russian envoy to the Qing. Roque de Oliveira (2012), 494–498. Like Verbiest, Thomas actively sought any information on overland routes from Saint Petersburg to Beijing controlled by Russia. One Manchu memorial, dated 1702 and only available in

Nerchinsk, Selenginsk, and Albazin.[102] On the map, Thomas delineates the Qing, Russian, Khalka ("Tartaria Halha") and Zunghar ("Tartaria Eluth," from "Ölöt") territories established in the treaty, as well as indicating a neutral "region which in the peace treaty with Muscovy was undecided."[103] Nerchinsk occupies a central position on one of the sheets comprising the larger map, labeled "place where the peace of Nerchinsk was reached."[104] This map's drawing therefore must reflect at least minimal input from either Gerbillon or Pereira. However, as a king's mathematician, trained to adhere to the *Académie*'s exacting standards, Gerbillon explicitly distanced himself from Thomas so far as concerned mapmaking:

> One should not rely on the maps, charts, observations and memoirs of F. Thomas, unless it is about places where he himself has been to, and about things he has seen; because he is too credulous in these matters, and easily draws maps on less certain information. He made one of Tartary based on information from some Russians and Chinese, which I have found to be so wrong, that there was not even one river marked in its right place, and that the Great Wall itself, and the land that surrounds it, is marked in an entirely different way from how it actually is.[105]

On January 24th, 1691, more than a year after the conclusion of negotiations, the emperor asked Gerbillon specific questions about positions along the Russian frontier, including Nerchinsk. In the passage from Gerbillon's diary

---

Chinese translation, reports Thomas had unsuccessfully attempted to obtain maps from the Russian mission in Beijing. An (2014), 282.

102 For the full story behind this map, see Roque de Oliveira (2012); Lo Sardo (2003). Other sheets comprising Thomas' map are of the Chinese provinces, the overland route to Moscow, and routes to Persia and India. They reside at ARSI, JS 105:5, 98 & 227; JS 110. Based on Lo Sardo's scholarship, it appears that two very similar maps by Thomas are in the library of the Archivio di Stato di Roma (ms.493). Antonucci (2008), n.19.

103 "Regio in Tractatu Pacis cum Moschis indecisa," in the upper right corner of the map.

104 "Locus conclusae pacis Nipchu."

105 "il ne faut pas compter sur les cartes, plans et observations et relations du P. Thomas, à moins que ce ne soit de ces lieux, où il a été lui-même, et des choses qu'il a vû; car il est assez crédule en cette matière; et fait facilement une Carte sur des rapports peu certains. Il en avoit fait une de la Tartarie sur le rapport de quelques Moscovites et Chinois, que j'ai trouvé tellement fausse, qu'il n'y a pas même une rivière marquée en sa place, et que la grande muraille même, et le paїs, qu'elle environne, y est marquée d'une toute autre manière, qu'elle n'est en effet." de Thomaz de Bossierre (1994), 39–40.

recording this event, it becomes clear that the emperor had launched further expeditions into the far north:

> The emperor asked me about the height of the pole in Nerchinsk, and of the other principal places in Tartary that I had traveled through in my two journeys. He told me on this occasion that he had sent people to the eastern part this year, towards the estuary of the Sahaliyan [Amur] River; and that they had reported that beyond the estuary, they had found the sea still frozen in the month of July; and that the land there is entirely empty.[106]

This is another indication that the emperor must have sent Gerbillon along to determine geographic positions on the new frontier, replacing, as it were, the late Verbiest, the emperor's personal expert in the techniques of land surveying as taught by Western learning, who had passed away just months before the first team's departure for Selinginsk. It also reflects the emperor's continued attention to the practical use of land surveying, this time in the shadow of peace negotiations with the Russians. On the other hand, by informing the emperor of strategic frontier locations in terms of latitudes and distances from the capital, the Nerchinsk negotiations appear to have spurred further mapmaking in the region, as mentioned in the excerpt above: in 1690, Manchu generals Sabsu 薩布素 and Bahai 巴海 divided their troops into nine groups, each of which followed a separate route through the Amur River borderlands. A map based on accounts and draft maps brought back from this expedition later appeared in an updated version of the *Descriptive Account of the Unified Great Qing*, covering the lower reaches of the Amur River, as well as its tributaries and estuary, plus the Lena and Anggara rivers, together covering the entire Qing-Russian frontier.[107] It is clear, however, that new techniques of land surveying as taught by the missionaries did not enter into the making of this map. We can therefore conclude that, in the late 1680s and early 1690s, the emperor upheld two different cartographic practices in his explorations of the new frontier with

106 "l'Empereur me demanda la hauteur du Pôle de Niptchou, & des autres principaux lieux de la Tartarie que j'avois parcourus dans mes deux voyages. Il me dit à cette occasion, qu'il avoit envoyé cette année des gens du côté de l'Orient vers l'embouchûre du fleuve Saghalien oula & qu'ils avoient rapporté qu'au-delà de cette embouchure, ils avoient trouvé la mer encore glacée au mois de Juillet, & que le pays y étoit tout-à-fait désert." du Halde (1735), 4:244. These further expeditions are confirmed in Cheng (2015), 11.

107 Fuchs (1933), 8–12; Cheng (2011); Cheng (2009), 151–158.

Russia. On the one hand, he asked his Jesuit experts to determine the position of Nerchinsk and to trace the route from Beijing in the face of conflict with the Russians, while on the other upholding inherited practices in the areas now demarcated in an international treaty as constituting Qing territory.

Soon after the conclusion of the Nerchinsk negotiations, in the early months of 1690, the emperor called on the missionaries to initiate a new round of tutoring, to provide him further instruction in practical geometry. Gerbillon and Bouvet, both of the French mission, formed one group, teaching the emperor with Manchu materials, while Thomas and Pereira, of the mission supported by the Portuguese, taught the emperor in Chinese. Each team was to compile a written version of the subject matter, in close cooperation with inner palace personnel, in preparation for these tutoring sessions. As Jami has noted, there "was a convergence between the scholastic *quadrivium* (i.e. arithmetic, geometry, music, and astronomy) on which Jesuit education in the mathematical sciences had been constructed, and the cluster of topics on which the court Jesuits wrote most extensively for the emperor."[108] As practical geometry and trigonometry were among these tutoring sessions' key subject matters, the lessons and the textual companions prepared in advance of them naturally covered the basic methods of geometric calculation necessary for land surveying.[109] The focus of the "French" team of Bouvet and Gerbillon centered on both practical geometry and familiarizing the emperor with the monarch they depicted as his European counterpart, Louis XIV. The French Jesuits informed the emperor, who for his part was apparently curious about the French king, of Louis' patronage of the sciences through the establishment of the Parisian *Académie*. This strategy of association, they believed, would facilitate the emperor's embracing the French model of supporting the sciences and, by extension, his conversion to Christianity.[110] Thomas and Pereira for their part appear also to have produced a number of tables and textbooks directly relevant to practical land surveying, such as the *Practice of Instruments for Measuring Heights and Distances* (*Celiang gaoyuan yiqi yongfa* 測量高遠儀器用法) and the *Tables of Distance between the Horizon and the Terrestrial Sphere* (*Dipingxian li diqiu mianbiao* 地平線離地球面表), both produced as aids to the practical surveyor.[111]

108 Jami (2010), 384.
109 Jami (2012), 139–159, esp. 142–143.
110 Jami (2014), 58.
111 Jami (2012), 191–195.

Jami, having studying these tutoring sessions closely, notes that lessons typically included practical exercises, supplying the emperor's wish to master the many practical uses of Western learning as taught by the missionaries.[112] This included conducting observations with some of the instruments the emperor kept in his apartments and performing the calculations involved in processing the resulting data. This attention to scientific practice through the use of precision instruments serves to remind us that surveying, including the techniques that permit mapmaking in the field, interrelated deeply with mathematics and astronomy. After all, geometric calculations enabled one to determine one's position on the earth following observations of the heavenly bodies. The development of Qing cartographic practice during the latter half of the seventeenth century therefore correlated strongly with the development of imperial mathematics. Land surveying was one of the scientific practices that followed, and in part motivated, the Kangxi emperor's sincere interest in Western learning.

Another practice closely related to land surveying, and for which we know the emperor to have commissioned his court missionaries as technical experts, was the drawing of river charts to facilitate effective flood-control measures. Yee has argued that water conservancy projects had been "stimuli for techniques of direct measurement," since Han times at the latest.[113] No wonder, then, that the emperor decided to employ his experts in practical surveying for such tasks. On April 26th, 1698, the governor of Zhili province, bannerman Yu Chenglong 于成龍 (1638–1700), who had charted the river courses in the area between Bazhou 霸州, Lilan 里瀾, and Yongqing 永清 just south of Beijing, with the assistance of Antoine Thomas, presented a chart to the emperor.[114] In the winter of 1700–1701, it appears, Thomas, Bouvet, Dominique Parrenin (1665–1741), and Jean-Baptiste Régis (1663–1738) produced another map of a district some forty-five kilometers to the south of Beijing in Zhili province.[115] Around 1705, finally, in what seems to have been an even larger project, the

112 Peng (1975), 424; Jami (2012), 149.

113 Yee (1994), 98.

114 Concentrated around today's Zhongting 中亭 and Yongding 永定 rivers. *Qingdai qijuzhu ce* (2009), 11:T06299-300; Fu (1966), 110. At the occasion, the Kangxi emperor decided to call the river, previously known as Hun he 渾河, by the name "Yongding he." Witek (2003), 102. Note that Thomas also had two "hydraulic machines" built for use in the imperial gardens at Changchunyuan 暢春園. Golvers (2014), 142.

115 To my knowledge, this is mentioned only in correspondence between Bouvet and Gottfried Wilhelm Leibniz (1646–1716). Widmaier (1990), 159. It is possible that Bouvet or the copyist of the letter made a mistake in the dating and in fact referred to the 1700–1701 survey.

emperor again asked several missionaries to join a team of experts, ranging from inner palace personnel to officials from the Directorate of Astronomy to artisans from the Imperial Workshops, to draw charts of the two rivers converging just to the west of Tianjin, a city plagued by severe inundations. This time, Thomas, Gerbillon, Bouvet, Parrenin, and Régis were all on the team and assisted in executing a chart of the two rivers. They completed their task, "perhaps the greatest one [compared to what] one has seen in Europe," within seventy days, proselytizing along the way.[116] The resulting chart formed the basis for the construction of levies and new gullies to divert excess water in the region.[117]

1.2.3 *The Qing Court's Appropriation of Paris-Made Instruments*

I have argued thus far that the Kangxi emperor, during the late 1680s and early 1690s, applied himself to the study of European-style mathematics and, together with Verbiest, took the first steps towards incorporating its practical techniques into surveying practices at the court and along the Qing's northern frontiers. Later, the emperor dispatched more of his missionary experts to map the frontier with Russia and to chart river courses near the capital, activities requiring the use of precision instruments tailored to land surveying, escalating the demand for more and better surveying instruments so that it soon outstripped the supply. To meet this need, the emperor both allowed missionaries to travel to Europe, whence they often returned bearing new instruments, both large and small, as well as sending trusted personnel of the inner palace to Guangzhou in a secret search for more instruments and other products from Europe.[118] At the same time, the palace's workshops began manufacturing their own surveying instruments and timepieces based on prototypes made in Paris. Court production of instruments, however, in truth significantly antedated the French missionaries' arrival. Verbiest's so-called Constantinople Manuscript, for example, published and annotated by Golvers and Nicolaidis, records the following passage:

> [The emperor] kept sending, during almost a full year, all kinds of craftsmen from the court to our residence, and saw to it that all which belonged to instrumental mathematics was skillfully constructed, after the

116 Recorded in a letter by Gerbillon, edited and published in du Halde et al. (1702–76), 10:413–415; and in de Thomaz de Bossierre (1994), 135–136.

117 de Thomaz de Bossierre (1994), 135–136 & 148.

118 Several Qing and European sources, including du Halde (1735), 4:241–242; *Acta Pekinensia* (2013), 36; Witek (1973), 528; *Kangxi chao manwen zhupi zouzhe quanyi* (1996), 476; etc. confirm this.

> model of those instruments which we kept in our *Musaeum*, for example: all kinds of quadrants, geometrical quadrants, pantometers, compasses, proportional rulers, etc., the array which I have drawn (here) in my Instrument Book. The Emperor frequently practices operating these and similar instruments, and he was not ashamed to accustom his right hand, which was used to bear the scepter of such a great empire, to the ruler and to the compass.[119]

The emperor's interest in European-style instruments continued to manifest after Verbiest's death, aided by the arrival of the French missionaries. Precisely one month after Gerbillon and Pereira's return from Nerchinsk, for instance, in the fall of 1689, the emperor called several missionaries to the inner palace's *Yangxindian* 養心殿, one of the large compounds that make up the imperial palace. Gerbillon's published diaries record the following anecdote for November 15th:

> We went to the palace, Fathers Pereira, Thomas and me, in the order that they had assigned us. We were led into one of the halls of the emperor, called *Yangxindian*, in which some of the most able artisans, painters, wood turners, silversmiths, coppersmiths, etc. work. There, we were shown mathematical instruments, which His Majesty had placed in boxes or cases, very nicely [executed] and specially made. There were no major instruments: the entirety consisted of some proportional compasses, almost all imperfect, several ordinary rulers, small and large and of different kinds, some squares and geometrical rulers, a divided circle of about one *pied* in diameter with its alidades… It was all rather roughly executed and quite removed from the neatness and precision with which the instruments we brought had been made.[120]

---

119 Translation from Latin by Noël Golvers. Golvers & Nicolaidis (2009), 208–211.

120 "nous allâmes au Palais les Pères Péréira, Thomas, & moi, selon l'ordre que l'on nous en avoit donné. Nous fûmes introduits dans l'un des l'appartemens de l'Empereur, nommé Yang sin tien, dans lequel travaillent une partie des plus habiles ouvriers, Peintres, Tourneurs, Orfèvres, ouvriers en cuivre, &c. Là on nous fit voiries instrumens de Mathématique, que Sa Majesté avoit fait lacer dans des boëtes ou layettes de carton assez propres, & faites exprès. Il n'y avoit pas d'instrumens fort considérables: tout consistoit en quelques compas de proportion, presque tous imparfaits, plusieurs compas ordinaires grands & petits de plusieurs sortes, quelques équerres & régles géométriques, un cercle divisé d'environ un demi pied de diamètre avec ses pinnules. Le tout fait assez grossièrement, & bien éloigné de la propreté, & de la justesse dont étoient faits les instrumens que nous avons apportez." du Halde (1735), 4:218.

On this occasion, the three missionaries were to explain clearly the uses of these instruments to the workers and artisans of the Imperial Workshops and the next day, to bring the instruments they kept at their residencies. Instruction in using both older and newer instruments went on for several days, Gerbillon explaining, in Manchu, the precise use of some of the instruments not only to the workers, but also to the emperor himself. Gerbillon states that he felt comfortable enough with the language to do this, but did not yet dare to teach mathematics in Manchu.[121] Manchu was the emperor's first language, through which he preferred to familiarize himself with the instruments, which included several state-of-the-art Paris-made specimens. In the somewhat more informal environment of the inner palace, the emperor thus personally engaged with the material and with the scientific practices they prescribed, undoubtedly learning from the French Jesuits about the superiority of Paris-made instruments compared to those gifted and produced by Verbiest and others. Gerbillon again confirms the emperor manipulating instruments himself and hints at missionary-assisted production of instrument parts at the *Yangxindian* later in his diaries. For March 23rd 1690, he recorded the following:

> The emperor, having again come to the *Yangxindian* apartment, tested together with us the divided circle of one *pied* in diameter, which he had ordered made during his absence in order to measure heights and medium-range distances: this circle also had a divided geometric square within it, so that one would not need to turn to [calculations involving] the sine for resolving triangles. Next, he tested in the courtyard of this hall a large semi-circle, which the late Father Verbiest had made, and which His Majesty had mended and placed on a good footing, made after [the footing of] the semi-circle that I have presented to his Majesty. He computed his operations on the spot on his *suanpan* [abacus], so fast that Father Thomas needed more time than him to calculate with our numbers.[122]

121 du Halde (1735), 4:218–219.

122 "L'Empereur étant encore venu à l'appartement d'Yang sin tien, éprouva avec nous un cercle divisé d'un pied de diamètre, qu'il avoit fait faire pendant son absence pour mesurer des hauteurs & des distances médiocres: ce cerclé avoit aussi un quarré géométrique divisé en dedans, afin de n'avoir pas besoin de recourir aux sinus pour résoudre les triangles. Il éprouva ensuite dans la cour de cet appartement un grand demi cercle, que le feu Père Verbiest avoit autrefois fait faire, & que Sa Majesté avoit fait raccommoder depuis peu & mettre sur un bon genou, fait à l'imitation de celui du demi cercle que j'avois présenté à Sa Majesté, & il supputa sur le champ cette opération fur son Souan pan si vîte, que le Père Thomas fut plus long que lui à le supputer par nos chiffres." du Halde (1735), 4:227–228; Jami (2012), 191–195.

Over the next three decades, as more French Jesuit missionaries came to the court, instrument production in the Imperial Workshops appears to have increased exponentially, aided by their expert knowledge. In 1691, a number of workshops moved to an area of the imperial palace situated just north of the *Wuyingdian* 武英殿, another hall of the inner palace, as the *Yangxindian* could not longer accomodate its many functions.[123] Moreover, the residence of the French mission, the Northern Church (*Beitang* 北堂), funded by the emperor as a gift to the French Jesuits in 1693 and located near the western entrances of the palace and workshops, soon included its own workshop for glassworks.[124] Around the same time, production of clocks and instruments within the *Yangxindian* became established on a more permanent and systematic basis. Bouvet discusses this in his book on the Kangxi emperor, finished during his stay in Europe and published in 1697:

> The Emperor of China began, about five years ago, to erect within his own palace a sort of academy of painters, engravers, sculptors and craftsmen in steel and copper for timepieces and other mathematical instruments. To stimulate them in their emulations, he most often proposes [to take] European pieces as a model, and among them, those that have been made in Paris… regularly, when he is in Beijing every day at a certain hour, or once every two days when he is at some of his houses of leisure, he has those [pieces] that come from the hands of his new academicians brought to him. He examines them himself, criticizes for each one that which is imperfect, approves that which merits praise, and keeps those of which he finds nothing to criticize and that surpass the ordinary.[125]

---

123 *Da Qing huidian shili* (1899), *juan* 1173; Zhang (2009), 1:261–302.

124 Pagani (1995); Curtis (2009).

125 "l'Empereur de la Chine commença, il y a environ cinq ans, d'ériger dans son propre Palais une espece d'Académie de Peintres, de Graveurs, de Sculpteurs & d'Ouvriers en acier & cuivre pour les Horloges & autres instrumens de Mathématiques. Pour piquer leur émulation, il leur propose le plus souvent pour modéles des Ouvrages d'Europe, & entr'autres de ceux qui ont été faits à Paris… il se fait apporter regulierement tous les jours à une certaine heure, lors qu'il est à Pé-king; ou de deux jours l'un, lors qu'il est dans quelqu'unes de ses Maisons de plaisance, ceux qui sortent des mains de ces nouveaux Académiciens. Il les examine luy-même; il reprend dans chacun ce qu'il y a de défectueux; il approuve ce qui merite de la loüange: il retient ceux, où il ne trouve rien à redire & qui passent l'ordinaire." Bouvet (1697), 198–200. This is confirmed in an account by Jean-François Foucquet (1665–1741). Witek (1973), 464–465.

Of course, the intention behind Bouvet's words was to provide European readers with a fascinating account of the Qing court, but these claims surely had some real basis. Published catalogues show many entries for instruments and timepieces produced at the Imperial Workshops during the late Kangxi period. In fact, the Palace Museum in Beijing keeps an impressive collection of instruments, consisting of European prototypes, several of them produced at the workshops of Butterfield, Chapotot, and Bion, alongside similar instruments produced by the Imperial Workshops.[126]

With regard to cartographic practice, instruments used in land surveying included compasses, timepieces, telescopes, and drawing tools, such as sectors and rulers. It is difficult to trace many of these smaller instruments because contemporary written accounts often make no mention of them. The explanation is simple: contemporary travelers, missionaries, and scientists considered such smaller tools so essential that their presence was a given. After all, most of them had broad ranges of applications, not limited to land surveying. Drawing tools, for example, often neatly tucked together in small cases, were essential for anyone practicing mathematics (in France, the technical term for such implements was actually "*nécessaire(s) de mathématiques*," or "necessities for mathematics").[127] Even though they constitute an excellent example of just how diligently the Imperial Workshops recreated instruments and tools made in Paris (figures 14 & 15),[128] we cannot trace them with regard to field surveying and cartographic practice. The same holds for sundials, astrolabes and compasses.[129]

Besides these smaller pieces, the instruments perhaps most tailored to land surveying, although by no means exclusive to that purpose, were the mobile surveyor's quadrant and semi-circle (*demi cercle mobile*). Certainly, these would have been the largest instruments carried about by any mapmaker of the time. As discussed above, the Parisian workshops of Butterfield and Chapotot were among their first renowned producers. At the Qing court,

126 Zhu (2002), 476–486; *Qinggong Xiyang yiqi* (2011), 115–165; Feng (2011), 76–77 & 145–152.

127 For several examples kept at the Palace Museum in Beijing, see *Qinggong Xiyang yiqi* (2011), 149–158. For proportional compasses, rulers and squares, see *Qinggong Xiyang yiqi* (2011), 64–91. Drawing tools of European origin also still survive at the National Palace Museum in Taipei and the Palace Museum in Beijing today, see Feng (2011), 77; *Qinggong Xiyang yiqi* (2011), 155–156; Guo (2015).

128 *Qinggong Xiyang yiqi* (2011), 152. Images of a similar set appear in Zhu (2002), 487. Both feature wooden casings.

129 A gilded brass semicircular sundial with compass produced at the Imperial Workshops and dated 1701 appears in Feng (2011), 150. For a stardial produced in 1681, see Feng (2011), 148–149. For stardials at the Qing court, see Feng (2002).

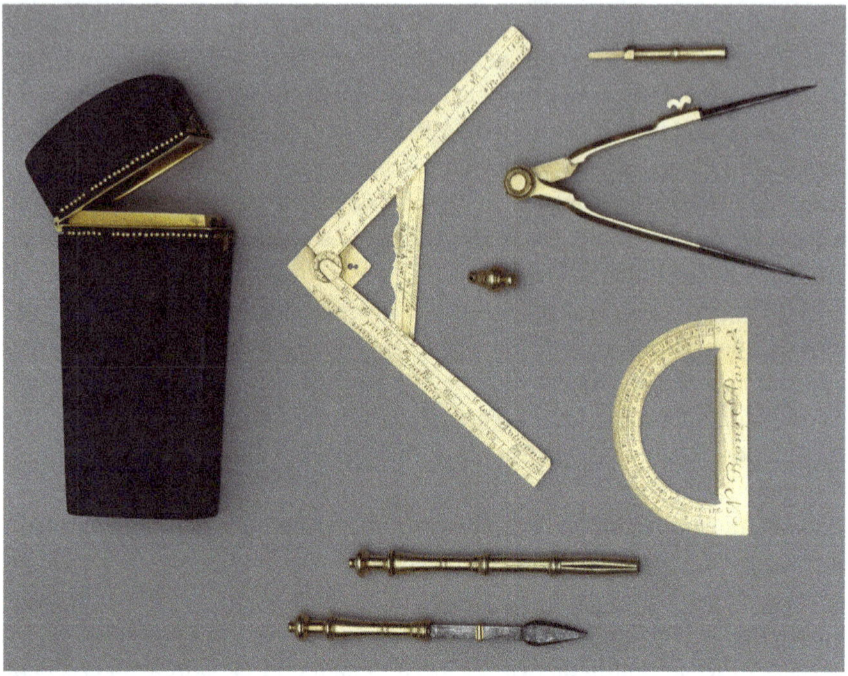

FIGURE 14 "Nécessaire de mathématiques" *with leather case, manufactured at the Bion workshop.*
LOUVRE MUSEUM.

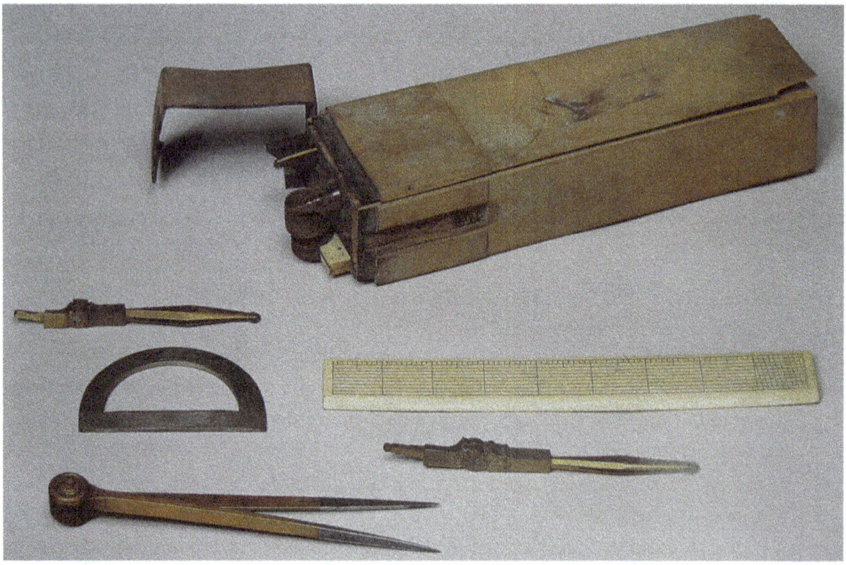

FIGURE 15 *A similar set of drawing tools produced by the Qing Imperial Workshops.*
PALACE MUSEUM.

these Picard-style mobile instruments also featured among those used as prototypes for emulation. One Palace Museum catalogue features several mobile quadrants and semi-circles produced by the Imperial Workshops and dating back to the Kangxi years (figure 16).[130] Some of these larger instruments rest on footings designed to facilitate both easy transport and vertical and horizontal angular measurements, precisely the characteristics Picard had prescribed in his treatise. Even so, one of the main issues we must address regarding these reproductions is their precision. After all, if surveying teams were to use both the Paris-made prototypes and their Qing adaptations side-by-side, both their technical production processes and their calibrations would have to accord. Unfortunately, due to lack of access to the instruments themselves or to full and detailed technical descriptions, it is not presently possible to compare the instruments meaningfully. What we can determine, however, is that telescopes, often featured on the Paris-made instruments, are absent from their Kangxi-era Qing reproductions still extant, presumably due to the technical

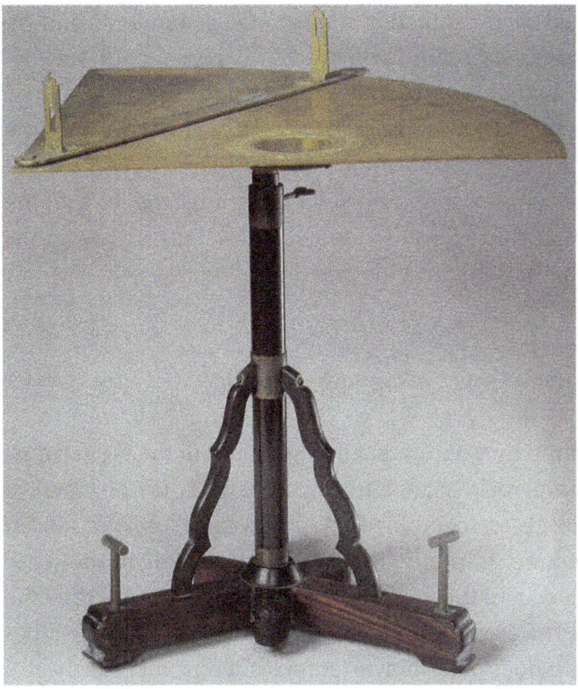

FIGURE 16
"Gilt-copper quadrant with the mark 'Kangxi yuzhi' (made by emperor Kangxi's order)," featuring alidades and built-in compass. Produced by the Qing Imperial Workshops.
PALACE MUSEUM.

130 *Qinggong Xiyang yiqi* (2011), 115–118 & 126–127. Guo (2015).

difficulty of their production. Thus, early Qing adaptations of larger instruments have alidades instead.

Overall, it seems clear that Paris-made instruments and the techniques of land surveying they served became an opportunity to fulfill the emperor's long-standing wish to improve imperial cartography. Given the impact of the French mission on the number of Jesuits working at court and the more specific and specialized knowledge they brought to the table, many scholars have called this a watershed relative to previous missions.[131] With regard to cartographic practice, however, we must also not exaggerate this watershed, as the French Jesuits' participation in Qing cartography largely built—and directly depended—upon the efforts of earlier missionaries like Verbiest and Thomas. Even so, with the successful conclusion of the Nerchinsk treaty, the Kangxi emperor turned his focus to stabilizing the frontier with Russia, inspiring him to push for new cartographies of these regions while also continuing his study of European-style mathematics and its practical applications. At the same time, as the need for precision instruments rose, the Qing workshops began, with the help of Jesuit experts, to emulate the Paris-made instruments. In so doing, they necessarily reinvented them, adjusting them to Qing standards and reproducing them through different production processes. Throughout, the emperor personally directed instrument production, judged the newly produced specimens' quality, and honed his own observational skills. This in turn led the emperor to go one step further in the mid-1690s, ordering land surveys carried out in his presence on imperial outings and military campaigns for the first time since 1683. The frontier thus continued to occupy a central place in the development of a new imperial cartographic practice.

1.3 Frontier Matters: New Qing Cartographic Practice

In the 1680s, the end of war in the southwest and the arrival of the French missionaries had given the emperor a chance to continue his study of practical geometry and to order various cartographic and astronomical observations in the far north. At the same time, court missionaries had been able to showcase their skills in determining geographic positions *vis-à-vis* Beijing, and thus the practical use of techniques of Western learning, during the Nerchinsk negotiations. I now turn to the emperor's persistent attempts, building on these beginnings, to enhance his knowledge of yet another new frontier, as Qing-Khalka

131 Du & Han (1992).

territories and peoples fell subject to the emperor at the end of the Qing-Zunghar wars (1687–1697).[132] In the wake of these events, the 1690s saw the gradual further evolution of imperial cartographic practice.

Discussing the events straddling the turn of the eighteenth century, I consider first the Qing-Zunghar wars and the Qing imperial excursions and military campaigns organized in its wake. On several occasions, the emperor asked one of the protagonists of the previous section, Jean-François Gerbillon, at times along with Joachim Bouvet or Antoine Thomas, to determine geographic positions in the shadow of the imperial excursions penetrating and skirting the Khalka frontier. Here, the emperor personally took part in, supervised, and evaluated the data obtained from observations and calculations. Next, I focus on a preliminary survey of Khalka territory conducted in late 1698 which followed the full integration of Khalka lands into Qing territories after the end of war with the Zunghar in 1697. Gerbillon and Thomas conducted the survey in the entourage of imperial envoys who presided over different assemblies of Khalka khans, one at Lake Buir, the other along the Tula River. In a final step, I reconstruct the story of the measurement of the distance covered by one degree of latitude, commissioned after a careful evaluation of the preliminary survey of 1698 and executed from about one degree south of Beijing.

I argue that the turn of the eighteenth century witnessed the crystallization of a new imperial cartographic practice, motivated by the need to explore the vast Khalka frontier. Aided by the emulation and adaptation to Qing standards of European surveying instruments in the court's Imperial Workshops, the Kangxi emperor personally directed and guided the gradual inclusion of select mapmaking techniques as practiced by the Parisian *Académie* into the preexisting framework of Qing cartographic practice. However, to accommodate these new techniques, the Qing needed to adjust their standards: when an imperial decree redefined the *li*, based on measurements taken during the preliminary Khalka survey, such that one degree of latitude equaled 200 rather than 250 *li*, the Qing court logically needed to re-standardize the *chi*, the most basic unit of length constituting the *li*, as well. The issuance of a new *chi* standard constituted the final step in establishing consistent Qing cartographic practice, allowing the Yongzheng 雍正 (r. 1722–1735) and Qianlong 乾隆 (r. 1735–1796) emperors to rely on these standards in mapmaking over later decades.

132 Perdue (2005), 133–208.

1.3.1 *Integrating the Khalka: Exploring a New Frontier*

The 1690s saw the full integration of Khalka lands into Qing territory after the Khalka recognized Qing suzerainty to shield themselves from the Zunghar, with whom they had been in violent conflict for decades. Following the Zunghar defeat in 1690 by a Qing army led by the emperor himself, Zunghar leader Galdan (1644–1697) quickly rebuilt alliances, among others with the sixth Dalai Lama of Tibet, and moved eastwards again in the mid-1690s. In this context, Jesuit missionaries participated in several imperial tours and military campaigns focused on stabilizing the Khalka frontier and thwarting Galdan's ambitions.[133] The missionaries' role was mostly determining positions along the routes linking Qing and Khalka territories. A first journey took place from May to June 1691, when the emperor himself set out for Lake Dolon, on the edge of the Mongolian steppe, some 300 km directly north of the capital (figure 18). There, three Khalka khans, having recently fled to Qing territories, pledged their allegiance on May 29th, 1691 and subsequently entered the Qing banner system.[134]

Gerbillon and Bouvet, then constituting one of the emperor's two tutoring teams for Western learning, joined the emperor on his journey to Lake Dolon. In the evening of May 10th, the day the imperial party had set out from the capital, the emperor called on them to help him review his knowledge of practical geometry. The emperor asked Gerbillon to determine the difference with Beijing in height of the pole star and to explain which corrections he ought to make in calculating the meridian shadow. On May 13th, in passing Gubeikou 古北口, a gate of the Great Wall, Gerbillon mentions that he had observed the height of the sun at noon "with the semi-circle of the Duke of Maine which I had offered to the emperor, who considered it an important matter and had it carried on the back of a horseman; he had a double casing made for it, so it could not run the risk of being damaged by transport; and he had two types of new footing made, one as well as the other featuring a joint."[135] A number of such Parisian semi-circle protractors indeed survive in the Palace Museum. One of these,

133 Perdue (2005), 174–208.

134 Gerbillon's account of this journey appears, edited, in du Halde (1735), 4:252–288.

135 "avec le demi cercle de Monseigneur le Duc du Maine, dont j'ai fait présent à l'Empereur, qui en fait très grand cas, & qui le fait porter sur le dos d'un Cavalier; il lui a donné un double étui, de manière qu'il ne peut courir aucun risque de se gâter par le transport, & il lui a fait faire deux sortes de pieds nouveaux, qui portent l'un & l'autre un genou." de Thomaz de Bossierre (1994), 44; du Halde (1735), 4:253–255.

only 17 cm in radius, produced by the Bion workshop and featuring a built-in compass, fits Gerbillon's description and is thus identifiable as the instrument gifted by the French Jesuits to the Kangxi emperor.[136] Executed in gilded copper, it seems quite suitable as a gift from the son of King Louis XIV (figure 17).[137] After completing his observations, Gerbillon carefully explained his calculations for determining latitude to the emperor. He makes no further mention of determining the difference in latitude with Beijing but throughout his diaries indicates the number of *li* traveled by the imperial party on a day-to-day basis. On the last day of the journey, only several *li* from the capital, the emperor ordered Gerbillon to assist him further with his geometric studies.[138]

Years later, Gerbillon joined the emperor on three different military campaigns. The first, launched during the spring of 1696, was a large-scale military campaign against Galdan, led by the emperor to stabilize Khalka lands and incorporate them into the Qing domain. The Qing armies were victorious, but Galdan managed to escape with only a few men. In the campaign's wake, in order to prevent Galdan from gaining support among the Ordos Mongols or to escape to Tibet, the emperor undertook two further campaigns, travelling respectively to Hohhot and to Ningxia (figure 18).[139] On all three campaigns, the emperor compared different cartographic practices: whereas missionaries Gerbillon and Thomas played their role of mathematician-observers, determining positions *vis-à-vis* Beijing in terms of latitude, other officials continued determining distances in *li* by "straightening out" road measurements. As mentioned above, thus reckoning road distances, employing measuring roads with graduated ropes followed by geometric calculations to account for unequal terrain and bending roads, a technique known in Chinese as *gougu*, dated back to Han times.[140] Han Qi and Catherine Jami have shown that the Kangxi emperor and one of his trusted mathematicians, Mei Wending (1633–1721), later used these *gougu* methods to argue that European-style geometry as taught by the missionaries had its origins in ancient Chinese knowledge.[141]

136 *Qinggong Xiyang yiqi* (2011), 128–131.

137 As mentioned above, the best instruments, reserved for the aristocratic and the wealthy, often contained silver or gilded copper.

138 In the summer of 1692, the emperor again asked to join him on one of his hunts at the Mulan imperial hunting grounds. Khalka lamas were also to meet with the emperor, but Gerbillon makes no mention of any scientific practices.

139 du Halde (1735), 4:304–355.

140 Yee (1994), 110–113; Feng (1998); Han (2016).

141 Jami (2012), 245–253; Han (2016).

FIGURE 17 *"Gilt-copper semi-circle protractor with a movable pointer" produced by the Bion workshop.*
PALACE MUSEUM.

During the 1696 military campaign against Galdan, at the border between Qing and Khalka lands, Gerbillon and Thomas determined the imperial party's position *vis-à-vis* Beijing by measuring the height of the pole star, whereas Qing officials measured the route from the gates of the Great Wall by rope. A few days later, on May 13th, 1696, the emperor compared the two techniques in the field, as recorded in the *Anthology of the Kangxi Emperor* or *Kangxi yuzhi wenji* 康熙御制文集:

> The road from Dushikou to the border, measured with a rope, is 800 *li* long, which is shorter when compared to what was measured by those who frequent [this road]. The road from Beijing to Dushikou is very short, roughly calculated no more than 423 *li*. The crown prince can send someone to measure it by rope. At the border post, instruments were used to observe the height of the pole star, it was 5 degrees higher than in Beijing. Measured like this, the distance in *li* is 1250.[142]

142 "自獨石口至喀倫，以繩量之有八百里，較向日行人所量之數日見短少。自京師至獨石口為路甚近，約計不過四百二十三里。皇太子可試使人量之。喀倫地方用儀器測驗北極高度，比京師高五度。以此

In determining the position of the Qing-Khalka border and the progress of the campaign relative to Beijing, the emperor here mistakenly (because not perfectly north-south oriented) compares the two different cartographic practices, showcasing, as he understood it, their near-compatibility: he first notes that the road from Dushikou 獨石口 to the border, measured by rope, extends 1223 *li* (800 + 423 *li*). He next gives the difference in altitude of the pole, observed to be five degrees. As Qing astronomical treatises at the time set one such degree to equal 250 *li* on the ground, the emperor then calculates the distance as equalling 1250 *li* (5 × 250 *li*). Just north of 45°N, this position constituted the Qing-Khalka border at the time, explicitly mentioned in the excerpt quoted above as well as in Gerbillon's diaries. Throughout the 1696 military campaign, Gerbillon and Thomas determined their positions again and again by observing the height of the pole star as well as of the sun at noon. This time, however, they carried out their observations using a large ring sundial from the Butterfield workshop that the emperor possessed.[143] Travelling northwest, the party eventually reached the Kherlen River, whence the main Qing army under the emperor drove the Zunghar troops westward into the arms of a second Qing army. This action completed, the emperor, disheartened by supply issues afflicting his troops, soon afterward returned to Beijing.

The emperor asked Gerbillon and Thomas to accompany him, in October that same year, on a second campaign, this one to the Ordos region, just south of the Khalka frontier. Their route passed through Hohhot, one of the gateways into Khalka lands and an important center for Qing-Khalka trade. The purpose of the tour was to prevent the defeated Galdan from forging an alliance with the Ordos Mongols or fleeing to Tibet.[144] Gerbillon mentions the latitudes of some of the places they passed through en route to Hohhot but provides little detail. He does mention that, on November 23rd, the emperor measured the width of the Yellow River "with his semi-circle."[145] After a Zunghar envoy met with the emperor near the Yellow River to negotiate the submission of the now-isolated Galdan, the emperor suddenly decided to return to the capital.[146]

度之，里數乃一千二百五十里…" SKQS (*Siku quanshu* 四庫全書 digital database), *Kangxi yuzhi wenji, juan* 2. Quoted in Qin (1997), 38. "*Kalun*" is a rendition in Chinese of the Manchu word "*karun*" (meaning "frontier" or "border"). Elliott (2014), 33–43. Since measurements appear in terms of *li*, I chose to translate this as "border post." du Halde (1735), 4:314; Elliott (2014).

143 de Thomaz de Bossierre (1994), 58–60; du Halde (1735), 4:314. A number of universal ring sundials survive in Beijing. *Qinggong Xiyang yiqi* (2011), 29–33. The Butterfield sundial in this collection seems too small for Gerbillon to have used it.

144 Perdue (2005), 174–208.

145 "Avec son demi-cercle." du Halde (1735), 4:345.

146 Perdue (2005), 197.

Immediately after reaching Beijing in mid-January 1697, the emperor launched a third campaign, planning first to travel all the way to Ningxia and from there to push a final military expedition into the Altai Mountains, where Galdan had set up camp. Thus, on February 26th, the emperor set out again with his armies, taking along Gerbillon and Thomas.[147] As before, Gerbillon determined latitudes and his diary mentions nearly every day the number of *li* travelled together with the general compass bearing. When they reached Shanxi and Shaanxi province, travelling alongside the Great Wall towards the southwest, Kangxi interviewed the son of Galdan. The Beg of Hami, a city in today's Xinjiang province, who sought Qing support against subsequent threats from another Zunghar faction, had recently captured him.[148] When the Beg of Hami and his territories gained incorporation into the Qing realm soon thereafter, the Turkic desert oasis was set to become the westernmost point surveyed in the comprehensive surveying of all Qing territories between 1708 and 1718 (Chapter 2).

When the imperial party finally reached Ningxia in mid-April 1697, the emperor occupied himself with the logistics of sending his armies out into the deserts.[149] Around that time, on April 21st, Gerbillon and Thomas observed a solar eclipse using Thomas' quadrant (more than thirty-two centimeters in radius), determining the start of the eclipse at 7:04 in the morning and the end at 9:10. Thomas notes that the emperor himself took part in these observations, using instruments "so that it could be admired by the princes and magnates who had gathered."[150] The *Anthology of the Kangxi Emperor* also mentions the observation of this eclipse:

> After my arrival, We have used instruments to observe the pole star, which, compared to the capital, stands 1°25′ lower. From east to west the difference is 2150 *li*. Today, Thomas calculated with [his] methods that the solar eclipse would describe 9′46″, whereas on the day of the eclipse we measured 9 minutes 30-something seconds and it didn't get dark enough

147 Gerbillon's account of this journey appears edited in du Halde (1735), 4:356–384.

148 The Beg system was a local political institution adopted by the Qing for its rule over its Muslim subjects in today's Xinjiang province.

149 Perdue (2005), 198–203.

150 ARSI, JS 149:643. Translation from Antonucci (2012), 117. A narrative left by Antoine Gaubil and written on the basis of notes and diaries kept at the French residence in the 1720s separately confirms all of this. Cordier (1915), 533; de Thomaz de Bossierre (1994), 79.

> to see the stars. Seen from Ningxia, the capital is located east and slightly to the north.[151]

Whereas the previous excerpt speaks of the position from the capital in terms of north-south distance in *li* or differences in altitude of the pole (latitude), this one mentions the position of Ningxia in terms of east-west distance in *li*. Gerbillon observed the latitude of Ningxia at 38°35′ and estimated the city's longitude at 10°20′ to 10°25′ west of the capital, the latter undoubtedly determined with the help of data obtained during the eclipse, one of the better ways of fixing a location's longitude at the time. He then checked this against the corrected road distance in *li*. Here he mentions, to my knowledge, the very first estimate for longitude employing observations in which the emperor took part, further indicating that Qing scientific practice was slowly incorporating missionary-taught techniques for determining latitude and longitude. It is important to stress that this combined practice included neither setting out base lines for precise triangulation nor observing the satelites of Jupiter, both practiced by the *Académie* in France. Rather, it utilized the inherited technique of geometrically correcting rope-measured road distances for estimating longitude.

Near the great bend of the Yellow River, northeast of Ningxia, the emperor watched his troops head into the deserts on their way to the Altai Mountains in pursuit of Galdan. The imperial party then returned to the capital, following the Yellow River back into the Ordos region, only later to discover that Galdan had already died in early April following dissension in his ranks.[152] When it was clear that the war with the Zunghar had ended, the Khalka and their lands could become an integral part of the Qing realm. Basic surveying by Gerbillon and others between 1691 and 1697, however, had for the most part only skirted the Qing-Khalka border, crossing it only once during the 1696 military campaign against Galdan (figure 18). As a result, although many Khalka had already resettled on the Qing side of the frontier, the emperor launched a large survey in 1698, the first one of its kind, to gather more data on these newly incorporated territories and so stabilize the northwestern parts of Khalka lands now firmly under Qing control.

151 "朕至此以儀器測量北極，較京師低一度二十分，東西相去二千一百五十里。今安多以法推算，言日食九分四十六秒，日食之日晴明測驗之，食九分三十幾秒，並未至昏暗見星。自寧夏視京師在正東而微北..." SKQS, *Kangxi yuzhi wenji, juan* 2; Qin (1997), 38.

152 Perdue (2005), 202.

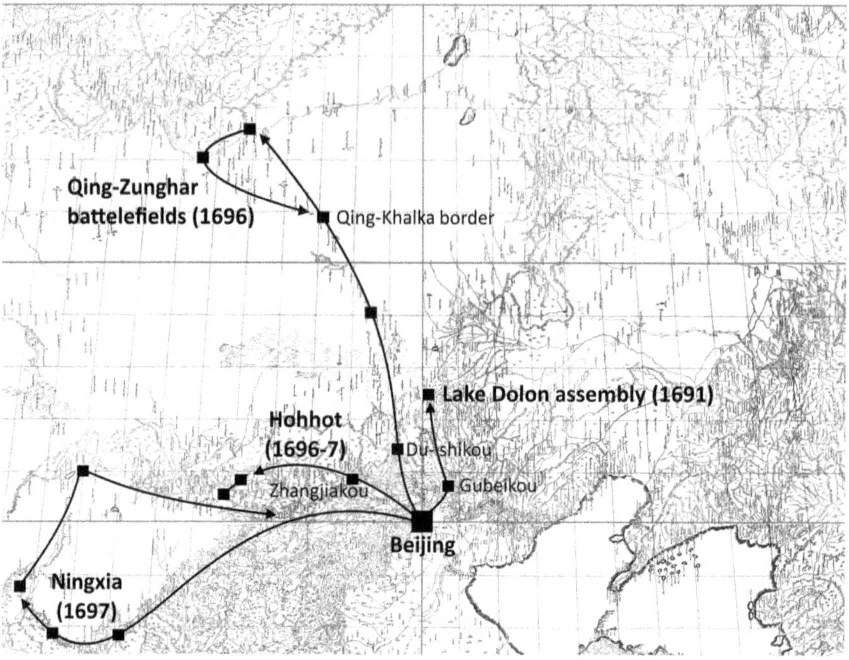

FIGURE 18 *The anti-Zunghar campaigns, marked on the 1719 edition of the Qing atlas (Chapter 3).*

1.3.2 *The 1698 Preliminary Survey*

On May 24th, 1698, according to an account left by Thomas, a team consisting of Gerbillon, Thomas, three Qing dignitaries, and a large number of officials and servants again set out from Beijing.[153] The three dignitaries did not preside immediately over the mapping of Khalka territory, but rather officially formed an imperial commission to oversee two different meetings of Khalka khans, the journey's primary objective (a follow-up of the Lake Dolon assembly of 1691). Their task was to work out in greater detail the integration of the Khalka clans into the Qing banner system and to discuss land control.[154] In other words, the 1698 journey constituted an overall effort to integrate Khalka peoples and territories into the Qing polity. In one of his letters, Gerbillon explicitly mentions the emperor having sent him

153 ARSI, JS 149:589–592; Bosmans (1926); Witek (2003). We find confirmation for this date of departure in de Thomaz de Bossierre (1994), 140. Gerbillon's account of the journey appears, edited, in du Halde (1735), 4:385–422.

154 du Halde (1735), 4:400–404.

along with Thomas to "make a map of the lands of the Khalka, his new subjects."[155]

After six days of travel, the party reached Xifengkou 喜峰口, another gate of the Great Wall, from which they started carefully measuring road distances using a rope 1/3 *li* in length. In addition to road distances in *li*, Gerbillon also meticulously notes the different compass bearings for each day of travel. He also seems to have observed the latitude nearly every day, weather permitting, by observing the height of the pole star or the sun at midday.[156] About one month into the journey, having covered some fifty to sixty *li* (c. 28–33 km) each day, they pitched their tents at Lake Buir.[157] The meeting of Khalka leaders lasted one week. The party then traveled on to a small lake north of the Kherlen River, determining latitudes as often as they could. Next, they followed the Kherlen River for 973 *li* upstream to the place Qing troops had forced Galdan to flee just two years earlier.[158] On August 3rd, they reached the Tula River, near present-day Ulan Bator, where Thomas makes explicit reference to the Qing-Zunghar wars: "we exactly put out tents where, in a hard battle, Galdan was defeated that very year [1696], by one of the two armies of the emperor."[159] Gerbillon, in his diaries for this stretch of the journey, also refers continually to the military campaign of 1696 while visiting places just to the southeast of present-day Ulan Bator where the decisive battles had occurred.[160] They followed the Tula River another 449 *li* through lush green hills until they reached the confluence with the Orkhon River, where they observed a latitude of 48°54′ (figure 21).

From there, they crossed the Orkhon River for another meeting with Khalka leaders between the 14th and 26th of August 1698. During their stay, Gerbillon and Thomas determined latitude by observing, due to bad weather during daytime, the height of the pole star. They also spoke with Russian merchants for information about the lands to the north and with a young Khalka boy in their service, who "traced a little map in the sand, on which he marked the rivers and their courses, the cities and the fortified towns, which were almost

155 "pour faire la carte de päys des Kalkas, ses nouveaux suiets." de Thomaz de Bossierre (1994), 140.

156 de Thomaz de Bossierre (1994), 84–92; du Halde (1735), 4:385–422.

157 du Halde (1735), 4:399.

158 Cordier (1915), 541.

159 "tentoria fiximus eo in loco, ubi Caldanus anno praedicto, praelio pertinaci, ab Imperatoris altero exercitu profligatus fuit." Bosmans (1926), 162–163.

160 du Halde (1735), 4:410–413.

all built on the banks of one of these rivers."[161] About the same time, Gerbillon fell ill and did not recover until they had returned to the capital. The party took a more southerly route, roughly that taken by the second Qing army in 1696 during the military campaign against Galdan, albeit in the opposite direction, reaching Hohhot on October 2nd. Passing the Great Wall at Zhangjiakou 張家口 on the 10th, they finally arrived back in Beijing on October 15th. Thus, throughout this survey of the new Khalka frontier, Qing surveyers again measured road distances in line with preexisting cartographic practice while Gerbillon and Thomas observed polar and solar altitudes to determine latitude and used recorded road distances to estimate longitudes. They used this data now not only to determine the positions of major cities or other relevant sites, as they had during the anti-Zunghar campaigns, but also to produce a map of a significant part of Qing territories.

In his diaries, Gerbillon mentions that during the 1698 preliminary survey they primarily used "a semi-circle which the emperor had provided, well-graduated and with a telescope… [mounted] above the alidade."[162] He also mentions that it featured a built-in compass. Catalogues include a gilded copper semi-circle with a radius of thirty-two centimeters, of uncertain European provenance, closely matching this description, and apparently emulated by the Imperial Workshops at some point, since the *Figures of Imperial Ritual Equipment* (*Huangchao liqi tushi* 皇朝禮器圖式), produced much later (in 1760), contains a similar instrument, named "semi-circle with movable telescope" or *Siyou qianli jing banyuan yi* 四遊千里鏡半圓儀, albeit smaller in radius (figures 19 & 20).[163] One such instrument, either the European prototype or a Qing adaptation, must have been the principal instrument used by Gerbillon and Thomas.

On December 8th, 1698, Gerbillon and Thomas presented the emperor with a map of Khalka lands. Thomas notes that, due to his familiarity with the terrain (having himself only recently led his armies in decisive battles for control over that area), as well as because the map undoubtedly symbolized that control, the emperor displayed great satisfaction with the map. He also immediately made plans for the missionaries to continue, mapping the Manchu

161 "il nous traça même sur le champ une petite carte, où il marqua les rivières avec leurs cours, les Villes, & les Bourgades, qui sont presque toutes bâties sur le rivage de quelques-unes de ces rivieres." du Halde (1735), 4:417–418; de Thomaz de Bossierre (1994), 91–92.

162 "Un demi-cercle que l'Empereur nous avoit prêté, fort bien divisé & à lunette… qui est sur l'alidade." du Halde (1735), 4:397.

163 SKQS, *Huangyu liqi tushi*, *juan* 3.

homelands, the frontier with Korea, and the shores of the northeast that following year. After studying the map carefully, however, the emperor called on Thomas to explain why the recorded road distances, including two stretches oriented almost perfectly north-south, did not correspond with north-south distances calculated on the basis of the observed difference in latitude. Thomas acknowledged the problem and explained that, based on their observations and measurements in the field, the actual *li* was likely only four-fifths of that used in producing the measuring ropes. In other words, the practice of combining land surveying techniques as practiced by the Parisian *Académie* with those inherited by the Qing court laid bare faulty standards in use in astronomy and cartography at the Qing court. As mentioned above, Matteo Ricci had first defined one degree of latitude as amounting to 250 *li*, which Qing astronomical treatises authored by Schall von Bell and Verbiest subsequently adopted. Now, based on measurements undertaken during the first preliminary survey of 1698, it had become clear that, roughly estimated, one degree of latitude only consisted of some 200 *li*. In Thomas' words, the emperor then "saw that, with accurate observations, many things could be understood, namely, how many Chinese *li*, invariable and defined in terms of *chi*, were contained in one terrestrial degree."[164] As a result, a new imperial decree officially and definitively redefined one degree of latitude to measure precisely 200 *li*, although efforts to re-standardize the *li* and the *chi*, the most basic unit of length, did not begin until 1702.[165]

Clearly, the gradual establishment of a new Qing cartographic practice, meaning the integration of selected new techniques of land surveying into preexisting cartographic practices, was inseparable from the vast Qing frontier: all land surveys and, as I will show in Chapter 2, the first stages of the comprehensive land surveys beginning in 1708, focused on the areas known in Chinese as *kouwai* 口外, meaning "beyond the gates of the Great Wall."[166] To be sure, the Qing did not actually consider all areas north of the Great Wall part

164 "cum accurate observari, plurimum conducere videret, quot scilicet stadia Sinica certa ac fixa mensura uniuspedis, continerentur in uno gradu terrestri" Bosmans (1926), 166. At the time, the earth was thought to be a perfect sphere with each degree of latitude covering the same distance.

165 The reason for this is unclear. A general preoccupation with other issues related to land control may be one possible explanation: as explored above, inundations in the capital province of Zhili necessitated smaller-scale surveys of the rivers there; in 1699, there were serious floods in Jiangsu province; and in 1701, Kangxi ordered the retaking of Dartsedo 打箭爐, today's Kangding 康定, a town on the Qing-Tibetan frontier. Dai (2010), 76–83.

166 Each of the expeditions mentioned in this chapter took a different route from the capital towards the frontier, passing the four different gates of the Great Wall that gave direct access to the capital. From west to east: Zhangjiakou, Dushikou, Gubeikou and Xifengkou.

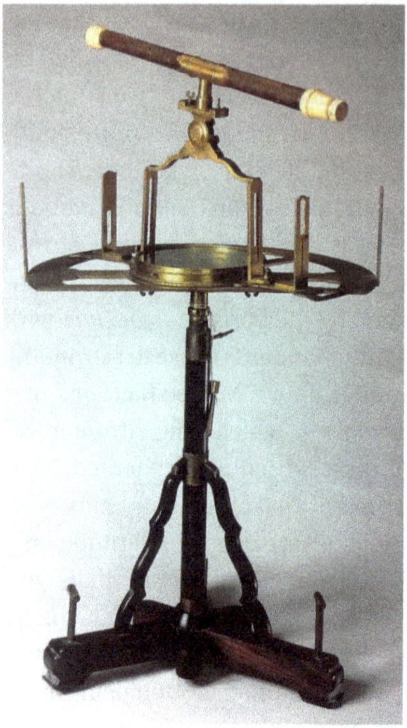

FIGURE 19 *"Semi-circle protractor with a telescope."*
PALACE MUSEUM.

FIGURE 20 *The "semi-circle with movable telescope" included in the* Figures of Imperial Ritual Equipment *(1760)*.
SIKU QUANSHU.

of their frontier (Manchu: *jecen*), most notably excluding the Manchu homelands. On the other hand, the frontier certainly did include the newly incorporated Khalka areas, the new frontier with Russia from just south of Lake Baikal to the estuary of the Amur River, and even the border with Korea, plagued by bandits, smugglers, and illegal crossings.[167] Only if the Qing court came truly to control these borderlands, if they could survey and "grasp" them via a consistent cartographic practice, could they make this new frontier a stable part of the Qing polity and no longer a threat to its heartland. Mapping these areas was therefore an absolute priority for the emperor. In total, I have argued that the evolution towards a new Qing cartographic practice, marrying newly introduced techniques of surveying to inherited practices, took place at the emperor's personal direction and initiative and through grappling with exploring and controling vast new frontiers. Having redefined the length of one degree of latitude from 250 *li* to 200, the court now needed only re-standardize the most basic unit of length making up the *li*, the *chi*, to implement this new Qing cartographic practice consistently and on a far grander scale.

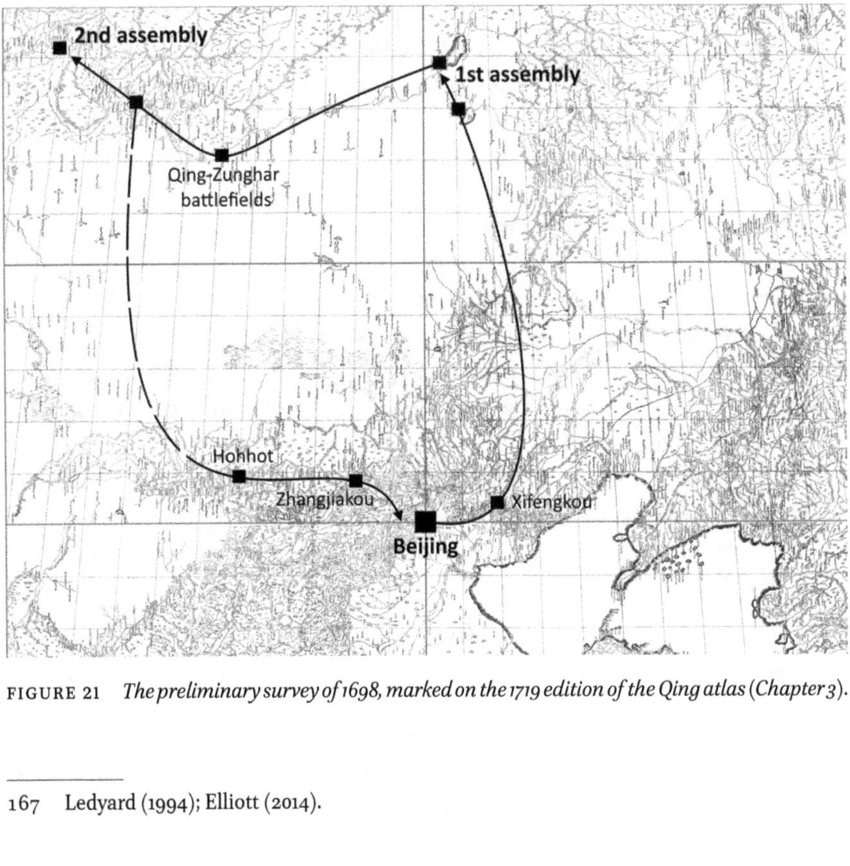

FIGURE 21 *The preliminary survey of 1698, marked on the 1719 edition of the Qing atlas (Chapter 3).*

167 Ledyard (1994); Elliott (2014).

1.3.3 *Re-Standardizing the Qing's Most Basic Unit of Length*

For each degree of latitude to equal precisely 200 *li*, as decreed by the emperor, the Qing court needed carefully to re-standardize their most basic unit of length. The way to achieve this was to measure, as accurately as possible, one degree of latitude on the ground. According to Thomas, on April 24th, 1702, the emperor decreed, at last, the necessity of doing exactly that, and then redefining the *li* as precisely 1/200th of that value.[168] He wanted to have it done at the end of autumn so that measurements would not interfere with the local harvest and vice versa.

Thomas' account, mentioned above, is the only comprehensive narrative of the measurements undertaken to determine the distance covered by one degree of latitude. He records that on October 20th the emperor viewed all the instruments chosen for conducting the measurements. On November 4th, he appointed his third son, prince Yinzhi 胤祉 (1677–1732), whom the emperor had personally instructed in geometry, to lead the expedition. Some days later, after the emperor had departed down the road from the capital towards Shandong province for his fourth imperial tour of the south, the prince sent out a team to search the region south of the capital for a vast stretch of suitably flat land. They soon found flat land, but dotted with too many trees and so unsuitable for such a survey. Finally, south from Bazhou, they reached an area "flat like the sea's surface, without even the slightest hill, and extending about 200 *li*."[169] After the prince received the report of this, he fixed the date of departure for the mission on November 28th.

On that day, all necessary instruments proceeded to Yinzhi's palace, as his own servants would be carrying all goods beyond the walls of the capital, after which local runners from each prefecture passed through would take on the responsibility for transporting necessities. Responsible for the actual measurements and thus ultimately for the management of all instruments, porters, and technical personnel, was the Manchu director of the Directorate of Astronomy Mingtu 明圖, assisted by seven subordinates. Before departure, one of the top officials of the Imperial Household Department (Chinese: *neiwufu* 内務府 / Manchu: *Dorgi baita be uheri kadalara yamun*), Hesihen 赫世亨 (d. 1708), gathered all necessities and selected able artisans from the Imperial

168 Bosmans (1926), 166. Note that the debate on the shape of the earth had only just begun in Europe. For mapmaking purposes, most assumed that the earth was not flattened at the poles, so that every degree of latitude has the same length.

169 "instar superficiei maris, sine ullis prorsus colliculis, ad 200 circiter stadia protensum." Bosmans (1926), 167; Bernard (1935), 455–458.

Workshops to join the expedition. The large team of men left the Beijing city gates as planned on November 28th, apart from prince Yinzhi himself, who left the next day escorted by his personal retinue. On November 30th, the prince's party caught up with the rest of the team and settled on the plan of starting measurements from the outskirts of the city of Bazhou, which they reached the very next day. Thomas apparently welcomed Yinzhi's presence, writing that the prince's knowledge of geometry accelerated the measurements, enabling him to give precise orders and to help out where he was most useful.[170]

After exploring the city's immediate vicinity, the team selected a stone monument in a field to the east of the city as their starting point. They used an "iron thread" measuring one *li* to trace a meridian from this point towards the south. To make sure it remained in the right position, every 1/3 *li* (or 120 *bu* 步) they placed a rod as a marker, held in place, in turn, by a tripod cast in iron. Personnel positioned along the line ensured that it remained in position at all times. Wherever trees stood in the way or blocked the view, an order from the local prefect, already directed to join the team and to assist however he could, immediately saw them cut down. With each *li* covered, personnel placed two instruments "of first choice, equipped with two telescopes and a large needle compass" at each end of the iron thread, permitting observation of the height of the pole star.[171] They employed another instrument in surveying all the rods already in place to make sure the line traced was perfectly straight. In this way, *li* by *li*, the team measured a line from the stone monument just east of Bazhou to the banks of the Zhang 漳 River near the town of Jiaohe 交河 (figure 22).[172]

Selected precisely for its flatness and sparse vegetation, the line between Bazhou and Jiaohe nonetheless crossed three villages whose names Thomas' account does not mention. The low and flat roofs of the houses in two of these villages left measurement unimpeded. The third village, however, at *li* 79, did not, forcing the team to lay out the iron thread parallel to the original line but one *li* to the east. Measurement continued from this parallel line until the team reached the shores of a sizable lake at *li* 97, after which they resumed measurements one *li* to the west in perfect parallel with the original line. Finally, on December 15th, when measurements reached the banks of the Zhang River, at *li* 199, they erected a gnomon or *guibiao* 圭表 about three meters high atop

170 Bosmans (1926), 169.

171 "bene probatum, instructum duobus tubis opticis et magna acu magnetica." Bosmans (1926), 168.

172 The Zhang River is known today as Laozhang he 老漳河.

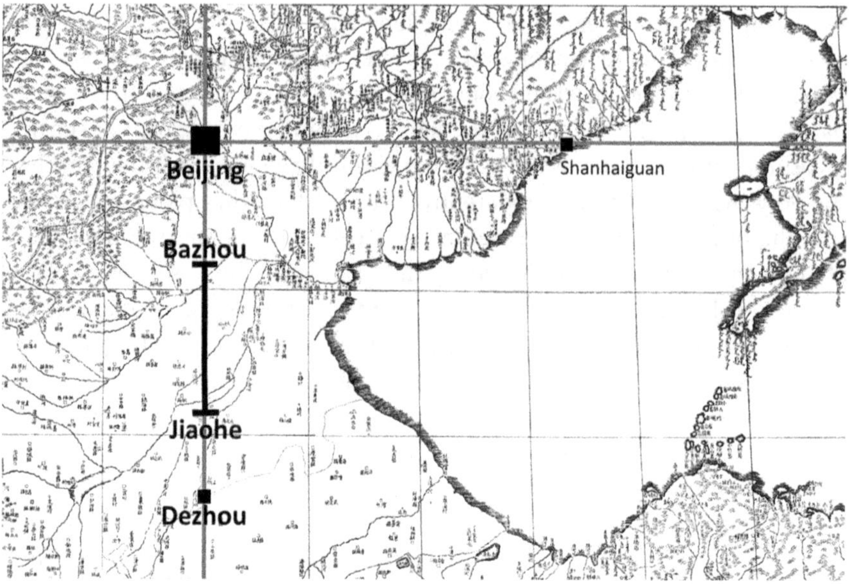

FIGURE 22 *The line traced for measuring the length of one degree of latitude in December 1702, marked on the 1719 edition of the Qing atlas (Chapter 3).*

a horizontal beam marking the meridian. The top of the gnomon formed a rectangular opening of about 30 cm. In this opening, they placed an iron sheet featuring a second opening. It left a copper ball of about 1 kg hanging, suspended from a thin copper wire, through this opening, and firmly fixed in position so the winds would not disturb it. On the horizontal beam, two grooves were filled with mercury to ensure its levelness.[173] For two days in a row prince Yinzhi personally observed the shadow cast on the horizontal beam to determine latitude.

Thomas, however, had doubts about the prince's observations. After all, with only a few days until the winter solstice, the sun did not even reach a midday height of 30°, and thus the phenomenon of refraction would make any readings of the height of the sun at midday unreliable at best.[174] For three consecutive nights, therefore, Thomas observed with a large quadrant the height of the pole star at 40°21′ above the horizon. Since the Zhang River prevented the team

173 Bosmans (1926), 171–172. Verbiest describes the similar use of the Beijing observatory's gnomon in Golvers & Nicolaidis (2009), 119.

174 Refraction occurs when the earth's atmosphere bends the rays of the sun, causing an unreliable observation of the height of the sun above the horizon. The lower the sun, the more its rays are bent.

from measuring one farther *li* to the south, however, they had only measured 199 *li* instead of the 200 the emperor had decreed. The team therefore decided to travel back to the other extremity of the line, at Bazhou, and to measure one more *li* to the north. When this was done, on December 21st, they installed the same large gnomon there, while Thomas again observed the height of the pole star with a quadrant, for two consecutive nights, with a result of 41°22′30″. Thomas also records the following calculation:

Height of the pole at the northern extremity	41°20′30″
Height of the pole at the southern extremity	40°21′00″
Difference	1°01′30″
Correction for refraction	02″
Corrected difference	1°01′32″

Thus, with 200 *li* or 72000 *bu* meticulously measured and covering an observed difference in latitude of 1°01′32″, one degree of latitude should cover a distance of 70206 *bu* or roughly 195 *li* according to the old standard. The emperor had the standard *chi*, or *yingzaochi* 營造尺, used to produce the iron thread of one *li* for conducting the 1702 measurements, cast in brass and kept at the imperial palace.[175] Therefore, since the emperor had decreed that each degree of latitude should measure precisely 200 *li* on the earth's surface, they had to re-standardize that old standard *chi* by shortening it by 195/200th of its former length. The recast standard, of which an even number of 360000 *chi* or 200 *li* now made up exactly one degree of latitude, ensured the profound simplification of the calculations for mapmaking (table 1).[176]

TABLE 1 *The different Qing units of length mentioned in this chapter with corresponding values in meters.*[177]

200 *li*	72 000 *bu*	360 000 *chi*	c. 110 880.000 m
1 *li*	360 *bu*	1800 *chi*	c. 554.400 m
	1 *bu*	5 *chi*	c. 1.540 m
		1 *chi*	c. 0.308 m

175 Witek (2003), note 39.

176 Régis confirms this in a manuscript account kept at the BNF written in the 1720s. BNF, Ms. français, 17242: 9-9v.

177 Based on a detailed study of different *chi*. Ferguson (1941). Note that Picard had estimated one degree of latitude to measure 110.46 km. *Supra*.

Qing sources confirm both the measurements and the casting of the new *chi*, but provide far less detail than does Thomas' account.[178] The edited writings of Li Guangdi 李光地 (1648–1718), then provincial governor of Zhili, where the measurement took place, include the following account:[179]

> The emperor sent officials to erect the gnomon and measure the land between the capital and Dezhou for one degree of difference in altitude of the pole and to have it coincide with 200 *li*. The theory of the ancients that one degree equals 250 *li* is correct, as one ancient *chi* begets 8 *cun* [0.8 *chi*] today. I took part in this. Observing horizontal and vertical graduations, connecting north to south, and obtaining degrees from the heaven, earth, sun, and moon was sufficient to rely on.[180]

A similar account in the *Preface to the Fundamentals of Astronomical Phenomena* (*Lixiang benyao* 曆象本要)—also considered largely written by Li—explicitly confirms the setting of the new standard *chi*:

> In the winter of *renwu* (1702), I joined the imperial tour of the south. The prince led the mathematician from the West from the capital to Dezhou, a distance of 700 *li*, to erect the gnomon and use the instruments to measure and observe. This took between two weeks and a month. They obtained the true angular height of heavenly bodies and land distance.A distance on land of 200 *li* was recorded for a difference of one degree of altitude of the pole. In times past, it was said that this should be 250 *li*, probably too rough an estimate. That is why we use the current *yingzaochi* of the Ministry of Works. It is still not known why the ancient *chi* now corresponds to eight *cun*. I learned about this on the scene as part of the retinue.[181]

---

178 Jami (2012), 245–246; Han (2014), 53–56. Jami notes rightly that use of different *chi* continued throughout the empire before and after standardization. Thus, when I speak in this chapter of the re-standardization of the *chi*, I do not claim consistent use of the new standard throughout Qing territories. Nevertheless, the imperial standard clearly had consequences for the court's involvement in mathematics, astronomy, and cartography.

179 Note that Li Guangdi was an important actor in the Kangxi emperor's engagement with Western learning, particularly with regard to mathematics. Han (1996); Jami (2012), 127–131 & 245–253.

180 "皇上遣官立表量地自京師至德州約極高移一度而地差二百里合之古人二百五十里而差一度之說，正為古尺得今之八寸也。夫既參之。縱橫之黍而合考之上下之聲而協而又有天地日月之度足為根據。" SKQS, *Rongshuji* 榕樹記, 29:19.

181 "壬午冬與南巡。命皇子領西洋籌人。自京城南至德州七百餘里立表施儀審加測望。淹歷旬月。乃得星度道里之真。計地距二百里而極高

Both excerpts mention both the practice of observing the meridian shadow with the help of the gnomon and the technique of observing the height of the sun at noon or that of the pole star. The Kangxi emperor here seems to be using the same strategy of combining inherited cartographic practice with new techniques as taught by the missionaries. In any case, with the length of one degree of latitude now precisely decreed in terms of *li* and a new standard *chi* defined to meet this objective, it was now possible to draw maps on which the observed latitudes and longitudes, or "heavenly" degrees, would match the measured roads, or "earthly" distances. This event therefore constituted the final step in the formal establishment of new Qing cartographic practice fully integrating selected techniques of land surveying as practiced by the Parisian *Académie* into the Qing court's inherited practices. Moreover, these new or hybrid practices formed the basis for the comprehensive land surveys of 1708–1718 (Chapter 2), allowing the increased cartographic precision seen in Qing imperial cartography upon which so many twentieth-century scholars have commented.

I have argued in this section that the measurement of the length of one degree of latitude led to the casting of a new standard for the *chi* and, in turn, to the formal establishment of a new Qing cartographic practice. The implications of this restandardization of the basic unit of length clearly reached far beyond the art of mapmaking. Han Qi, Catherine Jami, and others have long since shown how, in the first decades of the eighteenth century, the Kangxi emperor attempted to trace the missionaries' teachings in astronomy and mathematics back to ancient works of Chinese knowledge and have illustrated how he even sought to prove the authority of these works.[182] Intentionally or not, the emperor thus created a situation whereby the court could more easily appropriate those teachings in the interest of Qing statecraft. In the case of cartographic practice, therefore, such insights regarding the emperor's reliance on historical justifications for the appropriation of new techniques remain relevant, because they enabled him to incorporate observational techniques originating in Europe alongside the traditional practice for measuring road distances. In other words, not only new scientific discourse and teachings, but also new scientific practice, became embedded within preexisting knowledge and practices. In the eyes of the emperor, the old and the new mutually reinforced and justified one another.

差一度。舊說云二百五十里者，大疎濶矣。然所用者今工部營造尺或古尺當今八寸。則未可知爾臣地實扈從與聞之。" *Lixiang benyao*, 44–45. My translation. See also Han (2014), 54.

182 E.g. Jami (2012), 245–252; Han (2016).

To recapitulate, I have argued that an imperial desire to explore and to control new frontiers, particularly during and after the Qing-Zunghar wars, led to the exploration of new cartographic practices. Imperial tours and military campaigns in the 1690s focused on the Qing-Khalka frontier provided an ideal occasion for the emperor personally to compare inherited practices with techniques practiced in Paris (with the assistance of his European experts). The defeat of Galdan in 1697 saw the full integration of Khalka lands, necessitating a large-scale preliminary survey of the area, with mapmakers carrying instruments into the field in service to imperial conquest and control over new frontiers. Roads they measured in accordance with inherited cartographic practice, and the geometrical calculations using those road distances to estimate longitude differed from contemporary practice at the Parisian *Académie*. At the same time, observations of the height of the sun and the pole star ensured precise determination of latitude in accordance with European practice. As such, only selected new techniques were effectively married with preexisting Qing practices. However, this combining of techniques from different traditions created a problem: the preliminary survey of Khalka lands revealed clearly that the measured road distances did not correspond with the calculated distances based on the assumed length of one degree of latitude. To resolve this, the emperor launched an expedition precisely to measure the length of one degree of latitude on the Beijing meridian, on the basis of which the most basic unit of length in use at the Qing court, the *chi*, was then re-standardized. The implications of creating this new standard were far-reaching but, in the case of cartographic practice, signaled the official establishment of a new, or reinvented, imperial cartographic practice. As such, the consistent and comprehensive mapping of the entirety of Qing territories now possessed a solid basis.

Conclusion

In three steps, this chapter has treated the build-up towards the Qing mapping project by reconstructing the trajectories of surveying instruments from Europe to Asia and from the Qing capital, where the Imperial Workshops reproduced and adapted them, out to the frontier. First, I have shown how the scientific endeavors of the *Académie royale des sciences* under Louis XIV's patronage and the emergence of a solid market for precision instruments promoted technical evolution in the production of surveying instruments and improved cartographic practice in France. At the same time, the *Académie* and the Society of Jesus found one another: a state-sponsored French Jesuit mission could provide academicians with Jesuit access to continental East Asia while the Jesuit order could benefit from the support of a powerful scientific institution, facilitating proselytizing efforts under an emperor interested in practical applications of mathematics and astronomy. As a result, in 1685 a French Jesuit mission with scientific inquiry, including the geographic exploration of continental East Asia, high on its agenda, departed France. Many state-of-the-art Paris-made surveying instruments journeyed to Asia along with the French Jesuit mission to the Qing, for use in gathering geographic data for the Parisian *Académie*, as well as to please the Qing emperor and his family. At the same time, the *Académie* was unable fully to instill their methods and standards in Beijing, as only some of its missionary-representatives were to reside in the capital, and those who did had to operate within the tradition of an earlier generation of Jesuits.

In this chapter's second part, I focused on the Qing context with regard to cartographic practice. After the conquest of the Chinese provinces, the Kangxi emperor pushed for the systematic mapping of Qing territories. Concerned with the resulting maps' inconsistencies and with the problem of frontier control, he began experimenting with new techniques of land surveying by asking the Jesuit missionary Ferdinand Verbiest to draw maps on two imperial tours north of the Great Wall. After Verbiest's death in 1688, the emperor had to rely on a new generation of missionaries, most of whom enjoyed systematic contact with the emerging European scientific community. The newly arrived French Jesuits were particularly well-placed in this regard. One, Jean-François Gerbillon, used instruments from Paris to conduct basic observations as part of a team that negotiated a peace treaty with the Russians. Immediately afterward, the emperor decided to take up his study of missionary-taught practical geometry and personally practiced the correct use of different surveying instruments. He also ordered his Imperial Workshops to start emulating Paris-made prototypes, which led to their systematic reproduction in the following decades. The workshops adapted these instruments to Qing standards and thereby introduced both specific observational techniques and the geometric calculations involved therein.

The chapter concluded by tracing the subsequent evolution of imperial cartographic practices during the 1690s, concentrated on the frontier and leading to the full absorption of selected techniques of land surveying as developed in Paris, mostly in their observational dimensions. The Kangxi emperor took the first step in this regard when he became personally invested in cartographic field practice. He also asked Jesuit missionaries Gerbillon and Thomas to perform observations and calculations as experts on imperial tours and military campaigns designed to end the Zunghar threat to the Qing frontier. When the Qing-Zunghar wars had finally come to an end, the emperor launched an all-encompassing effort to integrate Khalka peoples and territory which included sending Gerbillon and Thomas along with his envoys to Khalka leaders to conduct the first large-scale survey. The emperor received the map of the Khalka lands and the routes connecting it to the Qing heartland enthusiastically, but also noted how the nominal distance for one degree of latitude did not correspond with actual road measurements. Therefore, the emperor entrusted one of his sons to lead an expedition to measure accurately the length of one degree of latitude south of Beijing, which imperial decree would redefine as covering precisely 200 *li*. On the basis of these measurements, the emperor then re-standardized the most basic unit of length, the *chi*, in a move that constituted the formal establishment of new Qing cartographic practice.

As such, this chapter's three parts each deal with a complex of factors, all of which shaped the eventual Qing mapping project, including but not limited to: the interests of two strong monarchs in cartographic practice, the technical evolution of surveying instruments, Jesuit mobility, education, and proselytizing strategies, instrument (re)production by the Imperial Workshops, the Kangxi emperor's determination to stabilise the Qing frontier, etc. This complex interplay of factors resists straightforward structural analysis, but one possible approach to understanding the many quantitative and qualitative shifts this chapter explores is to view this process as a converging of interests among three main forces: the Kangxi emperor, the Jesuit order, and the French *Académie royale des sciences*. The emperor sought to improve cartographic practice as a matter of statecraft and frontier control; the Jesuit order sought to please the emperor with regard to his interest in Western learning to procure greater protection for their religious mission; and the *Académie* wished to gather observational data from continental East Asia in its push for scientific exploration. All three of these actors thus pushed actively for change and benefited directly from the outcome. This view accounts for the main forces that shaped new Qing cartographic practice and touches on the interplay between individual and institutional factors, a point I will develop further in Chapter 2.

These converging interests led to the establishment of a new Qing cartographic practice from two divergent cartographic traditions, more precisely through the insertion of specific observational and geometrical techniques into a preexisting framework. This calls into question the idea that one type of cartographic practice, often termed "scientific," "early-modern," "Jesuit," or "European," simply displaced another, often refered to as "traditional," "premodern," "indigenous," or even "cultural." This chapter shows how, instead, a preexisting practice gradually absorbed select elements of another mapmaking tradition, with the expectation of improving the effective control of territories and peoples, which strong monarchs on both sides of the Eurasian continent desired greatly. This insight accords well with the view that the mapping project resulted from a convergence of interests among the emperor, the Jesuits, and the *Académie*.[183] After all, new Qing cartographic practice did not belong fully to any of them, but rather emerged through circulation made possible by a complex interplay of factors, mediated in part by surveying instruments, in an active effort to refine and improve existing practices.

183 Hostetler also used this phrase. Hostetler (2009), 99.

INTERMISSION 1

Missionaries or Mapmakers? The Mapping Project and Its Place in the Mission

Until recently, scholars have approached the Kangxi mapping project exclusively from the Jesuit missionaries' perspective.[184] The article most quoted, by Theodore Foss, speaks unequivocally of "Jesuit cartography," reinforcing the consensus that the Jesuit missionaries took it upon themselves to map Qing territories for the Kangxi emperor, exclusively employing "European" techniques of land surveying developed by the *Académie royale des sciences*. My argument in Chapter 1 challenges this consensus, showing that the emperor was eagerly mapping his domain well before the arrival of the French missionaries and that, rather than replacing existing cartographic practices wholesale, the Qing court instead appropriated specific land surveying techniques so as to enhance their own. In light of this, the question arises as to precisely how invested were the Jesuits, French or otherwise, in the mapping of Qing territories. In other words, how did the mapping project fit within the larger framework and objectives of their mission?

To be sure, scientific fieldwork was not something to take on lightly: these mapping expeditions required the missionaries to traverse vast expanses of land—often in remote deserts far to the north—for months on end as part of large surveying teams. These commitments curtailed their ability to carry out missionary work, particularly pastoral activities entailing major investments in local communities. On the other hand, travelling through different areas of Qing territory, including its vast frontiers, provided opportunities to learn more about the more remote Qing territories with an eye toward establishing new mission stations. This intermission, from the perspective of the Jesuit missionaries themselves, focuses on the place of cartographic practice within the larger framework of the Catholic missions to the Qing: did missionaries see the mapping project as a necessary evil, a means to securing greater support for their missionary activities? Were they eager to explore these lands in the hope of opening up new areas for proselytization, or did they squarely oppose the interference with their ministerial goals or personal ambitions?

184 Fuchs (1943); Foss (1988); Foss (1994); Yee (1994); Han (2014); Ribeiro & O'Malley (2014); Bai & Li (2014); etc.

© KONINKLIJKE BRILL NV, LEIDEN, 2017 | DOI 10.1163/9789004345362_004

To answer these questions, I temporarily leave behind surveying instruments and turn instead to a selection of sources that place the Jesuits' involvement in cartographic practice within the wider context of their lives at court at the turn of the eighteenth century. First, I explore what a number of the emperor's mapmakers themselves had to say about their involvement in the sciences at the Qing court (including how they justified it) in their private correspondence. Next, I turn to evidence suggesting that not all missionaries were eager to comply with the emperor's wishes, illustrated by the well-documented case of Joachim Bouvet and his unauthorized return to the capital during the Great Wall survey of 1708.[185] Several original accounts recording the events show clearly that Bouvet harbored strong reservations even before the start of the land surveys. The result is twofold: first, the court Jesuits emerge as the diverse group of individuals they were, with sometimes radically differing opinions and approaches; and second we contextualize the mapping project within the general framework of the Jesuits' lives as missionaries, from which perspective the mapping of Qing territories was merely a side-activity, even when viewed as a capital opportunity to procure court protection for their missionary work or when fully coincident with their personal interests.

Justifying Missionary Involvement

As mentioned in the introduction to this study, Jesuit missionaries to Ming and Qing China had long been active in cartography in that the sciences and their many practical applications as taught in Jesuit colleges throughout Europe served to attract the attention and interest of Chinese elites. As a tool for evangelization, Western learning went on to make an imprint on imperial policy, first and foremost in the field of calendrical astronomy but also later in mathematics, medicine, cartographic practice, etc.[186] Particularly with the onset of the French Jesuit mission, this strategy became more focused, as missionaries found employment in the Imperial Household Department as highly specialised technical experts in addition to at the Directorate of Astronomy. The Jesuits thus followed a general strategy of providing the court with scientific know-how to procure imperial goodwill, gaining some leverage in support

185 Claudia von Collani has done much of the work in this regard. von Collani (1999). The Great Wall survey marked the beginning of a period of more intensive surveying, discussed more fully in Chapter 2.

186 Standaert (2001), 689–710; Jami (2012).

of their missionary work.[187] They also hoped that association with Western learning and its useful applications would convince the emperor of the validity of their religious beliefs.

Antoine Thomas, who emerged as one of the main actors in Chapter 1, clearly considered his involvement in imperial learning, and therefore also his contribution to Qing cartography, within this general strategy:

> In China, without the favor of the court, it is hardly possible to preach the Divine Law and for it to bear fruit. It is necessary that this proud country is compelled to acknowledge that not all that comes from outside its borders is barbarian. The human sciences and its practical examples must be placed under their eyes, so that authority is given to the preachers of the Gospel and to win over the minds of those who always despise what comes from outside just as the European despises the American savage. This is why every now and then it is necessary to abandon our apostolic work, so that the nascent Chinese church may firmly weather the dangerous storms, and sooner to secure the strong support from the emperor and important persons.[188]

Thomas here reiterates the importance of using the sciences to promote Christian beliefs. He explains two ways this could help missionary work: first, it would inspire elites not to dismiss the missionary's apostolic teachings as barbarian; and second, as a result, it would gain the mission imperial and elite protection. Thomas thus subscribes to the idea of using the sciences as a tool to facilitate, by association, proselytizing efforts and everyday pastoral work. This opinion, in one form or another, recurs widely both in contemporary published accounts and in more recent scholarship. Jean-François Gerbillon, the

187 The Jesuits often sought this leverage more through imperial favours or honours than through the titles they held in the administration. After having assisted the emperor in the measurement of the length of one degree of latitude (Chapter 1), Antoine Thomas received a four-character calligraphy carrying the imperial seal. Saldanha (2012), 177.

188 "In regno Sinarum, sine favore curiae, stare vix potest et fructum ferre Divinae Legis praedicatio; cumque sit superbissima natio eam necesse est cogere ad agnoscendum quod non sit barbarum id totum, quod ab exteris regionibus venit. Scientias humanas et earum insignia specimina coram oculis illorum adhibere oportet, ut praedicatoribus Evangelicis authoritatem ac benevolos animos conciliemus eorum, qui semper externos omnes, non secus ac in Europa barbaros Americae, despexerunt. Quare necesse est, nos identidem hic ab Apostolicis muneribus avelli, ut stet firmius nascentis Ecclesiae Sinensis structura citra pericula tempestatum, et nixa authoritate ac benevolentia Imperatoris atque magnatum securius celeriusque assurgat." Bosmans (1926), 177.

other main actor of Chapter 1, who often worked together with Thomas, wrote the following in one of his letters:

> Our position as missionary exposes us to other difficult occupations, and those who are at court much more so than the others. This is because of the great dependence one necessarily has on the will of the emperor and [other] great men for the continuation of their protection of the Christian religion.[189]

Gerbillon and Thomas therefore strode in the footsteps of earlier generations of Jesuit missionaries, men such as Adam Schall von Bell and Ferdinand Verbiest, who had used their practical and technical skills to win imperial acceptance and protection for their religious mission. For these missionaries, this often meant building close personal relationships with the emperor and other high officials. When Gerbillon passed away in 1707 while Thomas was hardly able to walk, however, a younger generation of missionaries took over scientific practice at the Qing court in a much larger group with whom the aging emperor enjoyed far less or no personal contact.[190] This group included Pierre Jartoux (1669–1720) and Jean-Baptiste Régis, both of whom engaged in scientific work at court and took the lead among the court missionaries in the more comprehensive period of land surveying that began in 1708.

Several sources suggest that Jartoux possessed remarkable skills in mathematics and mechanics and that he especially enjoyed solving mathematical problems.[191] Following his arrival at Guangzhou in 1701 as one of the eight missionaries brought back from France by *Mathématicien du Roy* Jean de Fontaney, the emperor soon called Jartoux to Beijing, where he started working at court in 1702 constructing clocks and instruments.[192] From 1708 until his death in 1720, he led the other Europeans involved in the mapping project. Régis, on the

---

189 "Nostre employ de missionaire nous expose à d'autres fatigues et ceux qui sont a la cour beaucoup plus que les autres à cause de la grande dépendance qu'il faut avoir des volontez du prince et des grands pour conserver leur protection a la religion chrestienne." Quoted from a letter written in Beijing on October 7th, 1701. de Thomaz de Bossierre (1994), 115.

190 Even though the court missionaries constituted only a small minority of missionaries in Qing China, their numbers at the turn of the eighteenth century reflected a general increase. The presence of European missionaries in the Chinese provinces peaked at around 140 individuals by 1700. Standaert (2001), 301–304.

191 ARSI, JS 182:393.

192 Jartoux left a short autobiography. ARSI JS 175:381–386v; Baldini (2012), 267 (n.91). For clock-making at the Imperial Workshops, see Pagani (1995); Curtis (2009).

other hand, was one of the missionaries recruited by the other *Mathématicien du Roy*, Joachim Bouvet, after the latter had left for Europe during the years 1693–1698. Upon arriving at Guangzhou in 1698, Régis received orders to go and meet with the emperor, who was on one of his tours of the Jiangnan area. Soon after, Régis assisted in the charting of river courses close to the capital and, together with Jartoux, would play a major role during the comprehensive surveying of Qing territories between 1708 and 1718.[193]

Régis apparently subscribed to Thomas' and Gerbillon's general justification for the missionaries' involvement in Qing cartography, but also contextualizes his own contribution further with regard to events of the early eighteenth century in an account written in the 1720s. Régis refers explicitly to the impact of the first papal embassy to the Qing, led by Charles Thomas Maillard de Tournon (1668–1710):[194]

> We were entirely convinced that [the emperor's] protection is absolutely necessary for the conservation of our holy religion. The hope found in being able to procure him with [the atlas] during the very troublesome times that followed the arrival of the Patriarch at this court was a sufficient motive to persevere in the midst of so many dangers that had to be negotiated; and amongst the hindrances that are inevitable when one has to deal with such different kinds of people.[195]

Gerbillon and Régis were among those selected to welcome the papal envoy to Tianjin in December 1705.[196] The Kangxi emperor received de Tournon in audience, but the encounter turned sour when the emperor learned one of the embassy's main goals: implementing a decision by Rome to forbid Chinese Christians performing ancestral rites and offerings to Confucius. The papal embassy constituted the culmination of the notorious Chinese rites controversy, a long-standing debate among missionaries and in Europe that defies

193 Fully discussed in Chapter 2.

194 McCune (1991).

195 "Et comme nous étions entièrement persuades que sa protection est absolument nécessaire pour la conservation de notre sainte religion, l'espérance de la lui pouvoir procurer dans des temps très fâcheux qui ont suivi l'arrivée de Mr. Le Patch [Patriarch] dans cette cour, a esté un motif suffisant pour nous soutenir au milieu de tant de dangers qu'il a fallut courir parmi tant de traverses comme inévitable quand on a affaires a tant de différentes espèces de gens…" BNF, Ms. français, 17242:7. Régis here writes of the "Patriarch," since de Tournon held the title of Patriarch of Antioch.

196 Fu (1966), 1:114.

easy dating or delineatation. Essentially, the controversy revolved around the issue of whether Chinese Christians could permissibly perform ancestral rites and offerings to Confucius, but identifying the correct Chinese word for God also spurred debate.[197]

A wide array of sources make it clear the papal embassy provoked general suspicion from the emperor with regard to the Catholic missionaries as he grew increasingly frustrated and impatient with the infighting and disagreements that had come to a head during de Tournon's visit at court.[198] Almost immediately, he introduced a system of permits (*piao* 票) for missionaries willing to vow adherence to the example set by Matteo Ricci, implying general toleration for Chinese ancestral rites, and exiling to Guangzhou or Macao those unwilling to do so. In light of these events, Régis argues, the mapping project provided the court missionaries an opportunity to safeguard the mission and regain some hard-won imperial goodwill. Régis' comments about having to deal with "all kinds of people" may refer to opposition from influential people at the Qing court as well as to de Tournon, and perhaps even to Joachim Bouvet, who harbored serious reservations regarding his own involvement in imperial cartography. In 1708, the emperor selected Bouvet to start what would eventually become the first stage of the comprehensive surveying of all Qing territories: with Gerbillon dead and Thomas barely able to walk,[199] the emperor turned to Bouvet, his last remaining expert tutor in Western learning, and ordered Régis and Jartoux, each from a younger generation of missionaries well-versed in practical geometry, to assist him.

The Unauthorized Return of Joachim Bouvet

In August 1708, more than two months into the first survey (which mapped the entire length of the Great Wall), Bouvet fell from his horse. The team's decision

197 For a detailed overview of events surrounding the Chinese rites controversy, see Mungello (1994), esp. 3–12.

198 E.g. Rule (2011), esp. 148–151 & 155–158; *Kangxi chao manwen zhupi zouzhe quanyi* (1996), 434–437. The emperor appears to have harbored particular frustration towards Gerbillon in the years 1706–1707 in light of the latter's role during de Tournon's stay in Beijing. de Thomaz de Bossierre (1994), 158. Foucquet even accused one missionary of dismantling telescopic sights from a Butterfield quadrant and replacing them with alidades, thus reducing the appeal of a new generation of Paris-made instruments. Witek (1973), 603. de Fontaney made similar claims. Jami (2012), 155.

199 A 1708 memorial mentions Thomas' deteriorating health. *Kangxi chao manwen zhupi zouzhe quanyi* (1996), 586.

was that Bouvet should return to Beijing while the others continued on without him.[200] The emperor strongly criticized this decision, which provoked complex negotiations among the throne, top Manchu courtiers, and the court Jesuits, who together eventually forced Bouvet to resume his mapmaking duties. In what follows I present my own analysis, building on the work of Claudia von Collani, of the aftermath of Bouvet leaving the team of mapmakers and returning to the capital.

Bouvet was one of Louis XIV's *Mathématiciens du Roy*, rather hastily ordained into the priesthood before he boarded the ship for Siam in 1685. When he wrote to his mother from Siam, he told her of the joy he had felt hearing some of the ancient stories of the Siamese, in which he had recognized hidden references to the coming of Christ and other vestiges of Christianity.[201] Such ideas would come to typify Bouvet's approach to Asia's ancient texts and stories.[202] At the Qing court, he quickly developed an interest in the *Book of Changes* or *Yijing* 易經, one of the most inaccessible of ancient Chinese classics. He became convinced that this text contained veiled references to the writings of the Church Fathers and reflected the common Christian origins of all mankind. Hence, all that remained for him to do was to crack the codes of Chinese characters and the hexagrams contained in the *Yijing*.[203] This search for Christian "figures" in different cultures's ancient texts came to be known as "figurism," which contemporary missionaries sometimes even mockingly called "Yijingism." Bouvet's reading of the classics grew ever more intense over these years, rendering him the most prominent figurist among the court missionaries. In 1707, he exchanged letters with fellow Jesuit missionary Joseph Henri Marie de Prémare (1666–1736), in which he claimed that, with more work, he would be able to demonstrate that "not only Chinese characters but all the major alphabets of the world stem from the hexagrams."[204]

---

200 On the complex history of the "Great Wall" as the cumulation of different structures, see Waldron (1994).

201 Swiderski (1980–1), 136.

202 The missionaries were generally interested in early Chinese historiography and many of them discussed or wrote on the potential parallels with early Christian history, but their approaches differed greatly. Standaert (2012); Standaert (2016).

203 As evidenced by his correspondence with Gottfried Wilhelm Leibniz (1646–1716) on the 64 hexagrams or *gua* (卦) contained in the *Yijing*. This inspired Leibniz to develop the system of binary numbers. Bouvet also believed that all Chinese wisdom and customs resided in the *Yijing*, but that the true Christian wisdom was lost to contemporary Chinese. Swiderski (1980–1), esp. 139. Note the remarkable parallel with those Qing scholars of the time who believed the true origins of Western learning originated in an obscured Chinese past.

204 Swiderski (1980–1), 149. For figurism, see Rule (1986), 154–182; Standaert (2001), 668–676.

Several months later, right before they set out to map the Great Wall on June 4th, 1708, the emperor received Jartoux, Régis, and Bouvet, and also perhaps the Qing officials who were part of their team, in audience and questioned them about applied geometry.[205] Bouvet wished to use the occasion to talk about his many years' labor on the *Yijing* and to express his desire to continue his research, hoping the emperor would relieve him of mapping the Great Wall. Bouvet, however, had not asked permission to discuss the subject, and so, when the emperor asked why he looked so pale, the missionary answered it was from reading so many books and the restless study it implied. Bouvet then drew his writings on the *Yijing* out from beneath his clothes and handed them to the emperor. By the end of the audience, however, he could only accept the emperor's orders, commend the plan to have a map made, and promise to execute it in due time. Von Collani has argued that, for those present at the audience, including the emperor, it was clear Bouvet wanted to stay in the capital to work on interpreting the *Yijing*.[206]

Around mid-August, more than two months into the Great Wall survey, the mapmakers reached the town of Shenmu 神木 in the very north of Shaanxi province, where Bouvet fell ill. The team decided he should return to Beijing. Von Collani has discovered more details surrounding the episode through a close reading of the *Acta Pekinensia* by Kilian Stumpf (1655–1720). She notes that Dominique Parrenin (1665–1741), the superior of the French Jesuit residence in Beijing who accompanied the emperor at Rehe 熱河 (Manchu: *R'e Ho*) in late August 1708, had written a letter to Bernard Rhodes (1646–1715) in Beijing, stating that Bouvet had fallen from his horse at more than 200 *lieues'* distance, closely matching the road distance between Rehe and Shenmu.[207] Further in his account, Stumpf recorded the following:

> During the geographic expedition, which had begun with two companions at the Eastern Sea end of the Great Wall, Bouvet fell dangerously from his horse during the month of August. Following the fall, [he suffered from] weakness in the chest and headaches from which he couldn't sleep all night. Therefore, a council was held with his companions and the mandarin leaders of the expedition. They were of the opinion that he should leave the concern for the map to the other Fathers, and inform

205 Chapter 2 will address the composition of the different teams of mapmakers.

206 ARSI, JS 138:913; von Collani (1999), 240.

207 The contemporary transcription of Rehe used by the missionaries was Jehol. Later, Rehe, initially named for the local river, would receive the new name "Chengde" 承德 or, in Manchu, "Erdemu be aliha fu."

> the Emperor of his return home, so as to recover from his ill-health. The emperor was informed of the case but only answered "saha," "I know." The permission to return home, however, had not been requested or granted in any way. So this Father, without effective permission from the emperor, was not allowed to return to Beijing, but after having recovered on the way, did so late in the evening on September 16th.[208]

Meanwhile, in Beijing, the Jesuit Claudio Filippo Grimaldi (1638–1712), who held the highest position among the missionaries at the Directorate of Astronomy, had reported a mistake made in predicting an occultation of the sun for September 14th. Hesihen, who held one of the most important offices within the Imperial Household Department, including responsibility for all missionaries at court, suggested Bouvet should join Grimaldi and other missionaries at the planned audience with the emperor on the 17th to discuss the matter.[209] During the audience, the emperor strongly reprimanded Bouvet for his unauthorized return and accused him of feigning illness.[210] A memorial, written by Hesihen and presented to the emperor during the audience, returned a few days later with the following reply:

> From the beginning, Bouvet did not want to go. If he seems to be in normal health, then he has faked his illness in order to return. If he recovered after his return, then, being a missionary, acting in such an undignified way can be called excessive. If he would have returned by himself the day he recovered, he could have been forgiven, even though his crime would have been severe. But that all Westerners are now thankful for this, We do not understand. [As for] you, who just quickly forwarded these memorials, you should be dismissed from your position and put on trial for your

208 "Bouvet in expeditione geographica, quam secundum murum magnum ab ora maris orientalis coeperat cum duobus sociis, periculose lapsus fuerat cum equo, prout mense Augusti. Innui lapsum secuta est infirmitas pectoris, dolor in capite cum vigilia tota nocte nunquam conniventibus oculis. quare inito consilio cum socis et manadrinis expeditioni presidentibus, visum is fuit, patrem relicta alius mappe cura possa subsistere, aut domum redire premonito Imperatora, ut valetudinem curaret. Monuerant quidem Imperatoram de casu, sed imperato solum respondit segha, cognovi. Licentia autem domum redeundi pro pre aut petita non fuerat, aut carre nullo modo concessa. Sic pater sine licentitia Imperatoris affective prehibita redit Pekinum, de itenere convalescens, atque in Regium parvenit die 16 septembris sero vespere." ARSI, JS 138:1035; von Collani (1999), 241.

209 Hesihen was a Christian, baptized as "Petrus" by José Soares (1656–1736) one year earlier. von Collani (1999), 241.

210 ARSI, JS 138:1035; 149:459–459v.

> crimes. When you open your mouth, you speak properly, but how is this consistent with what's written in the memorial?[211]

This reply reached Hesihen a few days later, on September 22nd, after which Hesihen and the missionaries who had been present at the audience worked to formulate a set of apologies giving explicit expression to the regret and sorrow each felt. Bouvet, however, was hesitant to express regret for what had happened, and so an argument erupted between him and the other court Jesuits as to whether he should do so. His colleagues were clearly unable to convince him, as the next set of memorials to the emperor included no separate, explicit apology from Bouvet. As a result, the palace's follow-up reply was unforgiving:

> You have not indicated whether or not Bouvet was sent back [to assist in mapmaking]. What is the use of writing so much?[212]

In the end, the threat of a court case against him and the order of the Jesuit visitor, his immediate superior, forced a deeply offended Bouvet to rejoin his fellow mapmakers.[213] He set out again from the capital on October 1st, although he could not catch up with the other mapmakers, who would reach the westernmost point of the Great Wall on October 16th. Nine days after Bouvet's departure, Hesihen unexpectedly passed away, following days on end spent in conflict with Bouvet and the emperor.

When Jartoux and Régis finally returned to the capital on January 10th, 1709, Bouvet was not with them but arrived a few days later. Although there is no

---

211 "博津[白晉]起初不願去。若其臉色照常，則裝病而還矣。若返家後痊愈，則伊為傳教之人，行如此不體面之事，可謂足矣。若是病愈之日自行趕回，則其罪雖大，尚可垂憫。對此眾西洋人謝恩，朕有所不解。爾如此碌碌轉奏應即革職，執送慎形司儀罪。爾張口講禮儀，不知此何書所載。" *Kangxi chao manwen zhupi zouzhe quanyi* (1996), 594. Note the publication only of the Chinese translation of the original Manchu text. All of this is consistent with the account put forward by Stumpf and presented by von Collani, as well as with a narrative of "the Bouvet case" by Thomas. A translation of the memorial including the imperial rescript, taken from European archives, appears in von Collani (1999), 242–243 & 248–250. Another account of the case, used by von Collani and written by Stumpf, survives in the Bibliotheca Nazionale Centrale Vittorio Emanuele II in Rome (FG 1254/17) and includes memorials and other related documents. Bouvet's case is also partly discussed in Chen (2011), 287–290. For Thomas' account, see ARSI, JS 149:459–459v.

212 "未寫遣返博津與否，多寫話何用" *Kangxi chao manwen zhupi zouzhe quanyi* (1996), 596.

213 von Collani (1999), 245–246.

explicit mention of personal conflicts among the European mapmakers,[214] the emperor never again called on Bouvet, who eventually resumed his work on the *Yijing*, to assist in mapmaking. In the spring of 1711, Bouvet had a Manchu courtier gradually transmit his writings on the subject to the emperor, who dismissed them as lacking logic or coherence. Soon afterwards, the emperor summoned Jean-François Foucquet (1665–1741), who had assisted Bouvet with his work on the *Yijing*, to the capital for an interview. Foucquet, who arrived there on August 7th, 1711, would go on to astronomical work at the Qing court while also collaborating with Bouvet on the *Yijing*.[215] Even, so, after the completion of the mapping project in the 1720s, it was the more pragmatic Régis who would produce the first-ever translation of the *Yijing* in a European language, showcasing a very different approach to the ancient classic of Chinese literature, even ridiculing Bouvet's ideas.[216] Other missionaries involved in the mapping project, particularly Pierre Vincent de Tartre and Joseph Marie Anne de Moyriac de Mailla (1669–1748), would later openly accuse Bouvet of over-preoccupation with the *Yijing*, even voicing their concerns in front of the emperor.[217]

Overall, the immediate impact of the Bouvet case for the Great Wall survey apparently amounted only to a short delay in Shaanxi province. At the same time, the wider implications are more difficult to gauge, although it certainly severed Bouvet's personal ties with a number of actors involved in the project and clearly left its imprint on relations between the missionaries and the palace.

Conclusion

In this intermission I have shown how different attitudes pertained among the Jesuit missionaries with regard to their involvement in imperial cartography,

214 Apart from Bouvet's personal stance in the discussion with other court Jesuits during his conflict with the palace, von Collani mentions that Bouvet felt he had received unjust treatment because he had the permission of Régis and Jartoux for returning to Beijing. von Collani (1999), 244.

215 *Kangxi chao manwen zhupi zouzhe quanyi* (1996), 720–736 [nr. 1716, 1719, 1724, 1725, 1731, 1743, 1738, 1741, 1752, 1755, 1759, 1760, 1764, 1766 & 1768]. For the order for Foucquet to come to the capital, see also Fu (1966), 1:116. For a French translation of official correspondence in this regard, see Witek (1973), 470–500.

216 McCune (1991), 246–247. Copies of Régis' translation reside at the Bibliothèque de l'Observatoire (BO) 1/12/152, 1 and AN, 2JJ90 (not consulted personally).

217 Rule (1986), 154–160.

highlighting their varying individual viewpoints. Much seems to have centered on questions of priorities. On the one hand, those who took the lead in the mission's involvement in Qing cartography saw it as a way to procure greater protection from the emperor and the Qing elites. As I have shown, particularly after the papal embassy of 1705–1706, the position of the court missionaries had weakened. In an attempt to alleviate the damage, several Jesuits employed at court attempted pragmatically to go quite literally the extra mile in their service to the emperor. For them, the mapping project constituted an opportunity to regain the emperor's trust and goodwill with regard to Christianity, thus safeguarding pastoral work and proselytizing efforts, the very essence of the mission. We also have the example of Bouvet, who felt he could better spend his time studying the *Yijing*, believing deeply that cracking the ancient text's code would reveal undeniably universal Christian truths within, leaving the emperor and his subjects no choice but to embrace Christianity. It thus made no sense to him to take part in the emperor's mapping project to regain or strengthen his goodwill.

Needless to say, I have only been able to scratch the surface of the divergent attitudes and opinions surrounding the missionaries' involvement in Qing cartography and scientific practice in general. There are signs, for example, that the rivalry between the "Portuguese" and the French missions also shaped missionary involvement. In her article on Bouvet's unauthorized return to the capital, von Collani argues that, before the first survey, Parrenin asked the emperor to replace one of the three French Jesuits with Kaspar Castner (1665–1709) or Xavier Ehrenbert Fridelli (1673–1743) because the French mission lacked personnel for missionary work. This suggests concern about the prospect of the French mission being underrepresented in the capital during the surveys and efforts to decrease that mission's burden.[218] At the same time, the source used by von Collani is one of the few explicitly connecting Parrenin to the mapping project, often commented on in the literature but entirely absent from those early-eighteenth-century sources I have been able to consult.[219]

---

218 von Collani (1999), 240. Taken from ARSI, JS 172:331v. Parrenin apparently had proposed this to one of the princes.

219 In the mid-eighteenth century, Gaubil wrote that Parrenin aroused the emperor's desire for a map of the Great Wall. Gaubil (1970), 214. Quoted in Bernard (1935), 458; Yee (1994), 180; Foss (1994), 142; von Collani (1999), 235. Elsewhere, Gaubil suggests that Parrenin corrected the emperor's idea that Mukden, today's Shenyang, lay to the east on the same latitude as Beijing. du Halde et al. (1702–76), 26:146–171. Gaubil claims that the emperor then sent Parrenin to make observations in Shandong and Liaodong. If correct, these events must have occurred in the years 1702–1708. Gaubil (1970), 541–542; Bernard (1935), 458.

Taken together, it is clear that we should not take the missionaries' involvement in the mapping project at face value. Certainly cartographic practice was not generally or necessarily of primary personal interest to the missionaries, who were mostly concerned with their pastoral duties and the mission's survival following the papal embassy of 1705–1706. Rather, it constituted an opportunity to bring the emperor and his people closer to tolerating their missionary work and Christian beliefs. In sum, the place of the mapping project within the Catholic missions is not as straightforward as often assumed, and much work remains to take into account the many factors and voices informing missionary involvement in the sciences as practiced at the Qing court.

CHAPTER 2

Of Instruments and Maps: The Land Surveys in Practice

> 天上度數、俱與地之寬大吻合。以周時之尺算之、天上一度、即有地下二百五十里。以今時之尺算之、天上一度、即有地下二百里。自古以來、繪輿圖者、俱不依照天上之度數、以推算地理之遠近、故差誤者多。朕前特差能算善畫之人、將東北一帶山川地里、俱照天上度數推算、詳加繪圖視之。[220]
>
> 康熙 50年5月4日
>
> The degrees of the skies correspond perfectly to the size of the earth. When we calculate them using the *chi* of the Zhou era, then for one degree in the skies, there are 250 *li* on the earth. When we calculate them using the contemporary *chi*, then for one degree in the skies, we have 200 *li* on the earth. Since antiquity, those who drew maps did not follow the degrees of the skies in order to calculate geographic distances, so that there were many mistakes. We have already purposely sent people who are proficient in calculating and good at drawing to [go and] calculate, in accordance with the degrees of the skies, the mountains, rivers, and distances in the northeast and to carefully draw maps that show this.
>
> June 19th, 1711

Focusing on the comprehensive land surveys carried out between 1708 and 1718, this chapter builds on the previous one by continuing to trace the surveying instruments from the Imperial Workshops deep into the Manchu homelands and the remote regions beyond the Great Wall, and to the many cities and towns of the densely populated Chinese provinces. I will then follow the maps drawn there, for the first time and on the basis of measurements and observations conducted with these instruments, back from the many regions and provinces of Qing China directly into the hands of the emperor and his

220 *Qing shilu* (1985), 6:440–441.

© KONINKLIJKE BRILL NV, LEIDEN, 2017 | DOI 10.1163/9789004345362_005

workshops. This chapter therefore ties together two families of material goods, namely instruments and maps, both chronologically and spatially so that they can serve as vehicles for exploring the events discussed in this study.

I first use mostly European sources as I zoom in on the very first stages of the comprehensive land surveys: the mapping of the Great Wall and the areas beyond its many gates and passes, including northern and northwestern frontier areas as well as Tibetan and Korean lands. This will not only paint a fuller picture of the methods and strategies used but also illustrate how selected data and information obtained during the surveys circulated separately from the resulting maps. In this chapter's second section, I turn to the available Qing sources in an attempt to gauge the logistics involved in the land surveys' later stages: the surveying of the Chinese provinces. I will discuss at length the practical organization behind the mapping of China's heartland, including timing, progress, team composition, communication with the center, and local support, prompting a closer look at the mapmakers, both Qing subjects and European missionaries, whom the emperor personally selected to serve in the surveys. Zooming in on their institutional backgrounds will reveal the Imperial Workshops as a pivotal locus for the duration of the land surveys, providing not only a necessary route book for the entirety of Qing territories, which would later become a textual companion for the resulting maps, but also the requisite manpower, material support, and coordinating back office for what became in the end an unprecedented cartographic endeavor.

This chapter touches on the individual ties and institutional forces that enabled the practical execution of the comprehensive land surveys over almost ten years. In so doing we develop an image fully encompassing the implications of scientific practice in a time when the vagaries of travel, climate, and other practical matters weighed heavily on fieldwork: surveyors needed to exercise patience and strive for precision in times of hardship. I argue that the imperial cartographers, both Qing officials and European missionaries, could not have achieved their task without cooperating with one another and with others outside their teams, both locally and in the capital. In what follows, I will discuss such cooperation between individuals and expose the occasional lack of it, all of which inevitably shaped the land surveys most directly.

2.1 Beyond the Passes: Observations and Calculations

The comprehensive land surveys occurred in several phases. After the mapping of the Great Wall in 1708, a team of mapmakers set out for the northeastern part of Qing territory in 1709 to conduct a survey of the Manchu homelands and outlying regions. The same men then surveyed Zhili province and the middle reaches of the Amur River in 1710, returning to the capital by the end of that year. In 1711, a second phase began, accelerating the surveys with the formation of two separate teams of surveyors, enabling simultaneous operations: the teams surveyed both the province of Shandong and the Khalka frontier that year, followed by Shanxi-Shaanxi and Henan-Jiangnan in 1712. In the third and last phase, no fewer than three teams of surveyors worked simultaneously to map the provinces of Zhejiang and Fujian; Jiangxi, Guangdong and Guangxi; and Sichuan, Yunnan, Guizhou, and Huguang, respectively. By late 1716, the maps of all of these Qing-controlled territories were complete. Smaller teams of mapmakers—not including European missionaries—also went to Korea (1711–1713) and to Tibet (1717–1718).

The priority given to mapping the vast lands beyond the passes of the Great Wall is explainable in different ways. Perhaps the most important factor was control and stability on the frontier. An expanding Russia had in earlier decades prompted the Qing to consider and delineate a northern border: Jesuit missionaries had helped negotiate the treaty of Nerchinsk, which prompted the early mapping of the borderlands of the far north. The strengthening of the Qing presence along the Amur River in the treaty's wake and the incorporation of Khalka lands in the northwest during the 1690s, all discussed in Chapter 1, made further knowledge of these regions all the more desirable.[221] Second, a nostalgic gaze towards the Manchu homelands and the Golmin Šanggiyan Mountains (Changbaishan 長白山) motivated the court to delineate more precisely the border with neighboring Chosŏn Korea. Finally, turmoil in Tibet and the renewed Zunghar threat beyond Hami had led the emperor to take a number of measures in the 1710s and inspired him to push for better maps of those regions as well.[222] In contrast to these frontier regions beyond the Great Wall, the geography of the Chinese provinces was well-documented in a strong tradition of route books, gazetteers, geographic compilations, and cartographic material, ultimately reducing the need to prioritize the mapping of these regions.

221 Elliott (2000). Maps of the (Inner-Mongolian) regions that were already part of Qing territories before the incorporation of Khalka lands had long existed, but not on the basis of new Qing cartographic practice as described at the end of Chapter 1. Ha-si-ba-gen & Cheng (2010).

222 Dai (2010), 76–83.

2.1.1 *New Qing Cartographic Practice along the Great Wall*

The *Acta Pekinensia*, a manuscript account by Jesuit missionary Kilian Stumpf of events in Beijing during the years 1705–1712, records that on May 21st, 1708, the missionaries at court first heard that some of them would soon start mapping the Great Wall, in close cooperation with a number of Qing officials. In the beginning of June, after several days of uncertainty as to selection, the emperor nominated fathers Joachim Bouvet, Jean-Baptiste Régis, and Pierre Jartoux, all of the French mission, and provided them with the necessary instruments at the imperial palace.[223] The actual survey began on June 4th, 1708, when a team consisting of three or four Qing officials and the three French Jesuits set out from the capital to map the length of the Great Wall.[224] This map reportedly measured nearly five meters long, covering the main branches of the wall as well as more than three hundred sites in its immediate vicinity, mostly forts, military installments, adjacent towns, gates, and locations where rivers big and small cross the wall.

An account written and sent to Europe by Antoine Gaubil in the late 1720s can aid our analysis of the precise techniques and methods used during this first survey. Gaubil composed it based on Jartoux's personal notes, including detailed technical information about the observations, measurements and calculations the mapmakers carried out in the field.[225] To begin with, Gaubil states that mapmakers used a mobile quadrant by the Chapotot workshop with a radius of 2 *pieds* and 2 *pouces* (c. 70 cm), which they calibrated at regular intervals so as to ensure ongoing precision.[226] Other instruments taken

223 ARSI, JS 138:907 & 913. No further details are available with regard to these instruments.

224 Régis, in his account, mistakenly writes of July 4th. BNF, Ms. français, 17242:2. du Halde's *Description* repeats this mistake. du Halde (1735), 1:xxix. Lunar calendar dates provided by several contemporary accounts help to rectify the mistake. von Collani (1999), 238 (n.48). Note that the Great Wall did and does not consist of one single structure and that mapmakers probably did not travel on the wall itself, but rather on roads along the wall. In an account written much later, for example, Antoine Gaubil mentions that the wall was in a "very bad state" in its more western sections. Gerbillon, in his diaries of Kangxi's 1697 tour to Ningxia, confirms this. du Halde (1735), 4:356–384. On the complex history of the Wall as a structure and its importance as a construct, see Waldron (2014).

225 Cordier (1915), 529–540. The original manuscript account is at APF, Brotier, 148:47–66. Elsewhere, Gaubil writes that Jartoux's diaries were damaged and incomplete: "J'ay quelques memoires, tout mutilés et epars, du feu P. Jartoux sur cette carte; c 'est luy qui la redigea et l 'envoya. Ces memoires mutilés du P. Jartoux en supposent d 'autres que je n 'ay pu trouver." Gaubil (1970), 668; Bernard (1946).

226 Cordier (1915), 536.

along included a number of compasses, a pendulum clock, and several smaller instruments.[227] With precisely divided ropes, each 1/10th of one *li*, they then started measuring the road from a Beijing city gate, via Tianjin and a gate of the Great Wall known as Shanhaiguan 山海關, to Chenghailou 澄海樓, the wall's easternmost extremity. On their way, mapmakers observed:

1. Compass bearings (direction), at all times;
2. The height of the sun at noon, weather permitting; and
3. The declination of the needle (the difference between the true north and the magnetic north), as a control measurement for (1).[228]

When mapmakers returned to Shanhaiguan, around June 20th, they had recorded the following measurements and observations (table 2):[229]

TABLE 2 *Observations and measurements recorded between Beijing and Chenghailou.*

Measurement/observation	Instrument	Result
Height of the upper edge of the sun at noon at Shanhaiguan	Mobile quadrant	73°43′
Declination of the needle	Mobile quadrant Compass	2°
Road distance from Beijing (including a list of compass bearings)	Divided ropes Compass	482 *li*
Road distance from Chenghailou (including a list of compass bearings)	Divided ropes Compass	10 *li*

227 Supposing Gaubil referred to the *pied de Paris*, this equals to about 70 cm. Régis notes that the instrument came from the Chapotot workshop in Paris. BNF, Ms. français 17242:9v. In his correspondence, Gaubil notes that as late as 1750, the French Jesuit residence still harbored an old damaged quadrant of 2 *pieds* and 2 *pouces* made by the Chapotot workshop in Paris. Gaubil (1970), 633. Elsewhere, he notes the use of instruments of the same type for the duration of the comprehensive land surveys. Pinot (1932), 185–186.

228 At the time, scientists believed that the declination of the needle could correct observed compass bearings and thus lead to better estimates of longitudes. Régis notes, however, that variations of the needle followed no logic and could not explain the magnitudes of variation, sometimes very divergent, on the same meridian at different times of the year. In the end, they did not use recorded declinations of the needle for correcting measurements. BNF, Ms. français 17242:11-11v; du Halde (1735), xxlvi–ii.

229 Gaubil also summarizes these methods elsewhere in his correspondence. Gaubil (1970), 214.

The Jesuits determined latitudes and longitudes for both Shanhaiguan and Chenghailou on the basis of these data. On the one hand, they calculated the latitude for Shanhaiguan after observing the height of the sun's upper edge at noon:

Height of the upper edge of the sun at noon	73°43′00″
Estimation of error (instrument + refraction[230])	1′17″
Corrected height	73°41′43″
Radius of the sun (in degrees)	15′49″
True height of the sun (at its center)	73°25′54″
Declination of the sun (from published tables)	23°26′45″
Calculated difference	49°59′09″
Latitude (difference with 90°)	40°00′51″

At the time, this method constituted one of the most precise ways of determining latitude, as the result clearly reflects.[231] On the other hand, the road was surveyed as a series of distances in combination with observed compass bearings (direction), as an itinerary traverse. Simply put, each section straight stretch of road is measured from a previously determined point (a) to a new point (b) (figure 23). If one imagines point (c), such that ACB is a plane right triangle, then the compass bearing is angle *cab*, so that the lengths AC and CB can be calculated trigonometrically. These are then converted directly to the difference in latitude (AC/200 *li*) and longitude (CB/200 *li*, with a correction for the poleward shortening of arcs of longitude) between (a) and (b). Do these calculations repeatedly, and the result is a series of small differences in latitude and longitude, from which the location of any point along the road can be determined with the road's origin, whether Beijing (taken to be 0° longitude) or some other site where repeated and stationary observations had established the longitude carefully *vis-à-vis* the Beijing meridian.[232]

230 They corrected refraction errors on the basis of published tables. Wang (1995), 145.

231 In comparison, satellite images on Google Maps record a latitude of 40°00′34″ for the gate's precise location.

232 Determining longitude was an important problem in scientific practice at the time. The French *Académie royale des sciences* had perfected the technique employing observation of the satellites of Jupiter (Chapter 1). It was the most precise method for determining longitude on land at the time, but Régis explains in detail that the technique was nearly impossible to implement in surveying Qing territories. BNF, Ms. français, 17242:6-6v; du

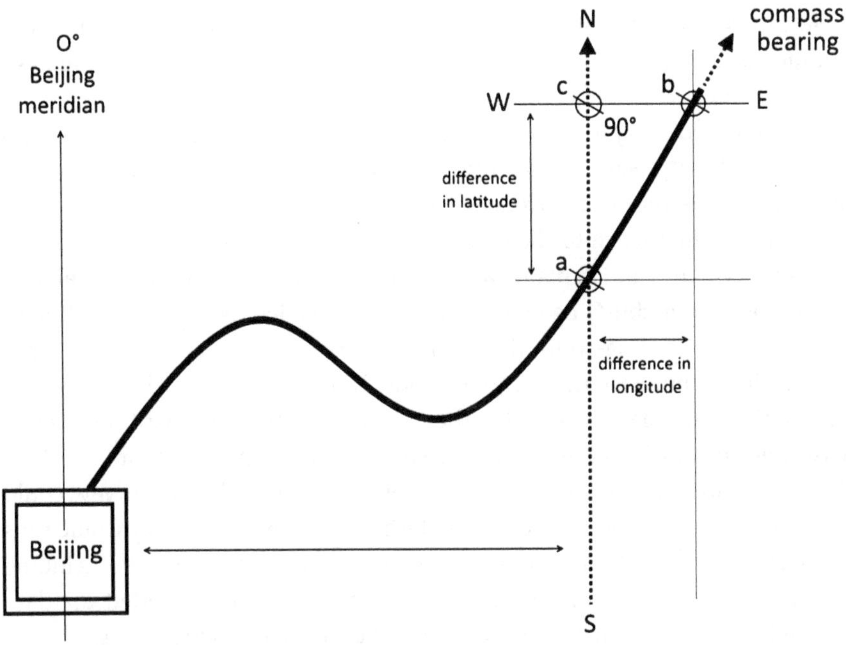

FIGURE 23 *New Qing cartographic practice and the method of determining longitude based on rope-measured road distances.*

We will best understand longitudes calculated in this way as estimates rather than precise determinations, although Régis staunchly defended this method as the most practical solution available. Considering the vastness of the territories to be surveyed and the density of towns and villages in the Chinese provinces, Régis argued, it was simply not feasible to widely use the method of observing Jupiter's satellites as developed at the Parisian *Académie*. He also points out that adhering to the methods described above permits the use of every observed latitude as a control for calculated longitudes by correcting the latitudinal distance as calculated between two points along the route with the distance inferred from the difference in latitudes as observed in those points.[233]

Halde (1735), xxxv–xxxvii. At the same sime, the missionaries had worked for decades on the longitude of Beijing *vis-à-vis* Paris, and de Fontaney had even performed observations of Jupiter's satellites in Beijing, according to the standards of the *Académie*. In his diaries, Ripa later confirmed the methods as explained by Régis and Gaubil, stating that they measured distances with "long chains" to obtain longitudes. Ripa (1991–96), 22:6. In several Qing sources, the terminology used is *zhanglian lucheng* 丈量路程. *Qingdai qijuzhu ce* (2009), 19:T10936-7.

233 BNF, Ms. français, 17242:6v; du Halde (1735), xxxvii.

One can correct an entire succession of route traverse sections in this way, ultimately also improving the estimates of longitudes calculated on the basis of trigonometry.

This method's major disadvantage is that it causes significant inaccuracies no matter how precise its execution: the farther from the Beijing meridian one travels, the more the errors accumulate with each traverse section. This is due in the first place to the earth's curvature, which this method does not take into account, and explains why mapmakers surveyed the Great Wall first even though Manchu-Qing expansion had superceded its ancient function as a boundary. With Qing-controlled territories stretching from the Amur River in the north to Hainan Island in the south and from the borders with Korea in the east to the desert oasis of Hami in the west, the Great Wall constituted a clear east–west axis with an average latitude not too divergent from that of Beijing. It was therefore useful as a correcting baseline to which all other surveys could reconnect; the longitudes determined during the Great Wall survey would thus serve as controls for future surveys. Several contemporary accounts, including one written by Régis, confirm that the Great Wall survey provided the data needed to "regulate" the longitudes of all other regions and provinces.[234]

In any case, the team mapping the Great Wall employed this method in estimating the longitude of Shanhaiguan at 3°09′ east of Beijing. Over the following years this estimate would adjust slightly, when another team of mapmakers passed through heading north. Thus, with both latitude and longitude for Shanhaiguan determined on June 20th, 1708, the mapmakers continued measuring road distances further along the Great Wall, constructing a succession of traverses on the basis of which they estimated longitudes. When mapmakers reached the next important pass and busy gate along the wall, Xifengkou, on June 26th, they again took their time in carefully determining latitude and using it in correcting the calculated difference in longitude from Shanhaiguan. According to Gaubil's narrative, they thus determined the latitude of Xifengkou as 40°26′00″ by observing the height of the sun's upper edge at noon, while estimating the longitudinal difference with Shanhaiguan at 1°13′, allowing them to infer a longitude of 1°56′ east of Beijing.

Shanhaiguan and Xifengkou thus became the first of a series of sites that would serve as necessary points of reference, the basis for the drawing of the map of the Great Wall. In his account, Gaubil notes that, at some of the less-important sites, or when the weather was unfavorable during daytime, they alternatively determined latitudes through directly observing the height of

234 BNF, Ms. français, 17242:2v-3; Cordier (1915), 538.

the pole star at night, which enabled a less precise measurement compared to observing the sun at midday.[235] Gaubil's narrative mentions many more of these key sites, as he notes their latitudes and longitudes and often specifies the dates associated with particular observations and calculations. An abstraction of this information lets us trace the mapmakers' progress step-by-step (figure 24 & table 3).[236]

This list further indicates the accuracy the mapmakers achieved in their calculations and observations: they recorded latitudes to the second, usually by observing the upper edge of the sun at noon with Paris-made precision instruments; whereas they recorded longitudes only to the minute, reflecting the much less precise method of calculating route traverses with the help of

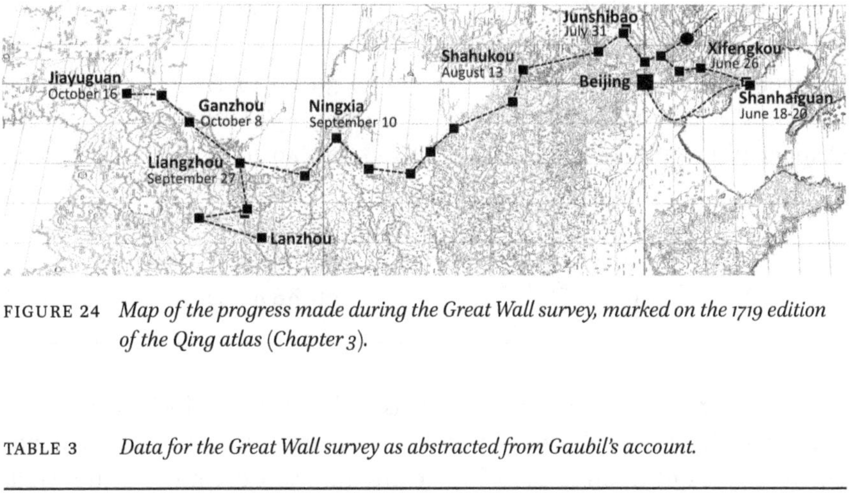

FIGURE 24 *Map of the progress made during the Great Wall survey, marked on the 1719 edition of the Qing atlas (Chapter 3).*

TABLE 3 *Data for the Great Wall survey as abstracted from Gaubil's account.*

June 18th	Shanhaiguan 山海關	40°00′51″ N	03°09′ E
June 20th	Chenghailou 澄海樓	40°00′10″ N	
June 26th	Xifengkou 喜峰口	40°26′00″ N	01°56′ E
June 29th	Tangquan 湯泉	40°15′20″ N	01°14′ E
July 6th	Gubeikou 古北口	40°43′15″ N	00°43′ E
	Sihaikou 四海口	40°35′ N	00°00′

235 For a more technical account of the two different ways to determine latitude, including a discussion of the necessary calculations and the inaccuracies involved in either method, see Wang (1995), 144–148.

236 de Moyriac de Mailla (1777–85), 11:314; Cordier (1915), 529–536.

TABLE 3 *Data for the Great Wall survey as abstracted from Gaubil's account.* (cont.)

July 31st	Junshibao 君示堡	41°13′24″ N	00°42′ W
	Dushikou 獨石口	41°19′20″ N[237]	
August 4th	Zhangjiakou張家口	40°45′15″ N	01°44′ W
August 13th	Shahukou 殺虎口	40°19′30″ N	04°10′ W
	[Side branch Great Wall]	39°37′ N	04°50′ W
	Shenmu 神木縣	38°55′20″ N	06°25′ W
	Yulinwei 榆林衞	38°15′08″ N	07°07′ W
	Qingpingbao 清平堡	37°42′40″ N	
	Huamachi 花馬池	37°52′45″ N	
September 10th	Ningxia 寧夏	38°32′40″ N	10′25′ W
	Zhongwei 中衞	37°40′45″ N	
September 27th	Liangzhou 涼州	37°59′00″ N	13°43′ W
October 8th	Ganzhou 甘州	39°00′40″ N	
	Suzhou 肅州	39°45′20″ N	
October 16th	Jiayuguan 嘉峪關	39°49′20″ N	17°56′ W
	Zhuanglangying 莊浪營	36°47′58″ N	
	Nandatongbao 南大同堡	36°40′20″ N	
	Lanzhou 蘭州	36°08′00″ N	
	Xining 西寧	36°39′00″ N	14°34′ W

compass bearings and road distances measured by rope.[238] The list also makes clear that what I have above dubbed "key sites" are further categorizable into different types of localities. First, there are the two extremities of the Wall, Shanhaiguan and Jiayuguan. Other key sites include important cities along the Great Wall, particularly along its more westerly reaches. Examples are Ningxia, Liangzhou, Ganzhou, Xining, and Lanzhou, which for obvious reasons they selected as major points of reference. For these cities, it is likely the mapmakers

237 They determined the latitude of Dushikou after measuring the road from nearby Junshibao, apparently due to time constraints.

238 For some of the key sites, Gaubil also includes road distances in *li* plus the number of extra ropes (1 rope equals 1/10 *li*). He sometimes also records declinations of the needle for places to the west of the Beijing meridian. Wang Qianjin has determined that the latitudinal values for important cities and towns as listed in du Halde (1735) are indeed more exact than the corresponding longitudinal values. du Halde (1935), 4:473–488; Wang (1995), 140–143. Baldini draws a similar conclusion. Baldini (2012), 271–272 (n.106).

had access to older but useful observational data, such as those obtained during the 1697 eclipse in Ningxia (Chapter 1), which would certainly have helped in double-checking calculated longitudes for these cities, all distant from the origin of longitude calculations at Beijing. In contrast, some of the minor key sites at first sight appear to have had no special status. Gaubil suggests that the selection of some reflected previous observations there that could help the mapmakers in verifying their observations and calculations. For example, Jean-François Gerbillon and Antoine Thomas had determined the position of Shahukou in 1697, and apparently those of other places between Xifengkou and Ningxia as well (Chapter 1). However, at least with regard to latitude, it seems no team ever adjusted the 1708 measurements. After all, Gaubil argues, the principal instrument used in determining latitudes was a mobile quadrant with a radius of seventy centimeters, which therefore delivered far more precise angular measurements than the instrument used by Gerbillon and Thomas, sporting a reported radius of only forty centimeters. Still other sites included in the list reflect perhaps more random selections, possibly dependent on whether the midday skies were clear enough for observation. In any case, the list of latitudes and longitudes at the end of du Halde's *Description de la Chine* (1735) contains most of these key sites but is not exhaustive, and thus does not constitute the full list of "anchor points" for the resulting maps, as claimed by much of the literature.[239]

Only once during their journey along the Great Wall, at Liangzhou, were mapmakers able to observe a lunar eclipse. This constituted a prime opportunity for precise determination of a place's longitude, provided they could compare the data with those of similar observations elsewhere.[240] Thus, they hoped the lunar eclipse would provide mapmakers a useful control for their determinations of longitude, which had already drawn them more than 1500 km from Beijing. Before the eclipse, just after midnight on September 30th, they prepared by calibrating a pendulum clock well in advance. Unfortunately,

239 This finds confirmation in (1) the title of du Halde's list: "Catalogue *d'une partie* des latitudes observées et des longitudes qui resultent des mesures geometriques…" [emphasis added]; (2) the fact that du Halde's list records longitudes to the second, suggesting they do not derive directly from the original data; (3) the pirated Dutch edition of du Halde's *Description*, which contains the claim that du Halde's list derived from Jean-Baptiste Bourguignon d'Anville's *Most General Map* (1734), a map only partially composed on the basis of data from the land surveys (Chapter 3). du Halde (1735), 4:473–488; du Halde (1736), 4:587–606. Therefore, it is a false assumption that latitudes and longitudes of precisely 641 sites formed the basis for the resulting Qing atlas. Several scholars claim this on the basis of the number of entries in du Halde's list. Baldini (2012), 271 (n.105); Bai & Li (2014), 104; Han (2014), 58; etc.

240 Wang (1995), 151–154.

the rain appears to have obstructed their efforts soon after the start of the eclipse, so the mapmakers were only able to observe and record the event's beginning. As it turned out, the observation would not be useful in correcting their estimates of longitude until several years later, when placed alongside data obtained by observing the same eclipse in Nuremberg, Marseilles, and Genoa. Gaubil claims that this comparison actually confirmed their estimate of Liangzhou's longitude, albeit some four years after they presented the map of the Great Wall to the emperor.[241] This episode hints at a wide circulation of specific data gathered during the surveys, separate from the resulting maps, a point on which I will elaborate below.

Finally, the dates included in the table above indicate that the surveyors spent nearly a month between two key sites just northeast of Beijing. The *Acta Pekinensia* helps to explain that a meeting between the mapmakers and the emperor in Kara Hoton, some two hundred kilometers north of the Great Wall, caused this holdup.[242] It was at least a fortnight before they continued their journey to the west. Von Collani, in her article on Bouvet, has demonstrated that, because of Jartoux's weak health, two Jesuit missionaries just arrived in Beijing, Xavier Ehrenbert Fridelli and Luigi Gonzaga (1673–1718), received orders to depart Beijing as potential substitutes, reaching Kara Hoton on July 18th. After eleven days of rest, however, no one replaced Jartoux and the same team of mapmakers continued mapping the Great Wall to the west of Gubeikou.[243] Even so, Jartoux's health issues were not the mapmakers' only holdup: as discussed in the intermission preceding this chapter, Bouvet fell ill in mid-August and had to return to the capital, marking the beginning of a remarkable incident in which Bouvet's personal reservations about mapmaking played out in unexpected ways. To make up for his illness, Bouvet's two personal servants supposedly "redoubled their efforts" to measure the road from Shenmu to Yulinwei, carefully checking compass bearings along the way and observing the exact latitudes of both places.[244] They recorded a road distance of 196 *li*, recalculated according to the methods described above as a difference of 42′ in longitude.[245] It is not clear whether the two servants returned to Beijing together with Bouvet, although it seems likely they did. After negotiations with

241 Souciet (1729–32), 1:35; Cordier (1915), 534.

242 According to the *Acta Pekinensia.* ARSI, JS 138 & 913. Kara Hoton is presumably present-day Harqin 喀喇沁, located north of the imperial retreat at Rehe and to the east of the Mulan imperial hunting grounds.

243 von Collani (1999), 241; ARSI, JS 149:460v & JS 175:382v.

244 "redoubler leurs efforts." Cordier (1915), 532.

245 Cordier (1915), 532.

the palace and his colleagues in the capital, Bouvet found himself forced to rejoin the mapmakers and only returned several months later, a few days after Jartoux, Régis, and the Qing officials had returned to the capital on January 10th, 1709.[246]

2.1.2 *Revisiting the Manchu Homelands and Northern Frontiers*

The Manchu homelands (1709). Only four months after the mapmakers' return to Beijing, they went to survey what the missionaries called "East Tartary," the Manchu homelands in the northeast of Qing territories roughly coinciding with today's Liaoning, Jilin, and Heilongjiang provinces. The emperor selected the same missionaries again for this project, with the exception of Bouvet, whom Fridelli had replaced for the reasons explored in intermission one. Furthermore, a set of recently-discovered Manchu documents, which Cheng Zhi has discussed, reveal the identities of some of the Qing officials on this team. It lists no fewer than eighteen individuals, from high-level officials to the clerks and craftsmen in their service. It mentions all the main team members by name: Colonels Dekjingge and Toofahaci, Investigating Censor Udari, Vice Director in the central administration Šereng, and one of the five directors of the Directorate of Astronomy, Cengde. Other team members included an inner palace clerk, Bursai, two messengers from the Escort Office, Nastai and Mingcing, another bannerman, Hemuyen, Director in the central administration Guwandung, two unidentified map-drawing experts, and three more low-ranking clerks.[247] This provides a rough idea of the size and composition of this team of mapmakers and indicates that several experienced bannermen led the mapmaking expeditions north of the Great Wall. During the the surveying of the densely populated Chinese provinces in the later stages of the land surveys, when team composition became more streamlined, both Bursai and Toofahaci again received orders to join the teams setting out for the purpose, discussed at length in the second part of this chapter.

On May 8th, 1709, the team of mapmakers set out to what is today Liaoning province. This region borders the Great Wall to the south, so that mapmakers could start from Shanhaiguan, one of the sites for which the Great Wall survey had carefully determined latitude and estimated longitude. They then followed the coastlines of the Liaodong bay before traveling to the border with Chosŏn Korea at Fengcheng 鳳城, eventually arriving at the Qing's other

246 von Collani (1999), 244.

247 Cheng (2015), 21–32. Cheng Zhi provides a Chinese translation, but I have not been able to consult the Manchu-language original personally. The *xiangdaochu* 嚮導處 or Escort Office provided the entourage for imperial outings or imperial envoys.

capital, Mukden, by way of Hetu ala (Xingjing 興京).[248] From there, they slowly made their way up towards Ningguta and reached the northernmost border with Korea at Hunchun 琿春 on July 25th.[249] The mapmakers then surveyed remote reaches of the Amur and Ussuri Rivers and ventured deep into the lands of the Yupi 魚皮 and Heijin 黑斤 tribes.[250] The Manchu documents translated by Cheng Zhi greatly emphazise the supply of horses, boats, and food in these remote regions as well as the need for guidance and security, all provided by local personnel and certified ginseng diggers familiar with the terrain. They also warn of the many insects during the summer and stress that the mapmakers should not risk their lives by carrying on with low supplies. On September 8th, 1709, the team reached the northernmost site of the 1709 survey, where the mapmakers determined latitude and longitude in the fishing village of Dondon on the banks of the Amur River.[251] They then returned to Ningguta, presumably by river, but their route back to Beijing remains a mystery. One source indicates they traveled through the very east of Mongol territories, crossing the Great Wall at Xifengkou, but no further evidence supports this.[252]

Just as data from the eclipse at Liangzhou during the Great Wall survey found its way across the globe and into the hands of European astronomers many years ahead of the resulting maps, so did surveying the Manchu homelands and outlying regions lead to an immediate, widespread, and separate circulation of particular information. In Europe, a letter written by Jartoux and published in the popular *Lettres edifiantes et curieuses* describing information and experiences this missionary-mapmaker gathered in the far north, contributed directly to greater appreciation and understanding of the ginseng root's qualities and uses:

> The map of Tartary, which we are now drawing by the emperor's orders, procured us an opportunity of seeing the famous plant *gin-seng*, so highly valued in China and so little known in Europe. About the end of July 1709,

248 Xingjing, today's Xinbin 新賓 autonomous county, was the first capital of the Late Jin or Hou Jin 後金 (1616–1636), precursor to the Qing (1636–1911).

249 The Manchu memorial confirms the planning of this route in advance. Cheng (2015), 21.

250 The Yupi, literally meaning "fish-skin," took their name from their clothing made of fish hides. The missionaries at the time transcribed "Heijin" as "*Ketching*." Nowadays, they are known as the Nanai or Hezhe 赫哲 people.

251 Just north of the river by the same name, known today as the Anyuy (Анюй) River in Russia. APF, Brotier 148:45.

252 BNF, Ms. français, 17242:3. No other sources confirm this.

> we arrived at a village distant but four short leagues from the Kingdom of Korea, and inhabited by tartars called Calca-tatse [Khalka]. One of those Tartars went and fetched, from the neighboring mountains, four plants of *gin-seng*, and brought 'em to us, unbroken, in a basket… [Later], finding myself so greatly fatigued and my spirits exhausted to such a degree, by our journey, that I could scarce sit on horseback; a Mandarin in our company observed it, and gave me one of these roots. I instantly took half of it, and, an hour later, my feebleness was quite gone.[253]

Jartoux's account would later inform published descriptions of the root's magical qualities, leading in turn to the quest for and the discovery of ginseng in North America.[254] The story of ginseng serves as another example of information gathered during the land surveys circulating widely over time and space separate from the resulting maps, informing dynamics extending far beyond cartography and this study's focus.

Scant evidence indicates that several mapmakers later faced charges of sabotaging the horse relay system during the land surveys' first stages. As a result, banner Colonel Dekjingge lost his title while his personal servant and the main culprit, Guan-bao 關保, received eighty lashes of the whip and had to carry a cangue[255] for no less than one month. In contrast, Fridelli and bannerman Hemuyen, both implicated in the matter as well, received pardons, even though Hemuyen had to relinquish his renumerations for three years.[256] It is unclear whether this case relates to the 1709 survey, however, since these three men could have been part of one or more other teams as well. In any case, this episode is strongly reminiscent of the controversy surrounding Bouvet's unauthorized return to the capital, providing yet another example of unforeseen circumstances shaping the land surveys' progress and practical execution.

Zhili province and the new Amur garrison (1709–1710). Scarcely had the mapmakers returned from the northeast before the emperor sent them out yet again, this time to map the capital province of Zhili, a barely-documented episode in the mapping project. Because of the large number of towns and

253 Narrative taken from an early English edition of the *Lettres édifiantes*. Lockman (1995), 2:424–437. The original is in du Halde et al. (1702–76), 10:149. Cheng (2015), 21. For more on ginseng in early modern times, see Nappi (2013).

254 Nappi (2013), 38.

255 French, from the Portuguese "canga," for "yoke." This device, like a mobile European stocks with abundant space to display information about the guilty party and his crime, served as public humiliation.

256 *Qingdai qijuzhu ce* (2009), 19:T10936-7.

villages in that province, especially compared to the regions in the northeast of Qing territories, mapmaking in this relatively small province lasted from December 10th, 1709 until June 22nd, 1710, interrupted only by the mapmakers' return to the capital for the Spring Festival.[257] Immediately afterwards, on July 22nd, the emperor asked the same team to map the area surrounding a remote garrison established in response to an increased Russian presence on the banks of the Amur River. Along the route thither, they passed through Cicahar and Mergen,[258] two towns established to supply that remote town (named Sahaliyan after the Manchu name for the Amur).

Régis notes that, on the way back from this survey in the remote far north, the mapmakers were able to precisely measure the distance of six degrees of latitude (between 47°N and 41°N): they found no houses, trees, or large rivers, only "plains as far as the eye can see."[259] The mapmakers thus gained an excellent opportunity to confirm or improve the 1702 measurement of the length of one degree of latitude discussed in Chapter 1. To their surprise, they found a difference in distance for each degree measured, no matter how careful their rope measurements and observations of latitude, and even after repeatedly checking the ropes and recalibrating their mobile quadrant. This perplexed Régis and Jartoux: the added combination of all possible inaccuracies from the use of instruments and ropes would not nearly explain a difference of 258 *chi* between 47°N and 46°N and all subsequent degrees.[260] It is unclear whether they discussed the matter with the emperor when they arrived back in the capital on December 14th, yet when Régis wrote his manuscript account on the geography of Tartary in the mid-1720s, now at the Bibliothèque nationale de France, he seems to have realized they had discovered divergences in the distance covered by different degrees of latitude, a phenomenon he called a "conjecture of Mr. Huygens."[261] Christiaan Huygens was a celebrated mathematician who had built the first reliable pendulum clock and, with Isaac Newton (1643–1727), had been the first to argue that the earth should flatten slightly at the poles, implying differences in the distances covered by different degrees of latitude.[262]

---

257 BNF, Ms. français 17242:3.

258 Known in Chinese as Qiqiha'er 齊齊哈爾 and Nenjiang 嫩江, respectively.

259 "que plaines à perte de vue, sans maison, sans arbre et mesme sans riviere bien considerable." BNF, Ms. français 17242:3v & 9v.

260 c. 80 m. Further discussion of this issue occurred later in the eighteenth century. Pinot (1932), 185–186.

261 The actual difference in distance between these degrees of latitude is on average 12.44 meters. Wang (1995), 154–155.

262 Pendulum experiments had shown a variation in effective gravity for different latitudes. Pinot (1932), 185–186; Levallois (1988); Greenberg (1995), 15–16.

In spite of this realization, the teams retained the fixed imperial standard, dating back more than a decade, of 200 *li* for one degree of latitude throughout the land surveys. According to Régis, the mapmakers were right to leave the "exact solution for this new problem to those who have more facilities and free time than we do," although the authority of the imperial standard must have played a much more significant, if not decisive, role.[263]

Shandong province and the Mongol frontier (1711). Mapping efforts gained momentum in 1711 after the surveying of what the emperor considered Manchu territory and its outlying regions, as well as the important frontier garrison and its supply route, was complete. Two teams went out at once, the first to the Mongol territories in the northwest, including the lands of the Khalka, the second to the Chinese province of Shandong. In order to deal with this, the only two missionaries who had been involved in every survey so far, Jartoux and Régis, finally split up. In spring, Régis went to Shandong with João Francisco Cardoso (1677–1723) of the Portuguese mission. Information on this episode is very scarce, with only the observation of a lunar eclipse by Régis in the town of Zhucheng 諸城 and the determination of latitudes and longitudes in the towns of Laizhou 萊州 and Tengzhou 滕州 recorded in a 1729 publication.[264] Meanwhile, Jartoux went to map Khalka territories on the frontier together with Fridelli and Guillaume Fabre-Bonjour (1669–1714), a French Augustinian known in Europe as a gifted intellectual.[265]

As described in Chapter 1, the first mapping of the Khalka frontier took place in 1698, but the re-standardization (Chapter 1) of the Qing's most basic unit of length, the *chi*, necessitated doing it all over again. Largely following the route taken during the 1698 survey, Jartoux, Fridelli, Fabre-Bonjour, and their Manchu companions traveled far north of the Great Wall from Dushikou to determine the position of several lakes, the biggest of which was Lake Hulun. They then followed the banks of the Kherlen River westwards to the exact location of Galdan's defeat in 1696, near present-day Ulan Bator, before continuing northwest towards the Selenga River. Somewhere along that river, Jartoux

263 "la resolution exacte de ce nouveau probleme à ceux qui auront plus de commodité et de loisir que nous en avons." BNF, Ms. français 17242:11. This passage appears, remarkably, heavily edited in du Halde (1735), xliv–xlvi. Gaubil notes that Jartoux later proposed to trace and precisely measure the Beijing meridian from north to south, as Picard and Cassini had done in Paris (Chapter 1), but the emperor refused. Gaubil (1970), 444. Gaubil believed the actual distance for one degree of latitude was closer to 193 *li*. Gaubil (1970), 735.

264 Souciet (1729–32), 1:36.

265 Baldini (2012).

and Bonjour supposedly held conversations on religion and world geography with several Buddhist lamas.[266] Continuing on their journey, mapmakers finally deviated from the 1698 route, heading southwest until they reached Hami, which was to become the westernmost site covered by the land surveys. The *Observations mathématiques, astronomiques, geographiques, chronologiques et physiques* (1729) records that the mapmakers measured the meridian altitude at Hami on September 16th, 1711 using an instrument with a radius of more than two French *pieds*.[267] The teams then made their way back down to Jiayuguan, the westernmost outpost of the Great Wall, where they could link up with sites determined in 1708. Observations of latitude in Jiayuguan and in the nearby towns of Suzhou and Ganzhou were similar to those made during the Great Wall survey, and mapmakers found to their surprise that, after calculating a long succession of road distances and compass bearings, they estimated the longitude at Jiayuguan as differing by only a few minutes from the 1708 estimate. The mapmakers then returned to Beijing through the Ordos region, following the course of the Yellow River north of the Great Wall from Ningxia and arriving back at the capital in January 1712.[268] Jartoux would later write of this journey as "extremely long and most difficult" and, more dramatically, that "I could have lost my life a hundred times."[269]

In all, studying the first stages of the land surveys executed between 1708 and 1711 makes it clear that mapmaking took place with remarkable speed, particularly considering the great distances involved. Several factors may account for this: first, all Qing territories beyond the Great Wall were remote compared to the Chinese provinces. Formal prohibition barred Han Chinese from migrating into the Manchu homelands and frontier areas, and the further north and west one traveled the fewer towns and villages one encountered. Second, and related to the previous point, several mapmakers' accounts mention that war and displacement over previous decades had left many towns in the northeast in ruins and deserted.[270] Third, the mapmakers followed established routes

266 Baldini (2012), 272–273.

267 Souciet (1729), 1:12. This instrument was likely the same quadrant used during the Great Wall survey, produced by the Chapotot workshop in Paris.

268 In his writings dating back to the 1720s, Gaubil suggests the mapmakers surveyed the course of the Yellow River in the Ordos region. Cordier (1915), 544. *Qingdai qijuzhu ce* (2009), 19:T10936-7 confirms mapmaking in the Ordos region during the fall of 1711.

269 ARSI, JS 175:381-86v. Quoted by Baldini (2012), 269 (n.99). Baldini also traces the route taken by the mapmakers. Baldini (2012), 270–271.

270 E.g. BNF, Ms. français 17242:3-3v. These are partly a reference to the Qing-Zunghar wars of the 1690s (Chapter 1).

across these regions and, particularly in remote frontier areas, did not have to interrupt their travels to determine the positions of nearby settlements. As a result, they only mapped routes across the region, allowing them to cover great distances in a relatively short time (figure 25 & table 4).

On the other hand, the fact that the mapmakers only charted certain routes across these territories, some of them very remote, meant little or no backtracking between surveyed sites was possible, which in turn entailed fewer control measurements and less precise determinations of longitude, a problem made worse by the distances involved.[271] Here the 1708 survey of the Great Wall came into play: traveling north from the capital and back, mapmakers inevitably had to pass at least twice through the gates of the Great Wall, allowing them to link up with sites located in 1708. In other words, longitudes and latitudes determined during the Great Wall survey served as controls, particularly useful considering the very large distances involved in surveying the northern frontiers. This was crucial since long distances exponentially magnified the inevitable errors in longitude, as explained above, in calculating route traverse sections with the help of road distances and compass bearings. The mapmakers could likewise link the first province surveyed, Zhili, up with the 1708 survey, in turn providing a link for the Shandong survey of 1711.

When we plot the routes followed during the surveying of areas north of the Great Wall in a schematic fashion on one of the resulting overview maps (figure 25), it becomes clear that the maps depict places the teams did not survey, including the region surrounding Lake Baikal, the Upper Amur, the estuary of the Amur, and the coastlines to the north of the Korean peninsula. Gaubil confirms that the depictions of areas beyond the surveying routes outlined above reflect very rough estimates derived from conversations with locals about routes and distances involved, travel accounts of other imperial envoys, and data missionaries had gathered during imperial outings over previous decades.[272] This explains why the shapes of Lake Baikal and of the coastlines of the northeast in the resulting maps are entirely off the mark. The fact the mapmakers did not actually survey all areas depicted relates intimately to Qing military and administrative control: they deemed certain areas safe enough for the mapmakers to visit, whereas other areas they simply did not. If anything, this speaks to the Qing court's projection of power over areas depicted but not surveyed.

---

271 Also discussed by Régis in BNF, Ms. français, 17242, f. 6v-7.

272 Cordier (1893), 42–43; Cordier (1915), 526–527 & 540–541.

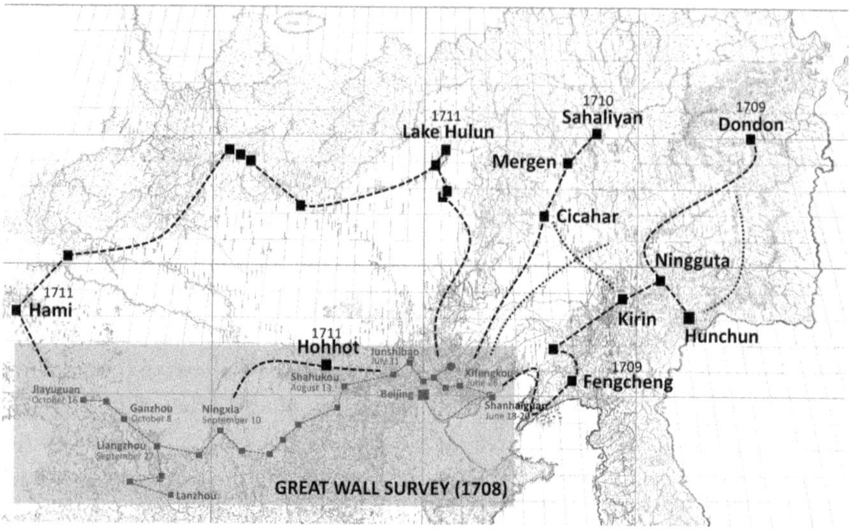

FIGURE 25 *Attempted schematic reconstruction of routes followed for the surveying of northern Qing territories (1708–1711), marked on the 1719 edition of the Qing atlas (Chapter 3). Dashed lines indicate confirmed routes taken by the mapmakers, dotted lines indicate likely routes.*

TABLE 4 *An overview of the first six surveys with the names of the European mapmakers involved. Only the Qing officials involved in the 1709 survey of Manchu homelands are identifiable.*

04/06/1708 – 10/01/1709	The Great Wall	Jartoux, Régis, Bouvet, & Qing officials
08/05/1709 – 11/1709	Manchu homelands	Jartoux, Régis, Fridelli, & Qing officials
10/12/1709 – 22/06/1710	Zhili province	Jartoux, Régis, Fridelli, & Qing officials
22/07/1710 – 14/12/1710	Amur garrison and supply route	Jartoux, Régis, Fridelli, & Qing officials
04/1711 – 10/1711	Shandong province	Régis, Cardoso & Qing officials
14/04/1711 – 01/1712	Khalka and Ordos territories	Jartoux, Fridelli, Fabre-Bonjour, & Qing officials

2.1.3 *Strategic Expeditions into Korea and Tibet*

Korea (1711–1713). In 1627 and 1636, two Manchu invasions forced the Chosŏn court to cease supporting the Ming and submit themselves, at least nominally, to Qing overlordship.[273] Athough this may well have contributed to the Kangxi emperor's interest in Korean geography, the peninsula's proximity to what the Qing considered the Manchu place of origin, the Golmin Šanggiyan Mountains or, in Korean, Paektu Mountain, certainly also played a decisive role. To the east, the mountains straddled the Qing-Chosŏn border, an area not easily accessible at the time. As early as 1677, the emperor ordered the eastern slopes of these mountains mapped, apparently causing the Korean court to debate the threat, as they saw it, to Korean settlements in the area.[274] In 1692, Korean officials refused to cooperate with a Qing team sent to inspect the headwaters of the Yalu 雅魯 and Tumen 圖們 rivers. In 1699, the Qing court, further pursuing its interest in Korean geography, demanded that Chosŏn envoys produce a map of their lands, which the envoys avoided only through careful negotiation. After sending a surveying team to map the Manchu homelands in 1709, however, the Kangxi emperor became more determined in pushing for a better map of the Qing-Chosŏn borderlands.[275]

For the 1709 survey, the mapmakers had carefully determined the positions of the border towns Fengcheng and Hunchun, where the Yalu and Tumen rivers meet the sea, constituting the Qing-Chosŏn border's two extremities. But this survey had not covered the entire length of the border between these cities, marked by the course of both rivers into the remote territories surrounding the Golmin Šanggiyan Mountains. Fully aware of this gap in his map, the Kangxi emperor soon found his pretext: In 1711, he dispatched two Qing officials from the capital to solve the case of Lee Wan-ki 李玩枝, a Korean ginseng smuggler who had murdered several Qing merchants on the Qing side of the Yalu River. Fully convinced of the importance of border stability in governing his territories, and with the blanks on the map in mind, the emperor decided to push ahead:

> I have dispatched two officials from the ministry, to go to Fengcheng to bring to trial the case of the Korean Li Wanzhi. I have also sent the general commander of Daxing Ula,[276] Mu-ke-deng, to go with them. When you ask [the Korean side] to execute the verdict, which We will secretly decree, you will go there and inspect the place. You will follow the river

273 Schmid (2007), 130. This offers some insight into the question of why all editions of the resulting Qing atlases (Chapter 3) include Korea.

274 Li (2006), 85–120.

275 Ledyard (1994), 298–299.

276 A post in Jilin (named "Butha Ula," in Manchu), established for controlling the ginseng trade and to collect ginseng for the state.

> upstream with the Korean official. If you can travel over land that belongs to China [*zhongguo* 中國], then travel over land that belongs to China with the Korean official. Alternatively, if there are obstacles on the land that belongs to China that do not allow passage, you will travel on land that belongs to Korea. Ride to the farthest place and observe carefully. When the task of clearly ascertaining the border [is finished], report back.[277]

Thus, in 1711, after supervising the smuggler's beheading on the Chosŏn side of the border in the village of Wiwon, a Manchu official named Mu-ke-deng was to inspect the border as far as possible upstream along the Yalu River.[278] A well-informed Chosŏn court ordered its officials to obstruct Mu-ke-deng's efforts to enter Chosŏn territory in any way possible but eventually had to yield when Mu-ke-deng produced the secret imperial edict. Even so, following many tribulations, the expedition came to a halt shortly after passing the Korean village of Manpo, due to the region's remoteness as well as because of Korean officials' efforts at obstruction.[279] A second expedition including Mu-ke-deng's party and their Korean escorts, more than fifty men in total, repeated the effort in the spring of 1712. A significant part of this group finally reached the crater lake of the Golmin Šanggiyan Mountains, approaching it from the east on June 18th, 1712 and erecting a stele to mark the border.[280] From the maps eventually included in the Qing atlas, it appears that Mu-ke-deng and his assistants determined the longitudes and latitudes of several places along the Yalu River, although the precise circumstances and methods used remain unclear. As Mu-ke-deng mapped the border, the team of Régis and Cardoso was surveying Shandong province. In his account, Régis mentioned Korea in the context of reaching the easternmost tip of the Shandong peninsula at a latitude of 37°24′

277 "前遣部員二人往鳳城會審朝鮮人李玩枝事。又派出打牲烏喇總管穆克登同往。伊等請訓旨時朕曾密諭云爾等此去並可查看地方同朝鮮官沿江而上。如中國所屬地方可行即同朝鮮官在中國所屬地方行。或中國所屬地方有阻隔不通處爾等俱在朝鮮所屬地方行。乘此便至極盡處詳加閱視。務將邊界查明來奏." *Kangxi chao shilu, juan* 246.

278 In his diaries, Ripa claims to have befriended him and that the Jesuits had trained him. Ripa (1991–96), 2:26–27.

279 Schmid (2007) includes the full narrative of these events as recorded by the Chosŏn court.

280 Schmid notes this date as May 21st, most likely referring to the lunar date. Schmid (2007), 145. See also Ledyard (1994), 301; Li (2006), 110–112; Ma (2009). Li records the inclusion of imperial bodyguard Su-lun 蘇倫, secretary of the Ministry of Rites E-Shi 鄂世, secretary Su-er-chang 蘇爾昌, and others.

and a longitude of 6°44′ east of Beijing: "[we] could not perceive any Korean lands, nor the islands that are at the mouth of the Han River."[281]

The emperor sent Mu-ke-deng to Korea once more in 1713, this time as a special envoy delivering an imperial edict to the court of the tributary state. No missionaries traveled with him, but several assistants and at least one mathematician were in his entourage. The mathematician was He Guozhu 何國柱, a minor official trained by the European missionaries and employed by the Directorate of Astronomy.[282] Chosŏn sources speak of men running up and down the road with ropes measuring every step along the way while mathematicians observed latitudes and compass bearings. In his diaries, Jesuit missionary Matteo Ripa (1682–1746) writes that Korean guards kept a close watch on the company, not letting the envoys out of their sight and taking notes of everything they did, even when "evacuating their bowels."[283]

Even before reaching Seoul, the Chosŏn court learned that Mu-ke-deng would demand sensitive cartographic data. On July 19th, 1713, Mu-ke-deng officially requested (1) hydrographic information on the border region, (2) a list of distances between Korean towns, and (3) a general map of the peninsula.[284] In return, Mu-ke-deng apparently offered a copy of the road map from Fengcheng to Seoul drawn by his team, as well as several surveying instruments I have been unable to identify.[285] Behind the scenes, the Koreans also discussed how best to acquaint themselves with the Qing surveying team so as to learn as much technical information as possible. In the meantime, the mapmakers observed the latitude of Seoul at 37°39′, which they would later use in drawing the map of Korea together with the road distances from Fengcheng and the newly obtained maps and distance data handed over by the Chosŏn court.[286]

---

281 "il ne put apercevoir aucun lieu de Corée, ni des isles qui sont vers l'embouchure de la Rivière." APF, Brotier 148:45v.

282 His title at the time, identified on the basis of Korean sources, was *wuguanzheng* 五官正, the official title for a director of one of the five offices in the Directorate of Astronomy. He was the cousin or younger brother of He Guozong 何國宗 and He Guodong 何國棟, all trained in mathematics. Hummel (1943), 1:285–286 & 330. Note that this was also an important episode with regard to exchanges in mathematics between Korea and the Qing. Jami (2012), 277–280.

283 Ripa (1991–96), 2:26–27. Régis' account in du Halde's *Description* confirms this nervousness of Korean guards. du Halde (1935), 4:423–430.

284 Ledyard (1994), 301–303. Note that the Korean court did not give their most elaborate maps to Mu-ke-deng. Instead, after careful deliberation, they handed over a less-accurate map.

285 Ledyard (1994), 301.

286 APF, Brotier 148:45v.

This explains why the resulting maps show Seoul positioned correctly in terms of latitude and longitude while the overall shape of Korea's coastlines, as well as the depiction of most of its towns, contain many inaccuracies of all sorts compared to the depictions of most of Qing territories. The party finally left Seoul on July 29th, 1713.

Tibet (1717–1718). As early as 1704, the emperor had sent two Qing officials, a Manchu by the name of Shu-lan 舒蘭 and a Mongolian named Raxi (La-xi 拉錫), to investigate the sources of the Yellow River. The fruits of this expedition were mostly textual, but apparently also included a map of the Kokonor region.[287] In 1709, he also sent Vice Minister of the Ministry of Personnel He-shou 赫壽 (d. 1718) to Lhasa to supervise the effective regent of Tibet, Lha-bzang Khan (d. 1717). This envoy had rough maps made of "all the countries immediately subject to the Dalai Lama" and brought back an integrated map. When the Jesuits examined it in 1711 they judged it rather badly executed, although later using it for the first editions of the resulting atlas.[288] Not until 1717, however, did the Kangxi emperor take steps to start mapping Kokonor and Tibet proper in a more consistent manner. In the spring of that year, a team led by two Tibetan lamas, Chu-er-qin 楚爾沁 and Zang-bu La-mu-zhan-ba 藏布喇木占巴, both of whom Jartoux had trained and instructed in calculating route traverse sections and observing latitudes, arrived in Lhasa. Also along was a secretary from the Lifanyuan 理藩院 (Manchu: *Tulergi golo be dasara jurgan*), a Manchu by the name of Šengju 勝住.[289] En route, the surveyors had observed compass bearings, measured road distances, determined latitudes using gnomons and, it seems, gathered the coordinates and altitudes of the primary mountains. Much as during the Korean mission, they obtained a number of maps and route books in Lhasa and local officials accompanied them.

The year 1717 marked the culmination of almost a decade of turmoil in Tibet. Following their defeat by the Qing armies in 1696, the Zunghar had been slowly rebuilding their armies and looked towards Tibet, which the question of allying with the Zunghar or the Qing had divided since the time of Galdan. In the wake of these tensions, the situation remained unstable for the

287 Known in Chinese as *Xihai* 西海 or *Qinghai* 青海.

288 Fuchs (1943), 1:12–18; Petech (1972), 19; Wang (1995), 135; Bai & Li (2014), 23–24. The title for Vice Minister was *shilang* 侍郎. For the quote, see du Halde (1735), 4:459.

289 Petech (1972), 25; SKQS, *Da Qing yitongzhi* 大清一統志, *juan* 413, 2b-3a. The Lifanyuan, sometimes translated as "Court of Colonial Affairs", was mainly responsible for Mongolian and Tibetan affairs. Fuchs (1943), 1:12–13. I have found no biographical information for these men. La-mu-zhan-ba possibly stands for the Tibetan academic title of Rabjamba.

first decade of the eighteenth century. From 1709 onwards, the well-informed Kangxi emperor, fearing Zunghar resistance to Qing influence over the Yellow sect, took a number of measures aimed at controlling the borders with Sichuan and Kokonor.[290] When the Zunghar suddenly attacked Hami in 1715, Qing forces pushed back and reacted by establishing outposts on the trade routes from Hami and sealing off the passes between Kokonor and Tibet proper. That same year, however, the Zunghar invaded Tibet with an army of six thousand men via a difficult backdoor route, hoping to escape Qing attention. Unaware of these developments, the emperor decided to send a team of mapmakers to survey the main route into Tibet, from Xining to Lhasa. It took the Zunghar one year to reach Lhasa, surprising the surveying team and leaving them no option but to return quietly to Qing territories in 1718 via another backdoor route, presumably into Sichuan. In the following years, as a reaction to the Zunghar sacking of Lhasa, the Kangxi emperor ordered several military campaigns into Tibet, undoubtedly aided by the new maps.[291]

Gaubil claims in his account that one of the lamas sent to survey Tibet kept a journal of his activities, on the basis of which he later drew a map that Jartoux corrected. He found the longitude of Lhasa to be to the west of Xining by twelve degrees. He also appears to have personally assured Gaubil that the length of the Sichuan route to Lhasa from Dartsedo (Dajianlu 打箭爐), should be 3000 *li* instead of the 3700 *li* recorded in route books, accounting for the many detours and high mountains.[292] It seems that this team also traveled beyond Lhasa to Mount Kailash, whence the mathematicians were not to continue together with the lamas and local officials towards the sources of the Ganges River. Gaubil notes that one should approach the Tibetan maps from the resulting atlas with more caution than those made with the assistance of the missionaries, although they included "at least the names of towns, rivers and big lakes, including many of the distances [between them]."[293]

Thus, these two surveys of lands under the influence but not under the direct control of the Qing court only included the primary roads into Korea and

290 The powerful local rulers of Kokonor had submitted to the Qing in 1697 after the Qing-Zunghar wars. Perdue (2005), 199.

291 Dai (2010), 76.

292 APF, Brotier 148:47v. Cordier (1915), 517. The distances recorded in the route books mentioned by Gaubil in this manuscript account also appear in Souciet (1729), 142. Gaubil mentions their uselessness in determining the longitude of Lhasa, since they do not mention corresponding compass bearings. Dartsedo is known today as Kangding 康定.

293 "[on sait] au moins le nom des villes, des rivieres, des grands lacs, avec beaucoup de distances." APF, Brotier 148, f. 47-47v; Cordier (1915), 516–517.

around Tibet, from Fengcheng to Seoul, from Xining to Lhasa and on to Mount Kailash. Moreover, the surveying of these roads lacked the level of precision and care of the surveys of Qing-controlled territories, due to suspicious hosts in the case of the Chosŏn court and because of the difficult environment and political turmoil in the case of Tibet. However, the inclusion of these maps in the resulting Qing atlas clearly illustrates the emperor's strategic interest in Korea and Tibet, concerned and inspired as he was by recent events on the Qing's vast frontiers.[294]

Through a case study of the first detailed survey of the Great Wall and an overview of the mapping of areas beyond its gates, including strategic expeditions into Korean and Tibetan lands, I have reconstructed the techniques used, the routes taken, and the progress made during the first stages of the land surveys. Most important in this regard is the fact that the mapmakers determined latitudes precisely by observing the height of the sun at midday, whereas longitudes they merely estimated after calculating route traverse sections with road distances and compass bearings. I have also argued that selected information gathered during the surveys sometimes circulated separately in addition to its uses in mapmaking: observing eclipses permitted comparison with European observations, whereas mapmaking in the remote far north allowed missionaries to describe the ginseng root for European audiences. In a last step, I have discussed excursions into Korea and Tibet, which resulted in maps differing greatly from those of Qing territories, based mostly on surveys of the main routes connecting these lands to Qing-controlled areas. Building on these insights, mostly derived from missionary sources, I now turn to Qing sources as the focus moves south, away from the Manchu homelands and the vast frontiers beyond the Great Wall and into the Chinese provinces.

2.2 The Logistics in Mapping the Chinese Provinces

While analysis of European sources has revealed the specific techniques used in mapmaking as well as the progress made and routes traveled in surveying regions outside the Great Wall, many questions remain unanswered. What were the precise backgrounds of the Qing officials in the surveying teams? How much did the emperor involve himself in the project, and what did he leave to

294 APF, Brotier 148:48–50v. Cordier (1915), 518–526. Note that these maps also eventually included other routes, from Hami to the ancient cities in today's Xinjiang province, but based entirely on textual descriptions.

other Qing institutions? More generally, what was the practical organization of the surveys and what non-technical strategies did mapmakers employ to cover such vast territories? Looking closely at Qing sources, I will now develop these hitherto unexplored aspects of the land surveys.

The main sources supporting the insights presented below are nine so-called secret palace memorials with vermillion endorsement. Belonging to the category of Palace Memorial (Chinese: *zouzhe* 奏摺 / Manchu: *wesimbure bukdari*) and typically reserved for important state affairs, this particular type of communication provided a way for the emperor to be in direct contact with his most senior officials, bypassing the many intervening levels of Qing bureaucracy. It was a highly efficient and secretive system: the palace's Office for Provincial Memorials (Chinese: *zoushichu* 奏事處 / Manchu: *Baita wesimbure ba*) received memorials to the imperial palace from the provinces, transported by private couriers and locked in special boxes, at any hour and presented them to the emperor the very next morning. The Kangxi emperor, known for his personal rule, spent a fair amount of time each day going through these memorials, thus keeping fully apprised of important matters.

Providing a direct channel of communication between the emperor and his most senior officials, palace memorials were the means chosen to transmit draft provincial maps to the capital in the land surveys' last stages, when mapmaking activities focused on the Chinese provinces. Many scholars have mentioned the memorials in passing when discussing the surveys, but a comprehensive analysis of all memorials related to the mapping project has not been undertaken.[295] Considered together, the memorials contain a wealth of information on the surveys' practical organization. In this section, therefore, I present a careful analysis of memorials and related documents. First, I explore the practical arrangements necessary for mapping the Chinese provinces by comparing data from all relevant palace memorials discussing timing, sequence, and strategy. Next, I discuss the role of the emperor and his administration during the land surveys by examining who was involved in this type of communication. Finally, our focus shifts to team formation and local support, with special attention to team composition and the tasks taken on by local administrators or their personnel. In all, by way of following the draft maps to the capital, aided by palace memorials and other scattered Qing documents, we discover many new details, all pertaining to the logistics involved in surveying the Chinese provinces.

295 Hostetler (2009), 102; Li (2012); Bai & Li (2014).

2.2.1 *Moving South: Sequence, Timing and Strategies*

With all northern areas and the two Chinese provinces of Zhili and Shandong mapped, from 1712 onwards the focus shifted fully to the areas south of the Great Wall. Jartoux, who together with Régis had so far partaken in every stage of the land surveys, was to remain in the capital to edit the draft maps, presumably due to his age and deteriorating health. With Jartoux thus unable to continue, the emperor sent Régis to the provinces of Henan and Jiangnan, while sending Cardoso, who had gained experience in Shandong as Régis' assistant, to Shanxi and Shaanxi province with Pierre Vincent de Tartre, who was stationed there. As a result, much like 1711, the year 1712 witnessed two teams of mapmakers surveying different provinces simultaneously. The order of the surveying of the Chinese provinces related directly to the fact that the surveyers would be able to link up with measurements from preceding surveys: Henan borders Zhili and Shandong provinces, while Shanxi and Shaanxi were the only remaining provinces directly abutting the Great Wall.

In the spring of 1713, the very last stage of the land surveys began with the simultaneous surveying of the remaining nine provinces by no fewer than three teams of mapmakers (figure 26): first, after Jesuits Romain Hinderer (1668–1744) and Joseph Marie Anne de Moyriac de Mailla went to assist Régis in Henan and Jiangnan,[296] the team did not return to Beijing but instead continued directly to survey Zhejiang and Fujian;[297] second, having returned from surveying Shanxi and Shaanxi in March 1713, Cardoso and de Tartre were to map Jiangxi, Guangdong, and Guangxi; and third, Fridelli and Fabre-Bonjour were to map Sichuan, Yunnan, Guizhou, and Huguang provinces.[298] By January 1717,

296 *Kangxi chao hanwen zhupi zouzhe huibian* (1984), 4:856–859.

297 Note that at the time Jiangnan province referred to the region covering today's Jiangsu and Anhui provinces. In 1667, the prefectures of Jiangning 江寧 and Suzhou 蘇州 gained some autonomy from Jiangnan, despite both conception and depiction as a unity.

298 In 1714, several students of mathematics went to different provincial capitals to measure the height of the pole star and the sun's shadow. It seems these observations related solely to the compilation of an imperial astronomical treatise, but it is possible the mapping project also used their observations as control measurements. *Kangxi chao shilu, juan* 260. See also Jami (2012), 279–280. One of these students in mathematics was He Guozhu, who went to Korea as part of Mu-ke-deng's party in 1713.

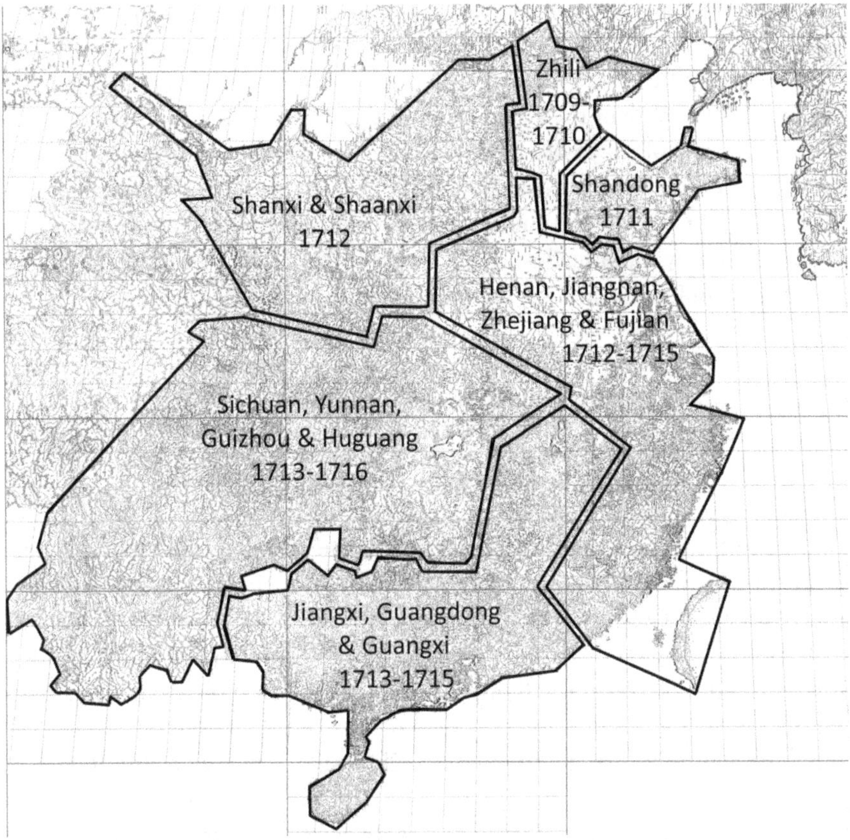

FIGURE 26 *Territories covered by different teams during the surveying of the Chinese provinces, marked on the 1719 edition of the Qing atlas (Chapter 3).*

all of these teams had returned to the capital, marking the virtual end of nearly a decade of intense mapmaking activity.

For these provincial surveys, one Manchu-language and eight Chinese-language palace memorials survive and have been published.[299] They divide naturally into two types: First, the majority transmit the draft maps for different provinces directly to the emperor. Second, two of the memorials were

299 I have separately discussed these memorials in Cams (2012). For the Chinese-language memorials, see *Kangxi chao hanwen zhupi zouzhe huibian* (1984), 4:856–859, 5:288–290, 5:328, 5:433, 5:674–676, 6:310–313, 6:605–606 & 6:786–787. For the Manchu-language memorial, see *Kangxi chao manwen zhupi zouzhe quanyi* (1996), 980.

follow-up replies to the vermillion endorsement of the emperor upon the original memorial. Structurally, however, all memorials follow the same pattern: (1) the official title and name of the memorialist, followed by a polite opening formula; (2) the date of the issuance of the instructions or endorsement; (3) a literal rendering of the instructions or endorsement; (4) confirmation and the date of the completion of instructed tasks or a reply to the imperial endorsement; (5) the name of the courier and a rendering of the instructions given him, followed by a polite closing formula; (6) repetition of the memorialist's official title and name; and (7) vermillion endorsement by the emperor. These

TABLE 5 *All palace memorials directly related to the land surveys.*

Author	Title of the author	Language	Content
Zhang Boxing 張伯行[300]	Governor of Jiangnan	Chinese	Completion of the map of Henan and Jiangnan
Wang Duzhao 王度昭[301]	Governor of Zhejiang	Chinese	Completion of the map of Zhejiang
Tong Guorang 佟國勷[302]	Governor of Jiangxi	Chinese	Completion of the map of Jiangxi
Tong Guorang 佟國勷	Governor of Jiangxi	Chinese	Reply to vermillion rescript of the emperor
Nian Gengyao 年羹堯[303]	Governor of Sichuan	Chinese	Completion of the map of Sichuan

300 Zhang Boxing (1652–1725) a native of Henan, became governor of Fujian and Jiangnan and later won promotion to serve in the capital as right vice-minister of revenue. He ended his career as minister of rites.

301 Wang Duzhao (d. 1724) was first in charge of the discipline of public functionaries. He went on to become governor of Hunan and Zhejiang province, and ended his career as vice-minister of public works.

302 Tong Guorang became governor of Jiangxi in 1712.

303 After his title as governor of Sichuan, Nian Gengyao (1679–1726) went on to become governor-general of Sichuan-Shaanxi and eventually received the title of Grand Guardian. He later suffered persecution under the Yongzheng emperor. His brother, Nian Xiyao 年希堯 (1671–1738), eventually became governor of Guangdong province and collaborated with the Jesuits in writing on the techniques behind perspective in painting. Mungello (2005), 41.

Author	Title of the author	Language	Content
Mambi 满丕[304]	Governor of Guangdong	Manchu	Completion of the map of Guangzhou
Gan Guobi 甘國璧[305]	Governor of Yunnan	Chinese	Completion of the map of Yunnan
Liu Yinshu 劉蔭樞[306]	Governor of Guizhou	Chinese	Completion of the map of Guizhou
Liu Yinshu 劉蔭樞	Governor of Guizhou	Chinese	Reply to vermillion rescript of the emperor

selected memorials, particularly considered together, therefore contain a wealth of information about the surveying of the Chinese provinces.

A list of these memorials (table 5) shows that they deal exclusively with the surveys' final phase, initiated in 1713. The table also makes clear that none of the memorials found originate from the last province surveyed by each of these teams, suggesting that the instructions prescribed the transmittal of provincial maps to the palace immediately upon completion, except the maps of those provinces each team surveyed last: Fujian, Guangxi, and Huguang. It is possible, therefore, that the teams personally carried these maps with them when they returned to the capital. Moreover, we should note that the mapmakers authored none of the memorials themselves: only officials with a rank of 4A or higher, which none of the mapmakers held, could present secret palace memorials.[307] In consequence, provincial governors (*xunfu* 巡撫) had to transmit the maps, appointing trusted household servants (*jiaren* 家人) for the purpose of carrying the sealed boxes, containing both the maps and the memorials, to the capital.

Examining the actual content of the memorials offers further clues with regard to the timing needed for completing the provincial surveys successfully. Analysis of the dates, for example, illustrates the mapmakers' progress (table 6):

304 Mambi became provincial governor in 1680 and won promotion into the Ministry of Works in 1717.

305 Gan Guobi (d. 1747) was a Han bannerman. Governor of Yunnan was the highest position he attained.

306 Liu Yinshu (1637–1723) was a Shaanxi native. He held sub-provincial posts before becoming governor of Guizhou in 1708.

307 Wu (1967), 25.

TABLE 6 *Dates extracted from the nine palace memorials.*

KX 52/V²/08[308]	30/06/1713	Henan & Jiangnan maps completed
KX 52/VI/09	30/07/1713	Jiangxi survey started
KX 52/VI/12	02/08/1713	Sichuan survey started
KX 52/XII/05	20/01/1714	Zhejiang map completed
KX 52/XII/26	09/02/1714	Jiangxi map completed
KX 53/VI/10	21/07/1714	Sichuan map completed
KX 53/X/05	11/11/1714	Guangdong map completed
KX 54/VI/24	24/07/1715	Yunnan map completed
KX 54/X/30	25/11/1715	Guizhou map completed

Taken together, these dates indicate that the provincial surveys were complete within timeframes ranging from a mere four months for Guizhou to more or less one year apiece for Sichuan and Yunnan. Apart from these provinces' varying sizes, we can also attribute the differences to the fact that in Guizhou, mapmakers were unable to survey the entire province. The resulting maps thus include blank spaces, sometimes labelled with the characters "*sheng miao*" 生苗, translatable as "barbarians" or, as on one extant copy kept in Brussels, "savage peoples" (*peuples sauvages*).[309] In contrast, the remoteness of large parts of Sichuan and Yunnan provinces at the time, as well as their difficult terrain, varying climate, and the complex ethnic situation, contributed to delays there. Another reason for the holdup in Yunnan was the sudden death of Fabre-Bonjour and of one of the Manchu officials in the same team, Bursai 布爾塞. Both died in the border town of Mengding 孟定 in late December 1714, presumably of malaria.[310] In the Yunnan memorial, governor Gan Guobi states that Régis replaced Fabre-Bonjour, while Cangboo 常保 replaced Bursai, all in accordance with instructions issued on April 17th, 1715.[311] Fabre-Bonjour's death notice, kept at the Jesuit archives in Rome, confirms this.[312]

Putting aside the differences in the times needed to complete the draft provincial maps, the dates indicate that, much as during the land surveys' first

308 Every second or third year in the Chinese lunar calendar has an intercalary month or *runyue* (閏月), which may follow any regular month. It has the same number as the preceding regular month and serves as a correction on the lunar calendar.

309 Chapter 3.

310 Mengding is on the border with Burma.

311 *Kangxi chao hanwen zhupi zouzhe huibian* (1984), 6: 310–313. Régis received orders to go to Yunnan on March 24th, 1715. du Halde (1935), 1:xliii. Cangboo held the same title as Bursai.

312 ARSI, JS 176:397. For a full account of Fabre-Bonjour's death and the controversy surrounding his funeral rites, see Baldini (2012), 277–282.

stages, work proceeded at a remarkable speed. The sheer number of towns and villages covered led one scholar to assume that "in most cases, the survey can only have touched the principal routes."[313] While this may well be true, the teams also employed strategies to speed up surveying, for example by splitting. Evidence or suggestions of such divisions of labor are apparent for at least four provinces. First, in Yunnan, Fabre-Bonjour and Bursai set out to map the southwestern part of the province while Fridelli and the two remaining Qing officials surveyed the southeastern part;[314] second, on the island of Taiwan, then part of Fujian province, de Mailla and one Qing official mapped southern Taiwan, while the others set out for the north of the island;[315] third, in Guangdong, only three of the five members sent to survey the province seem to have determined the southernmost point of Hainan island;[316] and fourth, in Shaanxi, two rather than five mapmakers apparently conducted observations and measurements in the city of Longzhou 隆州.[317] This strategy seems to have been common and may therefore have occurred in other provinces as well.

2.2.2 *Directed from the Center: The Emperor and His Administration*

Perhaps the only direct communication between the mapmakers and the center for the duration of the land surveys, the palace memorials also allow us to gauge the role of central administration in directing and controlling mapmaking efforts in the provinces. First and foremost, the emperor, for whom the maps and the accompanying memorials were intended, often wrote comments or endorsements (*zhupi* 硃批) in red ink directly on the memorials, either taking note (Chinese: *zhidao le* 知道了 / Manchu: *saha*) of what he had just read or adding instructions, questions, or comments for the memorialist.[318] Processing by the palace took only a few days, after which the memorial was copied,

313 Baddeley (2006), 1:clxxxvii.

314 ARSI, JS 176:397.

315 Recorded in a letter by de Mailla, published in *Lettres édifiantes et curieuses.* du Halde et al. (1702–76), 14:424.

316 Recorded in a gazetteer for the town of Yazhou 崖州 dating from the Guangxu period (1875–1908). However, the dates given in the gazetteer do not correspond with dates from the memorials. *Zhongguo guojia tianwen* (2013), 9. Hainan Island at the time was part of Guangdong province.

317 Recorded in a 1713 Kangxi-era gazetteer for Longzhou. ZFK (*Zhongguo fangzhi ku* 中國方志庫 digital database), *Longzhou zhi*, *juan* 2.

318 When a memorial required follow-up, the emperor simply turned down one corner, indicating he wished to discuss the matter the next morning in audience with his Grand Council. Hence the name 'folded memorial' or *zouzhe*. Elliott (2001), 47; Fairbank & Teng (1940), 24–32.

archived and sent back to the memorialist by private courier. This careful secrecy reflects the importance attached to matters handled through this type of communication and the fact that the Kangxi emperor clearly thought it necessary to keep in full control of developments.[319]

Four of the memorials are unique in this regard: two include vermillion endorsements in which the emperor goes beyond mere note-taking; the other two are follow-up replies by the memorialists to these longer endorsements. Upon receiving the first Jiangxi memorial in February 1714, for instance, the emperor not only took note of the completion of the Jiangxi map, but also inquired how the surveyors had gone about their business in that province.[320] Exactly fifty days later, the governor of Jiangxi replied to the emperor in a second memorial, an indication of the time needed for official correspondence over great distances, even considering the efficient handling of such documents in the palace.[321] The governor replied immediately:

> Imperial envoy Chuo-er-dai and others crossed the border into Jiangxi province on the 9th of the 6th month of last year.[322] They started drawing [the map] in the area subordinate to Jiujiang prefecture, and came to the provincial capital after that.[323] Your servant [I] and others have offered them wine and food. After drawing the provisional maps of the two districts of Nan and Xin, they immediately left for Raozhou, Guangxin, and other prefectures and subordinate areas in order to finish drawing [the map].[324] It was completed in the twelfth month. They then went on to Guangdong. Your servant has learned that in each prefecture and district, they have been offered wine and food and that there were no disturbances in these places. The imperial envoys and mapmakers have

319 With the exception of so-called "greeting memorials," also communicated through this system. Elliott (2001).

320 The full endorsement reads: "知道了。畫圖人員行事如何？具實奏来。" ("Noted. How about the behavior of the mapmakers? Report back truthfully.") *Kangxi chao hanwen zhupi zouzhe huibian* (1984), 5:328.

321 *Kangxi chao hanwen zhupi zouzhe huibian* (1984), 5:433. For more information on the time needed for the transmission of documents throughout the provinces, see Fairbank & Teng (1939).

322 July 30th, 1713.

323 Jiujiang is a prefecture in the north of Jiangxi.

324 Nan and Xin refer to the districts of Nanchang 南昌 and Xinjian 新建, comprising the prefectural city of Nanchang 南昌, the capital of Jiangxi. Raozhou and Guangxin prefectures lie in northeastern Jiangxi.

> been very respectful of the Emperor's kingly ways and went about their business in the most peaceful and quiet way.[325]

Other than reporting on the route taken by the surveyors in the province, the memorial stresses that the mapmakers operated in a peaceful and quiet manner and received hospitable treatment. Indirectly, the governor's reply thus articulates the emperor's true concern, whether the mapmakers had caused any local disturbances. This goes to show that the emperor was well-informed: he demanded direct intelligence about progress made and how his surveyors comported themselves in the provinces.

The governor of Guizhou province, Liu Yinshu, sent an equally revealing memorial to the throne on November 27th, 1715. In listing the surveyors, the governor rather ambiguously used the term "*daren*" 大人, precipitating confusion in the palace. From the vermillion endorsement, it seems the emperor interpreted this as "*dachen*" 大臣, meaning "Grand Minister:"

> Noted. There was no *dachen* sent. This is not consistent with the content of the memorial. Who is this *daren*? Ascertain clearly and report back.[326]

Roughly three months later, the same memorialist replied:

> In the outer provinces, generally, when we meet imperial envoys, we do not discuss [their] high or low rank. [Instead,] we call all of them *daren*. I thought this practice was common. Your servant is honest and had the best intentions. Thirty years have passed since entering officialdom, but I still have the manners of the countryside. I lack assistants and aides, so there is no one who practices the prevalent customs. Personally composing letters, I do not know the rules. I have absurdly spoken of *daren*,

325 "欽差綽爾代等於上年六月初九日入江西境。由九江府屬畫起後到省城。奴才等魯請過酒席送過吃食。畫完南新兩縣圖式即往饒州廣信等府屬繪畫去訖。十二月内畫完。隨往廣東省去。奴才知各府縣内俱送過酒席吃食地方並無騷擾在。欽差畫圖人員狠尊主子王法。行事甚是安靜。" *Kangxi chao hanwen zhupi zouzhe huibian* (1984), 5:433. "*Wangfa*" 王法, here translated as "kingly ways," is an expression referring to an emperor's virtuous way of ruling the country.

326 "知道了。並無差大臣去。摺内程大人不合。大人是誰? 察明再奏。" *Kangxi chao hanwen zhupi zouzhe huibian* (1984), 6:605–606.

> a crime I cannot take back. I beg your Majesty to exceptionally forgive me.[327]

Here, the Kangxi emperor clearly shows his thoroughness and eye for detail in dealing with the land surveys: confused or perhaps concerned that someone appears to be part of the group of surveyors without his knowledge, he demands to know the identity of the unspecified "*daren*." Taken together, the vermilion endorsements on the Jiangxi and Guizhou memorials show an emperor intent on reinforcing his patronage of the mapping project by gathering specific intelligence on the practical execution of the land surveys, including progress, team composition, and the behavior of his mapmakers.

Besides the imperial endorsements, the memorials mention several institutions of central administration that contributed to the logistics behind the provincial surveys. Most memorials indicate, for example, that instructions generally traveled by way of official communication from the Ministry of War. This ministry consisted of four departments or bureaus, one of which, the Remount Department (*chejia si* 車駕司) was in charge of all communications between the central administration and provincial governments, and responsible for the horse relay system enabling such interregional communication.[328] We can therefore assume that issuing instructions to provincial governors entailed standard procedures. Initial instructions by the ministry went to all governors and included ensuring that, upon completion, the mapmakers immediately handed over or sent the maps to the provincial governors, who would appoint trusted household servants to deliver them to the throne.

Some provinces received more specific instructions, as in the case of Zhejiang, Fujian, and Yunnan. The governor of Yunnan received follow-up instructions with regard to the replacements for the mapmakers who had passed away in 1714 as discussed above. Likewise, the Jiangnan memorial warned the team that would eventually continue on to survey Zhejiang and Fujian province of the dangers of crossing the seas to some of the islands, including Taiwan. Zhang Boxing, the governor of Jiangnan, writes of having received the following instructions on June 9th, 1713:

---

327 "在外省凡遇欽差不論官品大小，俱呼大人。習以為常。臣朴直踈率入任三十年尚是草野之氣。署乏幕客替理，無人習於流俗。任手成書，不諳規矩。謬稱大人罪愆莫贖伏祈格外恩宥。" *Kangxi chao hanwen zhupi zouzhe huibian* (1984), 6:786–787.

328 Liu (1990), 40; Fairbank & Teng (1939), 15.

> Officials, *baitang'a*, and Westerners who travelled to Henan and Jiangnan for drawing the map should not return. From there, they may continue drawing [the maps of] Zhejiang's Zhoushan and other regions, as well as Fujian's Taiwan and other regions. But when crossing the seas, we urge you and your men to [first] carefully consider weather and timing. There is no need to hurry. Please pass these instructions on to them.[329]

The Ministry of War, however, was not the only ministry issuing or forwarding instructions with regard to the land surveys: the first Jiangxi memorial mentions an official communication from the Ministry of Works alongside that of the Ministry of War.[330] A complementary document published in *Historical Materials of the Ming and Qing Dynasties* sheds more light on this ministry's involvement.[331] It consists of two separate documents, quite similar in general content but with substantial differences in wording.[332] The first renders instructions originating from the Remount Department of the Ministry of War, whereas the second contains similar stipulations issued by the Department of Waterways (*shuisi* 水司) of the Ministry of Works.[333] Judging from the strong similarities with ministry instructions recorded in the memorials, the combined documents must constitute a draft version or perhaps even an exact copy of the instructions transmitted to the governors.[334] Thus, it seems that two different ministries issued separate instructions for traveling, respectively, by land and over water. The extra document also indicates that an imperial decree dispatching the last two teams of surveyors formed the basis for these

329 "往河南江南畫輿圖去的官拜唐阿西洋人不必回來，就從彼處往浙江周山等處福建臺灣等處畫去。但走海時著你等謹慎看好天色時節。行走不必急了，須要仔細。再下旨與他們伊等。" *Kangxi chao hanwen zhupi zouzhe huibian* (1984), 4: 856–859.

330 *Kangxi chao hanwen zhupi zouzhe huibian* (1984), 5:328.

331 *Ming Qing shiliao ding bian*, 2:768–9. An annotated but incomplete translation appears in Fu (1966), 1:118–119.

332 Clearly translated from Manchu. In listing team members, for example, one of the documents speaks of *ga-zha-er-qi zhang-jing* 嘎渣爾氣章京, renderings in Chinese of the Manchu words "*gajarci janggin*" or "Escort Office banner Colonel" (*xiangdao canling* 向導參領). Many of the characters used to render the names of mapmakers differ as well, but are homonyms of those included in the memorials.

333 For more information on the Department of Waterways or *shuisi* 水司, see *Lidai zhiguan biao*, *juan* 14. The Manchu documents related to the 1709 survey of the Manchu homelands also mention the Ministry of Works. Cheng (2015), 21–32.

334 Certain passages from the Jiangxi and Sichuan memorials, for example, seem literal copies from this circular.

instructions and that it contained stipulations regarding team composition, the number of horses granted to each official, and other practicalities. It even included a warning for those travelling to the jungles of Yunnan, Guizhou and Guangxi:

> The miasma in Yunnan, Guizhou and Guangxi is very bad. From the 4th month until the 8th month, when the miasma can be encountered, personnel going there should temporarily go elsewhere to draw [the map] and wait until the miasma has dispersed before returning.[335]

Memorials indicate that, starting from May 1713, most of the instructions discussed above gradually reached the different governors, obliged to forward them to every prefecture and district within their jurisdictions and to have the Provincial Administration Commissioner or *buzhengshi* 布政使 of each province see to their swift and correct implementation.[336] Thus, provincial hierarchies were clearly key to the practical arrangements surrounding the land surveys: on the one hand, governors received and forwarded what we may regard as both arrival notices and detailed instructions, while on the other hand providing the quickest and safest links for transmitting the resulting draft maps directly to the throne.

2.2.3 *Team Composition and Local Support*

With the help of these selected palace memorials, I have so far reconstructed sequence, timing, and communication with the center, but the most striking information extracted is undoubtedly the composition and sizes of the three teams of surveyors active during the last stage of the land surveys. First, the Jiangxi and Sichuan memorials record the most clear and seemingly complete lists of team members, including official titles:

> Li Bingzhong, Westerners de Tartre and Cardoso… *langzhong* Chuo-er-dai and *wuguanzheng* Miao Shou.[337]

---

335 "雲南貴州廣西瘴氣甚惡。所去的人自四月起至八月止遇瘴氣暫在別處畫矣。瘴氣散時再去畫。" *Ming Qing shiliao ding bian* (1972), 2:768.

336 Recorded in the Jiangnan memorial. *Kangxi chao hanwen zhupi zouzhe huibian* (1984), 4: 856–859. The *buzhengshi* was an important provincial official, subordinate to the governor and alternatively translated as "Lieutenant-Governor" or "Financial Commissioner."

337 "李秉忠西洋人麥大成湯尚賢… 郎中綽尔代五官正苗受。" *Kangxi chao hanwen zhupi zouzhe huibian* (1984), 5:328. *Langzhong* 郎中 is the title for a director in a ministry or other important administrative body. According to the Manchu-language

Jianshi of the *Wuyingdian* Bursai, Westerners Fridelli and Fabre-Bonjour, banner Colonel of the Imperial Guard's Escort Office Ying-zhu, *langzhong* of the Ministry of Personnel Lang-gu-li, and *jianfu* of the Directorate of Astronomy Suwangde.[338]

When we put these lists next to the more scattered information on the team members included in other memorials and in the complementary ministry instructions mentioned above, certain similarities stand out with regard to the composition of the surveying teams, as well as the order in which individual names appear.[339] Compared to the list of team members for the surveying of the Manchu homelands in 1709, the composition of the teams was apparently fixed by the time the surveys moved south of the Great Wall, each team consisting of:

(1) A member of the Imperial Household Department with ties to the European missionaries at court, representing the inner palace and linked to the emperor's personal patronage of the project;
(2) two or three European missionaries, essential for their expertise in applied mathematics and instrument manipulation;

memorial, Chuo-er-dai worked for the Lifanyuan and the character "*deng*" 鄧 replaces "*zheng*" 正, presumably due to translation difficulties.

338 "武英殿監視布爾賽西洋人費隱單爻占向導護軍參領英柱吏部郎中朗古禮欽天監監副雙德。" *Kangxi chao hanwen zhupi zouzhe huibian* (1984), 5: 674–676. "*Jianshi*" 監視 was the official title for those overseeing production at the Imperial Workshops. Fabre-Bonjour here goes by "Shan Yaozhan" 單爻占, but his gravestone carries a homophone name in different characters: Shan Yaozhan 山遙瞻. Malatesta & Gao (1995), 166. *Hujun canling* 護軍參領 is the title for a banner Colonel in the Imperial Guard: an elite group within the banner system responsible for protecting the palace. Elliott (2001), 81 & 366. The prefix "*xiangdao*" 嚮導 here refers to the banner Colonel's ties to the *xiangdaochu* 嚮導處 or Escort Office, which provided the military entourage for imperial outings or imperial envoys. Lang-gu-li is otherwise written as "Lang-wu-li" 朗烏禮. "*Jianfu*" 監副 was the official title for a vice-supervisor in the Directorate of Astronomy. *Kangxi chao hanwen zhupi zouzhe huibian* (1984), 5:674–676.

339 The Jiangnan memorial mentions unspecified workshop personnel (*baitang'a*) and *hujun canling* Toofahaci 陶蕃齊 as being part of this team. *Kangxi chao hanwen zhupi zouzhe huibian* (1984), 4:856–859. According to de Mailla, four officials accompanied the European mapmakers in this team. du Halde et al. (1702–76), 14:8. Ministry instructions confirm that the team surveying Jiangxi, Guangdong and Guangxi included the same men earlier sent to Shanxi and Shaanxi provinces. *Ming Qing shiliao ding bian* (1972), 2:768–769.

(3) a secretary or director from an important administrative body, such as the Six Ministries or the Lifanyuan, indicating the involvement of different segments of Qing central administration;
(4) staff from the Directorate of Astronomy, strongly linked to the Jesuits' presence at court and with expertise in astronomy;[340] and
(5) a banner Colonel from the Imperial Guard working for the imperial Escort Office, serving as a military escort for the mapmakers.

When a list includes all team members (table 7) identified on the basis of data from the memorials as well as from ministry instructions, it becomes apparent that the teams were quite large, consisting of at least five to seven core members and unknown numbers of assistants and personal servants.[341] One notable exception to the seemingly fixed team composition was the group of surveyors for Jiangxi, Guangdong, and Guangxi provinces. As the excerpt quoted above makes clear, the first Jiangxi memorial did not list an elite banner Colonel as part of this team. In fact, the circular indicates that a bannerman named Lang-tu 朗圖 was originally part of this team, but in the event not sent along after all. The excerpt further implies that military escorts also took part in road measurements:

> *Langzhong* Chuo-er-dai and *wuguanzheng* Miao Shou will be sent to Jiangxi, Guangdong and Guangxi provinces. It is not necessary to send the Escort Office's banner Colonel. Their original task was to pull the rope. Enough people have been sent. The less people to pull the rope, the better.[342]

Ministry instructions further indicate that the selection of mapmakers did not all take place at the same time. Instead, for the very last stage of the surveys, an imperial edict selected both personnel from the Imperial Household

340 Note that several of the staff at the Directorate of Astronomy, part of which the Jesuits controlled, were Christians. du Halde et al. (1702–76), 3:74.

341 The complementary ministry instructions mention that imperial household personnel, led by Bursai and Li Bingzhong, included eight men, presumably referring to assistants or personal servants. For the 1709 survey of Manchu homelands, during the first stages of the land surveys, we know the names of several of these assistants and clerks, as discussed above and mentioned in Cheng (2015), 21–32.

342 "即中綽爾代五官正苗壽著派往江西廣東廣西省。嘎渣爾氣章京不必派出。伊等原為拉繩所派之人盡足。拉繩之用人少愈好。" *Ming Qing shiliao ding bian* (1972), 2:769.

TABLE 7 *Team members for the Chinese provinces.*

	Shanxi & Shaanxi 山西-陝西 **(1712)**	**Henan, Jiangnan, Zhejiang & Fujian** 河南-江南-浙江-福建 **(1712–1714)**	**Jiangxi, Guangdong & Guangxi** 江西-廣東-廣西 **(1713–1715)**	**Sichuan, Yunnan, Guizhou & Huguang** 四川-雲南-貴州-湖廣 **(1713–1716)**
Jianshi—inner palace's *Wuyingdian* or *Yangxindian* [343] 武英殿監視 養心殿監視	Foboo 佛保	*baitangga* 拜唐阿	Li Bingzhong 李秉忠	Bursai 布尔塞† < Cangboo 常保
Westeners 西洋曆法	1. Pierre Vincent de Tartre 2. Joao Cardoso	1. Jean-Baptiste Régis 2. Joseph M.A. de Mailla 3. Romain Hinderer	1. Pierre Vincent de Tartre 2. Joao Cardoso	1. Xavier Fridelli 2. Guillaume Fabre-Bonjour† < Jean-Baptiste Régis
Banner Colonel—Imperial Guard's Escort Office 嚮導護軍參令	Ying-zhu 英柱	Toofahaci 陶蓍齊	~~Lang-tu (朗圖)~~	Ying-zhu 英柱

343 The third part of this chapter further discusses the involvement of these men.

TABLE 7 *Team members for the Chinese provinces.* (cont.)

	Shanxi & Shaanxi 山西-陝西 **(1712)**	**Henan, Jiangnan, Zhejiang & Fujian** 河南-江南-浙江-福建 **(1712–1714)**	**Jiangxi, Guangdong & Guangxi** 江西-廣東-廣西 **(1713–1715)**	**Sichuan, Yunnan, Guizhou & Huguang** 四川-雲南-貴州-湖廣 **(1713–1716)**
Langzhong—Lifanyuan OR the Ministry of Personnel 郎中 (理藩院 / 吏部)	Lang-gu-li 朗古禮	(unknown)	Chuo-er-dai 綽爾代	Lang-gu-li 朗古禮
Wuguanzheng or *jianfu*—Directorate of Astronomy 五官正 / 監副 (欽天監)	Suwangde 雙德	(unknown)	Miao Shou 苗受	Suwangde 雙德

† indicates death during the surveys.

Department and European missionaries, on May 23rd, 1713. Next, the ministries appointed the other team members, selected four days later, on May 27th. This suggests that the household officials, together with the European missionaries, took the lead over the surveys or were otherwise the more crucial team members, which I will elaborate on in this chapter's last section.

Regardless of their time of appointment, all these men carried the title of Imperial Envoy (*qinchai* 欽差) a status that carried many privileges. First, a messenger traveled a few days ahead of the envoys, so that local authorities in each prefecture had ample time to prepare for their arrival. After all, wherever imperial envoys set foot, they received offers of sedan chairs, food, drinks, and the best lodgings available, often within the walls of the local official residence, the *yamen* 衙門. On water, yellow imperial banners standing two meters tall and bearing the words "envoys from the imperial household" (*qinchai* or *neiwufu qinchai* 內務府欽差) distinguished well-protected boats transporting *qinchai*, while at night they carried lanterns decorated with the same characters.[344] Prefectural or district magistrates were even to welcome them at the borders of their jurisdictions. In proof of their status, envoys received investigatory licenses known as *kanhe* 勘合 from the Ministry of War in Manchu and Chinese, confirming the purpose of their visit. This document also specified their right to a certain number of horses, amounting to four or five for each of the officials listed, depending on their rank, and to protection and local guides, all of which relay stations along the way were to provide.[345] When entering a province, they exchanged *kanhe* for *kanpai* 勘牌, separate sheets of paper on which local administrators had to sign off after having completed their duties as hosts.[346]

As the mapmakers journeyed into the Chinese provinces, their investigatory licenses as well as ministry instructions to the different governors, as discussed above, specified the necessary local preparations. The latter also included instructions to subordinate prefecture and district officials specific to the land surveys. These stipulated, among other things, the preparation of military horses in remote areas without relay stations, with one horse reserved for carrying instruments for each team. The Sichuan memorial also records that local administrators were to prepare route books, gazetteers, ropes, gongs, woodblock carvers and runners:

344 Described by Bouvet in his diary. von Collani (2005), 160 & 196.

345 *Ming Qing shiliao ding bian* (1972), 2:768.

346 *Kangxi chao hanwen zhupi zouzhe huibian* (1984), 4:856–859.

> Respectfully following [the orders], your servant has immediately contacted all the subordinate prefectures, departments, districts, military camps, military posts, etc. The officials have obeyed the texts of the ministry by arranging horses, local gazetteers, books, flags, gongs, and other items needed to serve [the surveyors].[347]

This excerpt illustrates the provision of local gazetteers and route books wherever the surveyors set foot.[348] The fact that the mapmakers consulted these works further indicates that mapmakers received geographic data from local gazetteers rather than relying solely on the compilations on geography kept at the imperial palace, allowing them to compare road distances recorded in different sources, which was crucial in determining the positions of smaller villages they lacked time to visit. A local gazetteer for Longzhou 陇州 in Shaanxi province records the following passage for December 1712, further illustrating the support provided on the prefectural level:[349]

> Imperial envoys, *jianfu* of the Directorate of Astronomy Suwangde and *jinshi* of the West Vincent de Tartre, arrived in Longzhou. They measured the road to draw a map. Department Magistrate Luo Zhangyi and Department Vice Magistrate Du Qingzhang indicated each *li* with flags, announced their passing with gongs, and compared measurements with ropes, determining the distance traveled by way of Western methods of calculation.[350]

From this excerpt, it appears that local administrators planted flags along the way to indicate road distances in *li,* took rope measurements, and warned passers-by of ongoing measurements to clear the way. This illustrates that local administrators were not merely to provide food and shelter but also to travel with the mapmakers through their jurisdictions and to participate in road

347 "欽尊臣即通行所屬各府州縣營驛等。官遵照部文預備馬匹並需用志書旗鑼等項。" *Kangxi chao hanwen zhupi zouzhe huibian* (1984), 5:674–676.

348 Ministry instructions confirm this.

349 Longzhou is a prefecture located just northwest of Baoji 寶雞.

350 "內欽差欽天監監副雙德西洋進士湯麥菰隴。丈量道路繪畫輿圖。知州羅彭齊州同督清丈用用旗以認道里，用鑼以明禁止，用繩以較丈尺，按西洋算法以定其里數跬步。" ZFK, *Longzhou zhi, juan* 2. A Department Magistrate was responsible for the lowest level of formal territorial administration, while a Department Vice Magistrate was his principal assistant.

measurements. As such, they cooperated closely with the team members and played a far more active role than hitherto appreciated. It is clear the mapmakers met with many local and provincial administrators, from governors to relay station staff and from prefects to district runners familiar with the terrain.[351]

Finally, there is evidence of the opportunities created by these numerous local contacts. On June 12th, 1712, de Tartre and Régis, both part of two different teams, happened to pass near one another on opposite banks of the Yellow River, with only a few days' travel between them. In a letter from de Tartre to Régis, now in the Jesuit archives in Paris, de Tartre writes of the hardships of traveling through the remote mountainous parts of Shanxi and complains that, due to the governor's negligence, local authorities had not implemented instructions in a sufficient manner.[352] De Tartre further notes that it would be useful for both teams to exchange and compare data for sites along the Yellow River, as it constitutes the common border between the provinces of Shanxi and Henan.[353] To facilitate this, he asks Régis to send his data to the military commander in Pinglu 平陸, a good friend who would certainly be willing to forward it.[354] Perhaps a rare occasion, the complex overlap of hierarchies and personal contacts at play here allowed direct communication and data comparison between different teams.

In this section, by tracing the finished draft maps from the provinces to the center, I have explored the logistics involved in mapping the more densely populated and well-documented Chinese provinces. An analyses of available Qing documents, the core of which included a set of secret palace memorials, has allowed me to examine sequencing, timing, communications with the center, supporting institutions, team composition, and local support. Many new

---

351 Confirmed by Manchu documents related to the 1709 survey of the Manchu homelands. Cheng (2015), 21–32. For other examples of local contacts, see the intermission immediately following this chapter.

352 De Tartre also mentions that mapmakers had to visit the Dizhu 底柱 rock that stands in the middle of the Yellow River near today's Sanmenxia. Both the emperor and his sons had visited the rock over the preceding decades. The rock held great symbolic significance dating back to ancient China, since Emperor Yu had allegedly partially destroyed it to let the river pass. This episode is strongly reminiscent of Gerbillon and Bouvet visiting the major battlefields of the Qing-Zunghar wars during the preliminary survey of Khalka lands in 1698 (Chapter 1). APF, *doss. pers.*, de Tartre.

353 APF, *doss. pers.*, de Tartre. Brucker (1889) publishes this letter in part.

354 Possibly a contact de Tartre had made earlier, since he had been stationed in Shanxi before working on the mapping project. The letter mentions that the official was a "tchihui," or *zhihui* 指揮, in Pinglu, a town in the very south of Shanxi province, located on the Yellow River.

details have emerged, all of which speak to the project's magnitude as well as to the involvement of individuals working at all levels of the Qing state apparatus: from the emperor himself to his ministries, and from the provincial governors to the runners and messengers in the service of local district magistrates. In the next section, I bring together the new insights this chapter has so far achieved to elaborate on one of the major missing links tying together the trajectories of maps and instruments: the Imperial Workshops.

2.3 The Imperial Workshops Connection

In closely following surveying instruments to the many sites where the mapmakers determined latitudes and longitudes, I have paid close attention to the techniques, progress, and separate circulation of selected resultant data. Likewise, I have traced the draft maps drawn on the spot and later transmitted to the capital, thereby examining the logistics involved in the surveying of the Chinese provinces. Geographically, I have thus covered the surveying of most Qing territories, stretching from the Great Wall to the remote frontier regions of the far north, into Korea and Tibet, and to all fifteen Chinese provinces. In this section, I build on my close analysis of the trajectories of the instruments and maps during the land surveys in order to explore the main institutional force binding them: the Imperial Workshops.

The workshops constitute a missing link emerging precisely from closely tracing the trajectories of instruments and maps during the land surveys. I begin my argument based on the insights regarding team composition distilled from the palace memorials in the previous section. As we can infer from the palace memorials and complementary documents, the emperor selected team leaders first during team formation and included personnel from the Imperial Household Department as well as a handful of European missionaries. In what follows, I discuss both groups at length in an attempt to gather biographical information on these men, most of them connected to the Imperial Workshops and working under the emperor's trusted Manchu courtiers. Based on their common institutional ties, I then explore the place of the workshops within the trajectories followed by maps and instruments. Temporary offices, the site for the composition of a textual companion for the resulting maps, also come into view, reminding us of the partial embedding of the mapping project within a Chinese tradition of textual geographic description. Overall, I argue for the logistical centrality of the workshops, at least for the duration of the land surveys, as I continue to stress the individual connections shaping the practical execution of the land surveys.

2.3.1 *Mapmakers from the Inner Palace*

As an institution, the Imperial Workshops changed greatly over time. During the Kangxi period, their functions and capacities multiplied so rapidly that in 1691 it was necessary to move a number of workshops from their original sites within the walls of the *Yangxindian*, one of the halls or compounds of the imperial palace, to an area of the palace situated just north of the *Wuyingdian*.[355]

Both clusters of workshops thus linked closely with these halls, sometimes simply called "*Yangxindian* Workshops" and "*Wuyingdian* Workshops" in Chinese-language documents, located to the left of the main axis of the imperial palace grounds. The Imperial Household Department, the main offices of which lay between these two halls, controlled the workshops and employed *booi* (litt. "of the house", *baoyi* 包衣 / *buyi* 布衣) or bondservants, mostly although not exclusively selected from the three banners personally controlled by the emperor.[356] Senior household personnel overseeing the workshops were highly trusted agents of the emperor who aided him in his control over matters he deemed important or essential to his personal rule. The term for lower-level personnel, on the other hand, was "*baitangga*" (*baitang'a* 拜唐阿 / 柏唐阿).[357] Of course, the workshops also employed highly skilled craftsmen from all corners of the land and beyond.

Those inner palace members of the surveying teams selected first, therefore apparently team leaders, were all such *booi* officials. In the list of team members presented above (table 7), their titles clearly refer to the *Wuyingdian* and *Yangxindian* within the palace, suggesting a connection to the Imperial Workshops, located precisely in and around these halls. At a time when the land surveys were well under way, Foucquet wrote the following about the significance of the two palace halls for both the emperor himself and the Europeans at court:

> Because [the emperor] has always loved the sciences and the arts, he has put, in order to perfect them, a large number of persons in both these palaces. Everything related to the composition and editing of books is done at the *Wuyingdian*: it is there that a very large library is [located], and he has it extended every day; it is there that he entertains many scholars and

355 *Da Qing huidian shili* (1899), *juan* 1173; Zhang (2009), 1:261–302. *Yangxindian* is translatable as "Hall of Mental Cultivation," the *Wuyingdian* as "Hall of Military Eminence."

356 Torbert (1977), 27; Xu (2012). The emperor personally controlled the bordered yellow, plain yellow, and plain white banners, also known as the "upper banners."

357 "*Baitangga*" commonly translates as "errand boy," "handyman," or "underling."

> other intellectuals who only occupy themselves with creating new works under his direction, or with renewing old ones, onto which he often adds commentaries. As for the *Yangxindian*, it is filled with workers who work tirelessly on all kinds of instruments, and with regard to mathematics, geometry, astronomy, music, physics, medicine and botany: it is there that [the emperor] is brought what arrives from Europe each year, and it is impossible to say anything of the quantity of rare and curious things that he gathered there from all parts [of the world]. In both palaces, there are mandarins who preside over them, and since the affairs of Europeans are generally related to the sciences and the arts, it is through the mediation and the agency of these men that they get dealt with at the emperor's quarters.[358]

As is apparent from their titles, the *Wuyingdian* and *Yangxindian* workshop officials chosen to lead the teams of mapmakers were such overseers as described by Foucquet, but what were their responsibilities within the workshops? A closer look at their appearances in official documents may help to answer this question and to clarify their role with regard to the land surveys.

First, for Shanxi and Shaanxi province, table 7 shows that Foboo went along with two Jesuits and three other Qing officials. We do not know Foboo's precise title at the time, but a Manchu memorial makes it apparent he was a measurement expert. The document reports that Foboo had measured and drawn up a plan of the *Wushu qingliang* 無暑清凉 gate at the imperial retreat in Rehe, so that the emperor could write down the characters for it in a way that would better fit the gate's size.[359] Therefore, Foboo must have worked in the

358 "Comme il a toujours aimé les sciences et les arts, il a mis dans ces deux Palais un grand nombre de personnes propres à les perfectionner. Tout ce qui regarde la composition et l'édition des livres se fait dans l'ou yn tien. C'est là qu'il a une très nombreuse Bibliothèque, et il l'augmente tous les jours. C'est là qu'il entretient beaucoup de docteurs et d'autres hommes sçavans uniquement occupés sous sa direction à faire de nouveaux ouvrages, ou à renouveler d'anciens, ausquels il ajoute quelque fois des commentaires. Pour l'yang sin tien, il est rempli d'ouvriers qui travaillent sans relâche à toutes sortes d'instrumens, par raport aux mathématique, à la géometrie, à l'astronomie, à la musique, à la physique, à la médecine, à la botanique. C'est là qu'on porte ce qui luy vient de l'Europe toutes les années, et on ne peut dire la quantité de choses rares et curieuses qu'il y ramasse de toutes parts. Il y a dans l'un et dans l'autre de ces palais des mandarins qui y président, et comme les affaires des Européans ont ordinairement raport aux sciences et aux arts, c'est par l'entremise et l'organe de ces messieurs qu'elles se traitent chez l'empereur." Witek (1973), 492–493.

359 *Kangxi chao manwen zhupi zouzhe quanyi* (1996), 767.

Wuyingdian workshops, his responsibilities during the land surveys possibly including drawing the first draft maps.[360] Years later, while surveying Shanxi and Shaanxi with his team in 1712, Foboo disobeyed orders to remain with the other mapmakers. Another team member, Imperial Guard Colonel Ying-zhu (table 7), sent a report via the Ministry of War, to the top officials of the *Wuyingdian*, Foboo's immediate superiors. They prepared plans to recall, impeach, and punish Foboo until another report arrived stating that he had rejoined the mapmakers. The emperor learned of all this through secret palace memorial and decided to give Foboo one last chance.[361] Put alongside the unauthorized return of Bouvet discussed in the intermission preceding this chapter, the comments of de Tartre about local authorities not sufficiently assisting mapping efforts in Shanxi (introduction), and the charges some mapmakers faced during the first stages of land surveying, this episode constitutes yet another example of unforeseen circumstances and unpredictable individual behavior shaping the land surveys and directly impacting progress and communication with the center.

The second workshop official in the list above is Bursai. He appears sporadically in European-language sources as Pursama or Purgama, a contraction of Bursai and *ama*, the Manchu word for "father," a form of address equivalent to the contemporary "*laoye*" 老爺 in the Chinese language.[362] As in many Qing institutions, officials working for the Imperial Household Department each had places within the two overlapping hierarchies of the bureaucracy and the banner system. This explains why Bursai was both an Overseer (*jianshi* 監視) at the *Wuyingdian* as well as a *booi* bannerman holding a hereditary title of non-imperial nobility (Chinese: *qiduwei* 騎都尉 / Manchu: *baitalabure hafan*).[363] He had previously been part of the team surveying the Manchu homelands as a clerk attached to the *Yangxindian*.[364] The fact that Bursai held positions in both halls indicates he may not have had a particular technical expertise. Rather, it seems Bursai had gained experience in dealing with the European missionaries at court. From extant documents, we know that he had been responsible for escorting Bouvet and Sabino Mariani (1665–1721) to Xiamen and

360 It appears that Foboo held the title of Director or *langzhong* by 1714. *Kangxi chao manwen zhupi zouzhe quanyi* (1996), 986.

361 *Kangxi chao manwen zhupi zouzhe quanyi* (1996), 835–836.

362 Witek (1973), 528.

363 This seems to have been relatively common for elite imperial household personnel. Chen (2011), 264.

364 As discussed above and mentioned in Cheng (2015), 21–24.

on to Guangzhou with gifts for the Pope in 1706. Afterwards, he scouted Guangzhou for missionaries trained in the sciences but eventually returned to Beijing with only a selection of European goods.[365] Soon after, in early October 1706, the emperor again sent him to Guangzhou in a great hurry to recall the gifts and to escort Bouvet and Mariani back to the capital, covering an estimated distance of about 2500 km in a remarkable twenty days.[366] Bursai thus certainly had experience in dealing with the European missionaries, and we should perhaps view his appointment to lead several surveys in that light.

Bursai was part of the team responsible for surveying the provinces of Sichuan, Yunnan, Guizhou, and Huguang. On the border with the Taungoo Empire (Burma), both Bursai and his fellow mapmaker Fabre-Bonjour perished, as mentioned above. While Fabre-Bonjour's body went back to Beijing, Bursai was cremated on the spot, although his ashes eventually also reached the capital.[367] Following his death, the court quickly appointed a replacement. His colleague at the workshops and overseer at the same *Wuyingdian*, Cangboo, went to Yunnan alongside Régis, whom the emperor had appointed as Fabre-Bonjour's replacement.[368] After the land surveys, Cangboo became salt commissioner for Guangdong and Guangxi.[369]

The fourth official is Li Bingzhong, Overseer at the *Yangxindian*. Li belonged to the Chinese plain blue banner and was part of the team of mapmakers surveying Jiangxi, Guangdong, and Guangxi. Like Bursai, Li appears in documents describing the transportation of European goods from Guangzhou to the capital in the fall of 1716. In the summer of 1720, he would escort papal envoy Carlo Ambrogio Mezzabarba (1685–1741) to Beijing and act as an intermediary between the emperor and the envoy during the latter's stay in the capital.[370] By that time, Li appears to have held one of the top positions within the Imperial

365 *Baqi manzhou shizu tongpu* (1989), 358; *Acta Pekinensia* (2013), 36; Witek (1973), 528.

366 Because of matters surrounding the Chinese Rites Controversy (intermission one). Rule rightly calls this "an extraordinary feat of horsemanship, even for a Manchu." *Acta Pekinensia* (2013), 36. Early in 1707, they arrived back at the capital. *Kangxi chao manwen zhupi zouzhe quanyi* (1996), 476.

367 ARSI, JS 176:397; Ripa (1991–96), 2:193 & 370–372. This was normal procedure for Manchus who passed away in the Chinese provinces.

368 A genealogy mentioning Cangboo is in the Harvard-Yenching Library (HYL), Rare Books, TMA 2748/9126. It is possible that Cangboo was the father of He-shen 和珅 (1750–1799), an official who played an important role during the Macartney embassy of 1793. Hummel (1943), 288.

369 Archives of the Grand Secretariat (AGS), 009120.

370 Recorded in a number of extant documents. *Qingzhong qianqi Xiyang tianzhujiao zaihua huodong dang'an shiliao* (2003), 1:15 & 27–47; Ripa (1991–96), 2:218 & 226–227.

Household Department and appears in documents alongside such top household officials as Zhao-chang 趙昌 and Li Guoping 李國屏. During the Yongzheng years, he continued to serve in different posts throughout the provinces, first as treasurer but later as provincial prosecutor in Henan and as prosecutor for the weaving industry in Hangzhou and Suzhou.[371]

In sum, all four men worked directly under the General Overseer of Works (*zongjianzao* 總監造) or Imperial Household Directors (*taijian* 太監), such as Hesihen ('Henk-ama'), Zhao-chang, Wang Daohua 王道化, Carki (Zhang Changzhu 張常住), and Li Guoping, most of whom had direct access to the emperor. These men appear sporadically in European-language accounts because they were responsible for the Europeans at court and, in that capacity, often acted as intermediaries in contacts with the emperor and his sons.[372] This combination of authority over the workshops and of issues pertaining to the court missionaries becomes explicit in the job description of Carki, Wang Daohua, and Zhao-chang taken from a 1717 memorial: "[officials who] manage production and deal with issues [pertaining to] the Westerners at the *Yangxindian*, *Wuyingdian*, and other places."[373] Dealing with European missionaries was not, however, a full-time occupation: reading through the Manchu memorials makes it clear that these officials dealt with and reported on a range of issues, of which the printing of books at the *Wuyingdian*'s Imperial Printing Office (*xiushuchu* 修書處) was perhaps most important,[374] although the memorials also report newly produced goods from the workshops, and the behavior and health issues of household personnel, to name only a few examples.

Having explored the backgrounds of workshop officials involved in the mapping project, we see clearly the important role they played as representatives of the inner palace, reflecting the emperor's personal patronage of, and investment in, the mapping project. However, at the time they took part in the surveys, the mapmakers mentioned above were all lower-ranking officials who had likely entered the imperial household as bondservants at a relatively young age and, according to Chen Guodong, often had a hard time working

371 *Kangxi chao manwen zhupi zouzhe quanyi* (1996), 1518–1519.

372 E.g. Witek (1973), 507–509. For the case of Hesihen, see Chen (2011), 257–290.

373 "養心殿武英殿等管制造帶西洋人事." *Qingzhong qianqi Xiyang tianzhujiao zaihua huodong dang'an shiliao* (2003), 1:53. Several Manchu memorials confirm this. For an overview of Manchu memorials featuring the missionaries, see Menegon (1997).

374 The formal establishment of the *xiushuchu* at the *Wuyingdian* occurred in 1680. Since this was the most important workshop there, it was simply called "*Wuyingdian zaobanchu.*" Chen (2011), 275.

their way up.[375] It seems that at least one of them, Foboo, received training as a measurement or drawing expert. Bursai, Cangboo, and Li Bingzhong were certainly overseeing production at the *Wuyingdian* and *Yangxindian* workshops as well, but we have no precise indication what their expertise may have been.[376] Apart from their skills, their appointments likely also depended on whether they had experience in dealing with the missionaries, so as more effectively to cooperate with or keep an eye on them during the surveys. After all, as we have seen, the emperor appointed the missionaries involved in the mapping of Qing territories together with these workshop officials by imperial decree several days before the ministries appointed the other team members.

2.3.2 *European Technical Experts and Assistants*

In Chapter 1, I argued that, from the 1690s onwards, court missionaries no longer worked exclusively at the Directorate of Astronomy. Instead, in their capacity as production experts for a number of material goods varying from glassworks to surveying instruments to clocks, several of them became attached to the palace's workshops, working directly under the Imperial Household Department. This is apparent from the fact that Jesuit missionaries tutored the Kangxi emperor on mathematics at the *Yangxindian*, which also kept, maintained, emulated, and adapted the emperor's clocks and other instruments in its workshops.[377] Moreover, I have argued that, much like the workshop officials in the teams of mapmakers, the missionaries were ultimately accountable to the same top household officials. Taken together, these elements help to explain why the emperor selected missionaries and workshop personnel first in forming the surveying teams, several days before the ministries appointed the other team members. They are also further evidence that the emperor tightly controlled the mapping project, which his most trusted courtiers managed, even though it also relied on a much wider base of central and local Qing administration.

Looking more closely at the European missionaries, we discover that some were involved in mapmaking throughout the duration of the land surveys,

375 Chen (2011), 274.

376 In the excerpt quoted at the beginning of this chapter, the Kangxi emperor speaks of "people who are proficient in calculating and good at drawing," perhaps another vague indication of their expertise.

377 The employment of foreign experts in the immediate and personal service of the Manchu emperor as part of his household was a practice dating back to the years of Nurhaci, see Torbert (1977), 15–16.

while others joined its later stages only briefly. The emperor initially appointed three French Jesuit missionaries to the first survey, and it was only after problems arose with the appointment of Bouvet (intermission one), that the emperor decided to appoint Fridelli, a newly arrived non-French missionary. Together with Fridelli, Régis and Jartoux formed a group that ended up working together on three of the earliest surveys. Not until the land surveys accelerated, sending two teams out simultaneously in 1711, did Régis and Jartoux finally split teams and begin cooperating with newly appointed missionaries. During the last stage of the land surveys, finally, when Jartoux remained in the capital to help in editing and integrating the draft maps, Régis continued mapmaking in Henan, Jiangnan, Zhejiang, and Fujian, and later went to Yunnan to replace Fabre-Bonjour, who had passed away while surveying in that province. We may thus safely assume that Régis and Jartoux, as the only missionaries involved in mapmaking for the entire duration of the land surveys, took the lead, at least among the European missionaries.

Further evidence supports this thesis. First, Régis authored the only comprehensive manuscript account on the surveys still extant and which Jean-Baptiste du Halde (1674–1743) used for his four-volume *Description de la Chine* (1735).[378] Moreover, during the land surveys' later stages, Régis appears to have been responsible for preparatory work before leaving the capital, even for teams other than his own. In an extant letter from de Tartre to Régis, de Tartre writes that Régis had selected, or perhaps translated, information on the routes taken by the team sent to Shanxi and Shaanxi provinces:

> We followed the line you have traced for us without any problem, because we only have had mountains there, as far as one can see, and without any fixed points except for what is marked in the gazetteers, [namely,] that their borders are distant from Henan by 6 *lieues*, 10 *lieues*, etc.[379]

We also know that Jartoux, on the other hand, instructed Qing officials, some of whom later traveled to survey the road into Tibet, in surveying techniques.[380] After taking an active part in the early stages of the land surveys, Jartoux took

378 BNF, Ms. français 17242; du Halde (1735), xxxv–xxlvij.

379 "nous suivrons sans tracerez, car nous n'avons eu de ce coté la que des montagnes a perte de vue sans point fixe que ce que marquent les Tchichu [*zhishu* 志書] que leur limites sont distantes du honan de six lieues, 10 lieux etc." APF, *doss. pers.*, de Tartre. "Lieues" here presumably refers to the *li*. Since the *lieue* is originally a French unit of length, another possibility is the so-called "*nouvelle lieue de Paris,*" in use from 1674 until 1793, corresponding to roughly 3.9 km.

380 Cordier (1915), 517.

the lead in reducing and editing the draft maps at the palace. He was apparently responsible for reducing the scale of the draft maps to one-half or one-fourth of the originals and combined all regional maps to form the first large atlas of the entirety of Qing territories.[381] This meant that he occupied himself with the projection of these maps and atlases, perhaps the most mathematical aspect of map production.[382] Jartoux's death notice, kept in the Jesuit archives in Rome, claims that he particularly enjoyed the more "subtle" mathematics of "infinity, the differential, the integral and (as it is called) the denominator."[383] Moreover, like Régis, Jartoux wrote his own memoirs about the surveys, although his account has not survived.[384]

In contrast to Régis and Jartoux, other missionary-mapmakers appear to have played somewhat less significant roles. In a first stage, when mapmaking efforts were still concentrated in the north, the emperor gradually appointed mathematically trained missionaries only just arrived in the capital, having been on the lookout for such men for at least a decade.[385] A letter written in 1714 by Jean Borghèse, physician of papal legate to the Qing court Charles-Thomas Maillard de Tournon, articulates the Kangxi court's desire for experts in a number of fields:

> The emperor makes continuous requests for having the people he asks for, that is, doctors, pharmacists, surgeons, musicians, painters, mathematicians, watchmakers, glassblowers, and it is on this that the conservation and the strengthening of the mission depends.[386]

---

381 Confirmed by several contemporary accounts and summarized in Pfister (1932), 1:584–586.

382 For projection methods, see Wang (1995), 159–168.

383 "subtiliore calculo" & "infinitus, differentialibus, integralibus et (ut vocant) denominatum." Quoted from Jartoux's death notice. ARSI, JS 182:393.

384 Gaubil (1970), 668 & 735.

385 Confirmed by several official documents. In a 1703 edict, issued only half a year after the 1702 measurements (Chapter 1), Kangxi explicitly states the following: "Pass this decree on to all Westerners: the use of foreign techniques should be seen as a very urgent need. Regardless of whether they are religious men or Macao residents, if you are able to recruit foreign experts, then do not linger and immediately make the necessary preparations. This is urgent." 傳諭西洋人：用外科甚屬緊要。無論其修道人或澳門地方人，若能得外科者，則當速找預備，勿致稍怠，關繫緊要. *Kangxi chao manwen zhupi zouzhe quanyi* (1996), 284; Witek (1973), 528.

386 "l'Empereur fait de continuelles instances pour avoir les gens qu'il demande, savoir, des medecins, des apoticaires, des chirurgiens, des musiciens, des peintres, des mathematiciens, des horlogeurs, vitrierеurs et c'est de la que depend la conservation et l'augmentation de la mission." AMEP 424:434–435.

Other sources explicitly state that the emperor had recruited some of the missionaries who took part in the surveys because of their background in mathematics. A 1727 manuscript biographical compendium prepared by Jean-Pierre Nicéron (1685–1738), for example, entitled *Biographies de Jésuites missionnaires en Chine*, contains short biographies of Fabre-Bonjour and Cardoso, stressing their training in mathematics:[387]

> [Fabre-Bonjour was sent] in the capacity of mathematician. He learned that it would be necessary to solve a few triangles and, since he found himself in the company of several trained geographers, learned to do like them.
> [Cardoso was] sent from Portugal to China as mathematician. It was in this capacity that he was presented to the emperor, but he knew nothing, or he knew very little, of mathematics. They did not fail to have him join one of the teams that made the maps. It was just a question of solving a triangle to have one's position on the earth calculated. This was not difficult and Father Cardoso learned it either before his departure or during the fieldwork in the countryside.[388]

The second selection of missionaries for the land surveys, consisting of Fridelli, Fabre-Bonjour, and Cardoso, none of them of the French Jesuit mission, thus appears to be the result of efforts by other Catholic orders to send more missionaries with backgrounds in mathematics. The emperor called each of these missionaries to the palace shortly after their arrival in Asia: Fridelli had been stationed in Jiangnan only two years when he received the order to proceed to the capital; Fabre-Bonjour, known as a gifted intellectual even before leaving Europe, received orders to proceed from Guangzhou to Beijing immediately

387 This manuscript constitutes the preparatory version of what would become Nicéron's 38-volume *Mémoires pour servir à l'histoire des hommes illustres de la république des lettres avec le catalogue raisonné de leurs ouvrages* (1728–38).

388 "[Fabre-Bonjour est envoyé] sous le titre de mathématicien. Il a apprit autant qu'il lui estoit nécessaire à résoudre quelques triangles et se trouvant en compagnie d'autres géographes exercés, il apprit à faire comme eux." & "[Cardoso est] envoyé de Portugal à la Chine comme mathématicien. Ce fut sous ce titre qu'il fut présenté à l'Empereur, mais il ne scavoit point ou il scavoit tres peu de mathématique. On ne laissa pas de le joindre à une des troupes qui faisoit les cartes. Il n'estoit question que d'apprendre à résoudre un triangle pour faire [calculer] son point sur le terre, comme les navigateurs le font à la mer. Cela n'estoit pas difficile et le P. Cardose le fit apprendre ou avant le départ au durant la marche de camp dans la campagne des quels il alloit." BNF, Ms. Français, 17670:1–85.

after arriving in 1710;[389] and Cardoso, carrying gifts from the Portuguese king, had also just arrived in Guangzhou when the emperor called him to Beijing together with Fabre-Bonjour.

When the emperor received both men in audience in February 1711, he asked Fabre-Bonjour three questions: (1) whether he had brought any new material on mathematics, to which he replied he had brought a book by Philippe de La Hire; (2) whether he was familiar with techniques of land surveying, to which he replied in the negative; and (3) whether he had brought any new instruments, after which he explained he had been obliged to leave behind in London a pantograph given by the pope.[390] The emperor then showed Fabre-Bonjour and Cardoso a book on trigonometry, likely the *Imperially Commissioned Treatise on the Derivation of Triangles*, and asked them a question regarding the measurement of distances.[391] Fabre-Bonjour's negative answer to the second question seems to confirm the biographical information recorded in the excerpts above, which states that Fabre-Bonjour and Cardoso still had to learn the art of land surveying during the land surveys and "in the company of several trained geographers," undoubtedly referring to Régis and Jartoux. Nonetheless, in late February both missionaries received unofficial information that they would soon start assisting in the land surveys. The emperor selected Cardoso to team up with Régis in mapping the province of Shandong, whereas he ordered Fabre-Bonjour to join Jartoux and Fridelli in surveying the vast Khalka frontier.[392] Other missionaries called to court, such as Franz Thilisch (1670–1716) and Jean-François Foucquet, did not find themselves selected.[393]

Thus, Régis and Jartoux, having served for years as imperial experts at the palace, certainly took the lead among the missionaries with regard to cartographic practice. On the other hand, the choices of Fridelli, Fabre-Bonjour, and

389 In contrast to the perception of Fabre-Bonjour as a gifted mathematician, Ugo Baldini convincingly argues mathematics was not actually among his main interests or competences. Baldini (2012), esp. 253.

390 Baldini (2012), 264–267.

391 Han (2016).

392 Baldini argues that the emperor possibly did not have a very favorable impression of Cardoso's talents and that he therefore decided to have Cardoso assist in mapmaking. If true, this would support the thesis that Cardoso was not one of the primary mapmakers, but instead acted as an assistant to Régis and Jartoux. Baldini (2012), 264–265 (n.83).

393 Witek (1973), 500–502; Baldini (2012), 266. Note that Foucquet had helped Bouvet with this work on the *Yijing*, perhaps influencing the emperor's decision (intermission one). The Kangxi emeperor later selected both men to explain the principles of astronomy to him. Jami (2012), 287.

Cardoso, who had spent little or no time at the Qing court, were specifically to assist Régis and Jartoux in the land surveys, based on their knowledge of mathematics and perhaps the fact they were not part of the French Jesuit mission.[394] From 1712 onwards, however, the land surveys would again accelerate, as the emperor wished to dispatch no fewer than three teams of mapmakers simultaneously into the Chinese provinces. This meant he urgently needed even more European missionaries with knowledge of mathematics. Upon advice from his court missionaries on the matter, the emperor decided to summon a number of missionaries stationed in the provinces. Not especially known for their knowledge of mathematics, de Tartre, Hinderer, and de Mailla received orders to leave their mission stations in, respectively, Shanxi, Jiangxi, and Zhejiang provinces to assist Régis, Fridelli, Cardoso, and Fabre-Bonjour, the latter three having received sufficient training in land surveying in the company of Régis and Jartoux.[395] It seems that, with the exception of Régis, de Mailla, and Fridelli, none of the missionary-mapmakers ever again participated in imperial cartographic practice.[396]

With Régis, Jartoux, and the workshop officials thus leading surveying teams as part of the Imperial Household Department and its workshops, while other missionaries assisted them, it appears that the workshops were far more than a place or a setting for material production or interaction between court

394 For more on Fridelli's life, see Zerlik (1962); Hsü (1986).

395 In the *Lettres édifiantes*, one finds a narrative of de Tartre correcting the emperor, who thought he had found a mistake on the map of Shanxi: "l'empereur prétendait que le cours d'une certaine rivière était mal placé… le P. du Tartre, voyant que l'empereur se trompait, soutint le sentiment de la vérité avec la modesté et le respect qui est dû à la majesté des princes, et il le fit d'une manière si claire que l'empereur fut obligé d'en convenir." du Halde et al. (1702–76), 14:v–vi.

396 Jartoux continued his work on the Qing atlas until his death in 1720, while the Yongzheng emperor later asked Fridelli, de Mailla, and Régis to contribute their cartographic knowledge to a revision of the Qing atlas in the 1720s (Chapter 3); De Mailla also wrote several religious treatises in Manchu and Chinese and composed the *Histoire générale de la Chine*. de Moyriac de Mailla (1777–85); Bouvet continued his work on the *Yijing* (intermission one); Fabre-Bonjour passed away during the surveys; Cardoso remained in Beijing until his death in 1723, but we know little about his activities during this time; De Tartre at first returned to his mission station in Shanxi but later became head of the French residence in Beijing; Hinderer allegedly refused politely the emperor's offer to stay at court, instead traveling through many of the Chinese provinces and focusing on missionary work. Chaney (1889). On the other hand, Hinderer does seem to have supplied local officials with maps. ARSI, JS 180:83r. I am indebted to Noël Golvers for this reference. For biographical information, see also Pfister (1932); Dehergne (1973).

missionaries and the throne. Instead, the appointment by imperial decree of its personnel, both European missionaries and Overseers of the *Wuyingdian* and *Yangxindian* workshops, illustrates the institutional involvement of the workshops in terms of manpower. It also reflects the strong need of the Kangxi emperor to extend his personal control, not just over the mapping of Qing territories as an imperial project but also over its practical execution: he personally selected men he knew and trusted to take the lead. When put next to the importance of the Imperial Workshops with regard to instrument production (Chapter 1), the institution emerges as an important logistical center for the entire mapping project, providing both manpower and material support. In further exploring this point, the next section shows that the workshops constituted no less than the logistical pivot for the duration of the land surveys, without which the practical execution of the mapping project would not have been feasible.

2.3.3 *The Logistical Centrality of the Imperial Workshops*

The Imperial Workshops consisted of different *zuo* 作, *chu* 處, *chang* 廠 and *guan* 館, sub-departments each with their own specialties. Material, labor-intensive workshops often carried the suffix *zuo* (from *zuofang* 作坊), such as the *yuzuo* 玉作 (jade workshop), *paoqiangzuo* 炮槍作 (cannons and firearms workshop), or *biaozuo* 表作 (watches workshop).[397] Large projects, on the other hand, such as the compilation of encyclopedias or important literary volumes, often required setting up more temporary offices (*guan* 館) to manage the projects. The mapping project, too, required establishing an *Office for the Overview of Imperial Territories* (*Huangyu quanlan guan* 皇輿全覽館), carrying a name similar to the atlas eventually produced based on the data from the land surveys (Chapter 3) and housed at the *Nanxundian* 南薰殿, just south of the *Wuyingdian*.[398] Presumably, these offices directed a large part of the preparations for the surveys and the processing of maps afterwards, since they appear in the records as early as 1713. One person working for the *Office for the Overview of Imperial Territories* was Zhang Wei 張位, an imperial student of whom we know little else.[399] Jartoux may also have worked in these offices,

397 This is apparent from a quick scan of the first volumes of *Qinggong neiwufu zaobanchu dang'an zonghui* (2005). Later in the eighteenth century, the workshops added a map-room (Chinese: *yutu fang* 輿圖房 / Manchu: *ba na I nirugan nirure boo*).

398 Jami (2012), 265; Chen (1990), 106. Note the establishment also in 1713 of an office for mathematics or *suanxueguan* 算學館 at the Changchunyuan. Jami (2012), 260–283.

399 Jami (2012), 265.

in which case cartographic work at the office would have included reducing all maps to corresponding scales, dealing with the problem of projection, preparing the original maps for publication, and combining all regional maps into one large overview map.

In any case, this newly established office certainly was responsible for the compilation of the *Imperially Commissioned Overview of Imperial Territories* or *Qinding huangyu quanlan* 钦定皇輿全覽. This impressively large work constitutes a textual description of routes throughout the surveyed territories, compiled with selected information from a wide range of gazetteers and historical works and arranged by main roads, starting from Beijing and slowly covering all the provinces and regions, including every prefecture (*fu* 府), department (*zhou* 州) and district (*xian* 縣) along the way. Each of these places has its own full section inserted, consisting mostly of several pages of quotations from a wide range of materials on sites and sights of interest. Therefore, the *Imperially Commissioned Overview of Imperial Territories* belongs not to the genre of gazetteers but to that of route books, as it constitutes the ultimate guide to the Qing road and waterway network.[400]

This large compilation, unlike the resulting atlases, was never published or distributed outside the palace (Chapter 3). One Manchu-language memorial, dated to late October 1711, announces the completion of the manuscript volumes for Zhili, Shandong, Jiangnan, and Zhejiang provinces, 42 *juan* in total, compiled by Hanlin scholars Qian Mingshi 錢名世, Li Qiyuan 李琪園, Wang Shihong 汪士鋐 (1658–1723), Peng Huiqi 彭會淇, Qiu Zhao'ao 仇兆鰲, and Gu Tuhe 顧圖河, identified as editors.[401] Others working on the project were Yang Kaiyuan 楊開沅, Yu Changce 俞長策, Gong Hongli 宮鴻曆, Guo Yuanyu 郭元釪, and Yang Shihui 楊士徽. Clearly, producing these volumes required forming a large team of scholars. Furthermore, I have been able to identify two

400 Besides the *Descriptive Account of the Unified Great Qing* (Chapter 1) and this *Imperially Commissioned Overview of Imperial Territories*, both large projects and the result of many years compiling textual information, there is a third similarly large geographic work compiled earlier during the Kangxi years (c. 1680–1705): the *Huangyu biao* 皇輿表. The precise relationship among these three major works remains unstudied. Cao (2007).

401 *Kangxi chao manwen zhupi zouzhe quanyi* (1996), 738; Chen (1990), 106; Library of Congress (LC), Chinese Rare Books Dept., B118 W19. I have not been able to consult the Sichuan copy. Other Manchu memorials dealing with the *Qinding huangyu quanlan* are also extant, with one mentioning the further editing of the Zhili volumes. *Kangxi chao manwen zhupi zouzhe quanyi* (1996), 434 & 741. The elite Hanlin scholars worked at the Hanlin Academy, an academic and administrative imperial institution dating back to Tang China (618–907).

versions of the *Imperially Commissioned Overview of Imperial Territories* in four extant copies, all of them constituting only small parts of the entire compilation. First, there is the draft version, also mentioned in the Manchu-language memorials and carrying the distinct title *Qinding fangyu lucheng kaolüe* 欽定方輿路程考略. It survives in two institutions: three bound parts (*ce* 冊) are at the Library of Congress, while another set of more than ten *ce* is at the Sichuan University Library.[402] Someone later crossed out the title on the Library of Congress copy in vermillion ink, typically reserved for the emperor or those writing in his name, and replaced it with "*Qinding huangyu quanlan*" (figure 27). It further includes many notes and comments, some in vermillion, and constitutes one single volume with descriptions covering the main roads of Shanxi as well as those connecting the province with Henan and Shaanxi. The Sichuan copy includes similar corrections to both the title and the main text and mainly covers roads in Shanxi and Sichuan.[403] On the other hand, a second, printed, version is at the National Library of China in Beijing and the National Central Library in Taipei in, respectively, three and eight *juan*.[404] The Beijing copy mainly covers roads into Zhejiang province, whereas the Taipei copy includes a large part of the roads leading to Shaanxi. It is possible that further parts of the printed edition survive in Beijing, Taipei and Shanghai.[405]

The Taipei copy illustrates the difference between both versions: it has a table of road distances inserted in the first *juan* of the province, between the table of contents and the actual geographic descriptions of towns and cities, arranged based on the Shaanxi road network.[406] These distances are in quantities of *li* plus extra ropes (*sheng* 繩) according to both the *zhouchi* 周尺 and *yingzaochi* 營造尺 standards. The fact that the text employs two standards, the Zhou-era *chi* and *chi* as re-standardized after the 1702 measurements (Chapter 1), must be the result of concerns about the authority of the *zhouchi* apparent in the emperor's words quoted at the beginning of this chapter.[407] At the same

402 LC, Chinese Rare Books Dept., B118 W19; Chen (1990).

403 Chen (1990), 106.

404 National Library of China (NLC), Rare Books Dept., SB 09779; National Central Library (NCL), Rare Books Dept., 000024.

405 Chen (1990), 105.

406 NCL, Rare Books Dept., 000024, 1A-27B [陝西九府州邊幅里數].

407 Chapter 1. Another work includes a passage with measurements for roads in Taiwan dating from the same survey. It must be based on the *Qinding huangyu quanlan*. SKQS, *Taihai shichai lu* 臺海使槎錄, *juan* 1, 22.

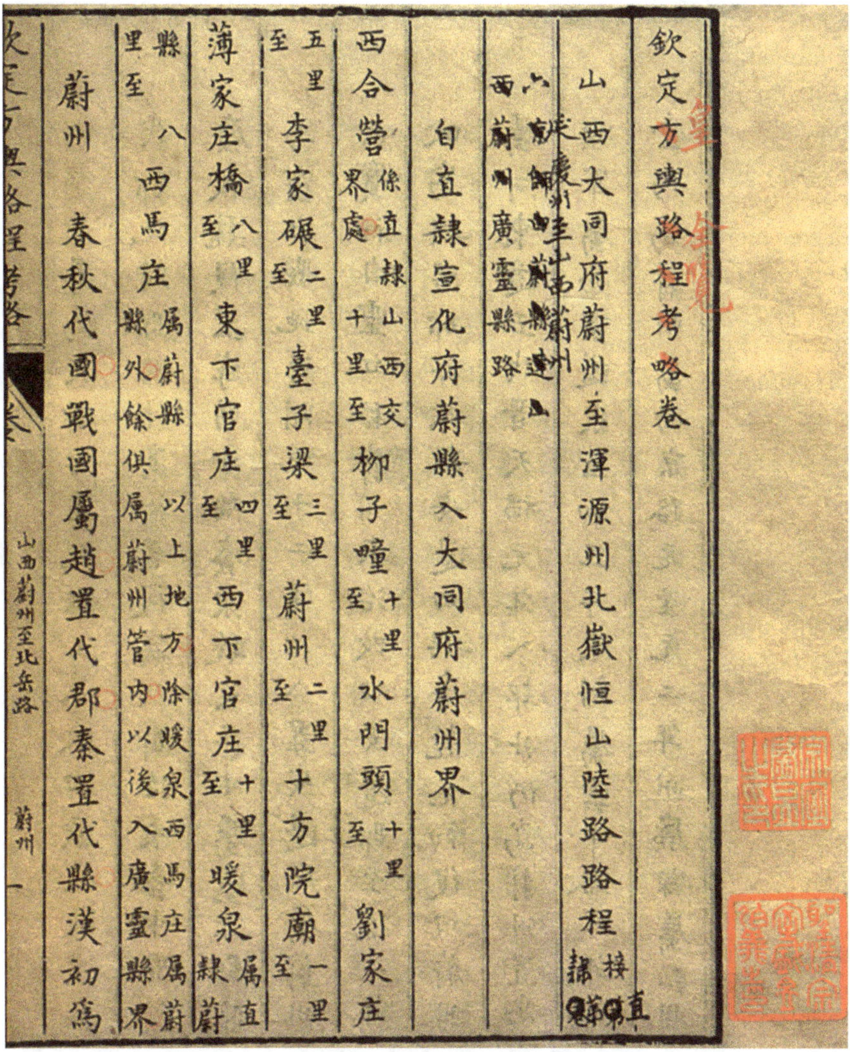
欽定方輿路程考略卷
山西大同府蔚州至渾源州北嶽恒山陸路路程
自直隸宣化府蔚縣入大同府蔚州界
西合營係直隸山西交界處十里至柳子疃十里至水門頭十里至劉家庄
五里至李家碾二里至臺子梁三里至蔚州二里至十方院廟一里至
薄家庄橋八里至東下官庄四里至西下官庄十里至暖泉屬直隸蔚
縣八里至西馬庄屬蔚縣以上地方除暖泉西馬庄屬蔚縣外餘俱屬蔚州管內以後入廣靈縣界
蔚州 春秋代國戰國屬趙置代郡秦置代縣漢初爲
山西蔚州至北岳路 蔚州 一

FIGURE 27 *The* Qinding fangyu lucheng kaolüe. *With vermillion ink, the draft title is corrected into* Qinding huangyu quanlan.
LIBRARY OF CONGRESS, CHINESE RARE BOOKS DEPT., B118 W19.

time, it confirms the project's embedding in inherited cartographic practice, by making sure readers could compare old and new data easily.

From the differences between both versions of the *Imperially Commissioned Overview of Imperial Territories*, we can presume the preparation of a first draft version for each region or province well before the emperor even formed the first teams, giving scholars and the palace plenty of time to make corrections at a later stage and giving team leaders the opportunity to plan their itineraries informed by centuries of accumulated knowledge. On the other hand, finalizing the printed edition after the surveys were complete involved inserting road distances for each province. This thesis finds further support in the dates put forward by Chen Li, who argues that the project to compile the *Imperially Commissioned Overview of Imperial Territories* began in 1704, not long after the degree measurements of one degree of latitude in 1702, and was not finished before 1721, the year of the completion of a revised version of the resulting atlas (Chapter 3).[408] Moreover, the need to compile this information in preparation for the actual land surveys partly explains the nearly six years between the measurements of one degree of latitude in 1702 and the first comprehensive survey in 1708. As such, the *Office for the Overview of Imperial Territories*' responsibility as an office for compiling textual material and perhaps even editing the draft maps is evidence of a third manner in which the workshops were involved in the mapping project's practical execution, besides material support and manpower.

In turn, viewing the Imperial Workshops as an institutional missing link binding together team leaders and producing the textual exponent of the mapping project sheds new light on the trajectories of instruments and maps during the surveys. First, the production of instruments large and small in the workshops over the decades leading up to the land surveys made mapping the vast lands of the Qing technically feasible. After all, in the mapping project's later stages, up to three teams surveying simultaneously necessitated production of more instruments, even if they mobilized the instruments kept at the Jesuit residences. The necessity of instrument production at the workshops, as discussed in Chapter 1, becomes even more apparent considering the possibility that the teams carried back-up instruments along, or that, more likely, different team members performed parallel observations for verification. We should also note that, when instruments did not go into the field, they remained within the imperial palace, either on display at the *Yangxindian* or in the workshops' storage rooms, contributing to the survival of some of these instruments until today.

Second, from 1709 onwards, maps slowly poured in from the various Qing territories. After the emperor's approval, the maps must have passed on to the

408 Chen (1990), 107. In a 1707 memorial, Li Guoping announces the compilation of a small part of the work. *Kangxi chao manwen zhupi zouzhe quanyi* (1996), 529. An extant document further shows the gathering of a great number of maps from different institutions, locations around the empire, the Jesuits in Beijing, and the imperial retreat in Rehe between 1708 and 1711. Qin & Liu (1997).

Office for the Overview of Imperial Territories, where another team worked on the draft maps and prepared them for printing.[409] After the editing, combining, and preparation for carving of all regional and provincial maps, specialists working in the *Wuyingdian* workshops printed different editions of the atlas that came to be known as the *Overview Maps of Imperial Territories* or *Huangyu quanlan tu* 皇輿全覽圖. Meanwhile, they also printed the textual route book companion discussed above, the *Imperially Commissioned Overview of Imperial Territories*.[410] All woodblocks and copperplates used in printing the maps and the textual companion remained in the workshops' storage rooms until at least the 1770s.[411] The workshops also produced reduced maps based on data from the surveys in a number of imperial compilations.[412] One example is the *Comprehensive Collection of Books and Pictures of the Past and Present* or *Gujin tushu jicheng* 古今圖書集成, also prepared, produced, and printed in the Imperial Workshops.

When we create an abstraction of these trajectories of instruments and maps, a highly similar pattern emerges within which the emperor and his workshops occupy a central position. First, the Jesuits carried instruments from Europe to Asia, some of which eventually reached the missionaries' residences in Beijing. Others were gifts for the emperor, who had his workshops reproduce and adapt them. In a later stage, European-made instruments and perhaps their Qing emulations as well, all kept at the workshops, went out into the field for mapmaking. On the other hand, when a regional or provincial map was complete, it proceeded immediately to the emperor and thence to his workshops for further editing and preparation for printing. This process produced different atlases, some of which made it back to the missionaries'

409 *Qingdai qijuzhu ce* (2009), 28:B013933, where it is mentioned that a map must proceed to the Wuyingdian having been checked by the Ministry of War. In the First Historical Archives in Beijing, it seems, a checklist from the workshops survives recording the dates of reception for most of the regional draft maps. The list remains unpublished but a rough summary is in Bai & Li (2014), 61–63. When we consider transmission times, the list corresponds to the dates recorded in table 6. This booklet also records information on manuscript maps kept at the First Historical Archives in Beijing, possibly constituting the draft maps in question. However, at present, the collection is not open to scholars. Bai & Li (2014), 74–90.

410 For the different atlas editions, see Chapter 3.

411 A memorial dating back from the 1770s mentions that all woodblocks and copperplates of the Kangxi emperor's *Huangyu quanlan tu* are kept at the workshops. *Qingzhong qianqi Xiyang tianzhujiao zaihua huodong dang'an shiliao* (2003), 4:248–249.

412 Chapter 3.

residences and on to Europe, as we shall discuss in Chapter 3, the circulation of material goods related to cartography between Europe and East Asia thereby coming full circle. The emperor thus remained fully in control during the land surveys: he gave the orders to produce the instruments, adjudged their quality and was the first to take a good look at the maps pouring in from the provinces. If the emperor was the driver, though, his workshops were the engine, producing, reproducing, and eventually preserving within its premises instruments, maps, and a route book companion. As such, the workshops as an institution were crucial throughout the duration of the land surveys, an aspect thus far unstudied.

Clearly, in terms of the daily operations and the precise functions performed by the workshops during the Kangxi years, for which sources are few and far between, there is tremendous room for further research. In this section, I have limited myself to the workshops' role during the Qing mapping project. More precisely, I have argued that the workshops shaped the project in two major ways: first, it was the place where the mapping project transmuted from an idea and a possibility into an organized project. After all, it was here that Jesuit missionaries served the Kangxi emperor as technical experts, and the entire project's planning and execution proceeded from offices attached to the Imperial Workshops and with the help of its personnel. Therefore, in providing institutional support for the land surveys, the workshops were in effect where the mapping project materialized. Second, the workshops also shaped the mapping of Qing territories in providing the necessary material support. European prototype instruments kept at the workshops and at the missionaries' residences permitted the production of more such instruments and the exploration of the techniques they stood for. This emulation and reproduction of Paris-made instruments made it practically feasible to survey such vast territories. At the same time, the workshops were also the site of the gathering, editing, and eventual carving and printing of all draft maps in different editions and on the basis of different techniques, as we will discuss in Chapter 3. Thus, by providing full institutional and material support, the Qing workshops and inner palace emerge as the main logistical pivot, under direct exhortation from the Kangxi emperor, for the practical execution of the mapping project.

Conclusion

Chapter 2 has focused attention on the trajectories of instruments and maps together during the land surveys. First, on the basis of mostly missionary sources, I have followed instruments to the many places in northern Qing territories and frontier areas where surveying teams conducted observations, took measurements, and made calculations. This uncovered the exact techniques they used, as well as progress made and routes followed. Teams also surveyed roads into the bordering territories of Korea and Tibet, albeit without the European missionaries' mathematical and observational expertise. This emphasis on mapping what lay beyond the passes and gates of the Great Wall reflected the emperor's desire for more effective control over the vast Qing frontier. At the same time, focusing on the precise techniques of mapmaking made clear that selected information gathered in the field circulated ahead of the resulting maps and as far afield as Europe, thus informing dynamics well beyond the Qing court's cartographic pursuits.

In a second step, I took on the available Qing sources, mostly Chinese-language documents replete with details about the surveying of the Chinese provinces. I gradually left instruments behind, my focus shifting to the draft provincial maps. When finished on the spot, maps proceeded directly to the emperor via local governors, who wrote memorials accompanying each map, aiding my reconstructon of details pertaining to the logistics involved in the land surveys. I have presented a range of new information on strategies and timing, communication with the center, supporting institutions, and local contacts. Perhaps most significant was the list of team members extracted from the memorials and other documents. In all, Qing sources confirmed the involvement of all layers of the Qing bureaucracy, both central and local, while also highlighting the connections between the many individuals involved in the land surveys, without whom and without which the practical execution of the mapping project would not have been feasible.

In the last section, I argued that the focus on instruments and maps with regard to the land surveys reveals a missing link: team leaders during the land surveys were institutionally bound to the Imperial Workshops, whereas two of the European mapmakers whom the Kangxi emperor chose to keep at court, Régis and Jartoux, also had a connection to this institution. At the same time, temporary offices established at the workhops supported the project, forming a back-office responsible for producing the resulting atlases as well as a textual route book companion. Combined with insights from Chapter 1, this led me to conclude that the Imperial Workshops occupied a logistical center for the duration of the land surveys, not only in institutional terms but also in taking on the responsibility of producing and storing surveying instruments and the resulting atlases.

With its attention to the connections between individuals, guided by the trajectories of instruments and maps, this chapter best exemplifies this study's overall approach. I have argued that, in spatial terms, the land surveys were central to these trajectories. More specifically, the surveys connect the global movements of both families of material objects studied in this book, as instruments and maps "met" in the many places where measurements and observations took place: in the case of instruments, the trajectory from Paris-made prototypes to their Qing adaptations to their test-of-concept use in the field (Chapter 1) extended to determining those sites' positions *vis-à-vis* Beijing; on the other hand, for the resultant maps, the trajectory from the first Qing atlas to its different editions and its European incorporation (Chapter 3) extends back in time to those same locations, where the mapmakers first put place names to paper and drew draft maps using the fruits of instrument use and geometrical calculations. Putting aside the more divergent trajectories, such as those of the instruments given to the emperor's sons, this abstraction has been fruitful in parsing the Imperial Workshops' logistical centrality.

In the past, conceptualizing the kind of cross-cultural circulation I describe has often distinguished between two conceptual levels, the "global" and the "local." Actors such as the Qing officials and local administrators involved in the land surveys, or institutions such as the Imperial Workshops, come to occupy problematic positions within this global–local dichotomy, constituting centralizing and connecting factors bypassing the weak links between the so-called local and global levels, in this case, the Jesuit missionaries. In other words, the global–local dichotomy to some extent obstructed efforts to appreciate the full range of actors and institutions involved in the mapping project. Likewise, as this chapter has shown, new data made its way across Eurasia ahead of the resulting maps, informing dynamics unbound by any constraints associated with the global–local dichotomy. Thus, rather than connecting abtract "global" and "local" levels, this chapter's true contribution lies in tracing out the most basic connections between individuals and taking into account the inherent complexities and contingencies that shaped, supported, and sustained the land surveys. It has put forward a long list of individuals involved in the project, and devoted attention to the institutions to which some of those individuals belonged. The picture that emerges as a result depicts a complex web of individual connections, mediated by different institutions and spatial factors, all of which shaped the production of a new body of cartographic knowledge.

INTERMISSION 2

Missionaries and Mapmakers: Missionary Activity during the Land Surveys

In Chapter 2, I have shown that extensive contacts with provincial and subordinate officials in the densely populated provinces of China facilitated the mapmakers' work.[413] In addition to their contacts with local officials, there are also strong indications the missionary-mapmakers met with Chinese Christians during the surveys. Jean-Baptiste du Halde, editor of the *Lettres édifiantes et curieuses*, a multi-volume publication presenting carefully edited letters of Jesuit missionaries to the European public, claimed the missionaries used the opportunity to travel throughout Qing territories to aid Christian communities wherever they could, as he described in his preface of the 1720 volume:

> What has been comforting for the missionaries during such a tough task of travelling around all the provinces of such a vast empire, is that, besides the imperial protection for the preachers of the Gospel that their work procured, they have been able, by themselves, to render great service to the religion... In several places they were able to recover, to the missionaries, churches that had been usurped by the infidels and devoted to profane uses. To others, they provided support and protection to undertake their task with more liberty. They consoled, instructed, and animated a great number of Christians, long since abandoned and deprived of a pastor. Everywhere they restored, to the religion and its ministers, the reputation and credit they had lost during these unfortunate times of division and troubles.[414]

413 This is a reworked version of an article published in *Frontiers of History in China*. Cams (2014).

414 "Ce qu'il y a eu de consolant pour les Missionnaires dans une occupation aussi pénible que celle de parcourir toutes les Provinces d'un si vaste Empire, c'est qu'outre la protection de l'Empereur que leur travail procuroit aux Prédicateurs de l'Evangile, ils ont esté en estat de rendre par eux-mêmes de grands services à la Religion... En plusieurs endroits ils ont fait restituer aux Missionnaires des Eglises qui avoient esté usurpées par les Infideles, & consacrées à des usages prophanes. Ils ont ménagé à d'autres de l'appui & de la protection pour faire leurs fonctions avec plus de liberté. Ils ont consolé, instruit, animé un grand nombre de Chrestiens abandonnez depuis long-temps, & destituez de Pasteurs. Par tout ils ont fait rendre à la Religion & à ses Ministres la réputation & le crédit qu'on avoit

© KONINKLIJKE BRILL NV, LEIDEN, 2017 | DOI 10.1163/9789004345362_006

We find a second example of such claims in a letter included in the same edition of the *Lettres édifiantes*. This letter is by the hand of Joseph Marie Anne de Moyriac de Mailla, one of the missionary-mapmakers, comprising a description of Taiwan based on his experiences drawing the map of Fujian province. He writes that, before setting sail from Xiamen, he learned of the presence of Christians on the island, whom he was later able to meet:

> Before departing from Xiamen, we were told that there were Christians in Formosa: we informed ourselves about it, and certainly there are none among the Chinese; but it seems that there had been among the islanders at the time when the Dutch were masters of the harbor... We have striven, as best we could, to teach them the most necessary truths of our holy religion: we advised in particular those who seemed best instructed to us, to instill these important truths in their fellow-countrymen, and above all to baptize the children as soon as they are born... We have left them the formula for baptism, and that is all that we were able to do.[415]

These short excerpts raise the question of whether missionary-mapmakers, passing through the cities, towns, and villages of the Chinese provinces, were truly able to further their religious agenda and impact positively the lives of Christians they encountered, as both du Halde and de Mailla maintain.[416] Might their comments perhaps have been carefully crafted rhetoric aimed at eliciting further support and legitimation for the French Jesuit mission? After all, at the time of publication, the Jesuits stood amidst increased scrutiny due to the so-called "Chinese rites controversy," which revolved in large part around the issue of whether Chinese Christians could perform ancestral

---

perdu dans ces malheureux temps de division & de trouble." du Halde et al. (1702–76), 14:xi.

415 "Avant que de partir d'Emoui, on nous avoit dit qu'il y avoit des Chrétiens dans Formose: nous nous en sommes informés, & certainement il n'y en a aucun parmi les Chinois ; mais il y a apparence qu'il y en a eu parmi les insulaires, du temps que les Hollandois étoient maîtres du port... Nous avons tâché, autant que nous le pouvions, de leur enseigner les vérités les plus nécessaires de notre sainte Religion : nous avons recommandé en particulier à ceux qui nous paroissoient le mieux instruits, d'inculquer à leurs concitoyens ces vérités importantes, & sur-tout de baptiser les enfans aussi-tôt qu'ils seroient nés... Nous leur avons laissé la formule du baptême, & c'est tout ce que nous avons pu faire." du Halde et al. (1702–76), 14:446–447.

416 Jesuit missionary Antoine Gaubil makes very similar claims in one of his letters, written in 1736, but does not speak of the restitution of churches. Gaubil (1970), 470–471; Bernard (1946), 199; Pfister (1932), 1: 590–592, 596–605, 607–609 & 611–617; Baldini (2012), 273–274.

rites and offerings to Confucius. In Europe and Asia, actors within the Catholic Church criticized the Jesuits' generally tolerant stance towards such rites. Outside the church, the Kangxi emperor reacted by establishing a system of permits (*piao* 票) for those missionaries in favor of Chinese rites, while exiling the rest.[417] By using the words "*division et troubles*," in the first excerpt quoted above, du Halde alludes to the difficulties experienced due to both of these developments. The Jesuits thus felt the need to defend their position in Asia as well as in Europe and often carefully edited their publications for this purpose.

Whether tainted or not by issues surrounding the controversy, the claims of du Halde and de Mailla seem to have had their impact on the literature. Some scholars have chosen not to question their claims directly, at least not explicitly or in writing. On the other hand, Chapter 2 has shown that the land surveys were conducted rather rapidly, especially considering the great distances and practical difficulties involved, following a strict schedule: mapmaking teams generally completed provincial maps within one year, which, given the limitations of traveling and surveying, probably meant the mapmakers had little leeway to pursue their religious agenda. These seemingly contrasting bits of information thus raise questions about the mapmakers' missionary activities during the land surveys: Was there ample time for numerous and widespread contacts with Christians in the provinces? Did European mapmakers actively support local Christian communities during the land surveys? If so, in what ways? How, then, did the power relations involved in the mapping project shape their contacts with others in the provinces?

The Restitution of Church Buildings

Perhaps the most concrete element in du Halde's claim is that the missionary-mapmakers had restituted confiscated churches to Christian communities. Achievements of this sort were discussed by members of the Paris Foreign Missions Society, the Missions étrangères de Paris (MEP), stationed in the Chinese provinces, some of whom wrote in their letters about the activities of the European mapmakers during the surveys. They based most of their accounts on hearsay, but presented forward a consistent narrative: the missionary-mapmaker actively using his status as imperial envoy to mediate between Christian communities and provincial or subordinate officials. In other words, it tells the story of missionary-mapmakers actively and consciously employing

417 When the surveys started, the rites controversy had come to a head. For a detailed overview of events surrounding the rites controversy (intermission one), see Mungello (1994); Huang (1997).

their imperial privileges to further the interests of their colleagues and fellow Christians in the provinces. In what follows, I will analyze these manuscript letters by MEP missionaries in an attempt to distill information helpful in reconstructing the local contacts that occurred as a result of the surveys.

Philibert Le Blanc (1644–1720), collaborator of Charles Maigrot (1652–1730) in Fujian province and vicar apostolic of Yunnan from 1696 onwards, authored the first set of excerpted passages, below. In a letter to Maigrot, written on October 15th, 1714, in the town of Yushan 玉山,[418] he writes about the impact of the surveys on the missionary work of himself and his colleagues:

> Reverend Father de Tartre who drew the map of this province with a Father Cardoso, a Portuguese, they [sic] have passed by Guangzhou since then and afterwards have to go and draw the map of Guangxi. Father de Tartre had the church of Guangxin prefecture[419] returned, where the mandarins had lodged entertainers, but as soon as he left, the mandarins put their entertainers back there… The maps of Nanjing, Zhejiang and Fujian have been drawn by the Reverent Fathers Hinderer, a Portuguese [sic], de Mailla, and Régis, both French. They had the [Church of the Holy Mother] in Quzhou[420] and the church of Jinhua prefecture[421] returned to the Dominican fathers, but these were retaken by the heathens two days after their departure. Reverend Father Fabre, together with two Jesuit fathers, is employed to draw the maps of the provinces of Sichuan, Yunnan, Guizhou and Huguang. Father Fabre attempted to return our church in Chengdu prefecture and that of Mr. Appiani[422] in Chongqing prefecture, but he could not succeed. The governor[423] says that he will not return our church unless an order arrives from the court. The mandarins have given fifty taels for the church in Chongqing to a Christian, but he has [used the money on] eating. I do not know what he has done in Yunnan to have our

418 Located in Guangxin 廣信 prefecture, east of Jiangxi province, near the border with Zhejiang.

419 Guangxin fu 廣信府, today's Xinzhou 信州 in Jiangxi province.

420 Quzhou 衢州, in Zhejiang province. "Xing mu tang" presumably refers to Shengmutang 聖母堂 or the Church of the Holy Mother. Most churches with this name were churches for women. Standaert (2001), 580–581.

421 Jinhua 金華 in Zhejiang province. A church erected by two Dominican fathers in 1656 is mentioned in Dehergne (1957), 18.

422 Lazarist priest Ludovico Appiani (1663–1726).

423 *Fuyuan* 撫院 or provincial governor. Hucker (1985), 221.

> church back. A long time ago, I learnt that since our departure, they have put a schoolmaster in it.[424]

It is remarkable that Le Blanc's letter mentions three different teams of surveyors all simultaneously mobilized by imperial edict during the mapping project's final stage: the surveying of (1) the provinces of Henan, Jiangnan, Zhejiang, and Fujian (1712–4); (2) of Jiangxi, Guangdong, and Guangxi (1713–4); and (3) of Sichuan, Yunnan, Guizhou, and Huguang (1713–6). The excerpt illustrates that the European mapmakers on all three of these teams inclined to aid Christian communities in the provinces, perhaps despite their entourage, which consisted mostly of Manchu officials from several Qing institutions as well as a Manchu military escort, usually a banner Colonel of the Imperial Guard, as Chapter 2 discussed. It remains possible, however, that some of these Manchu officials included in the teams of mapmakers were Christians, since we know some of them had close connections with the Jesuits at court.[425]

The basis for providing support to these Christians lay in the mapmakers' status as imperial envoys. As I have shown in Chapter 2, they carried with them imperial documents confirming their status and messengers traveled ahead to inform officials of different levels of their coming in advance, undoubtedly lending them a certain standing in the eyes of local administrators. It must have strengthened their negotiations with prefectural and district officials as well: Le Blanc reports that Jesuit mapmakers in both the first and the second teams of surveyors were able to arrange the return of the non-Jesuit churches of Guangxin, Quzhou and Jinhua to their Christian communities. Nevertheless,

---

424 "Le R.P. De tartre qui faisoit la carte de cette Province avec un Pere cardose Portuguais, ils ont depuis ce temps passé a Kanton, et ensuite ils doivent aller faire la carte de Kuang Sy. Le Pere de tartre avoit fait rendre l'eglise de Kuang Sin Fu ou les mandarins faisoient loger des comediens, mais aussi tost qu'il fut parti, les mandarins y remirent leurs comediens ... Les cartes de nan King, che Kiang et fo Kien ont esté faites par les R.R. Pere hinderer Portuguais, de mailla et Régis francois. Ils avoient fait rendre aux Peres Dominicains le xing mu tang de Kiu cheu et l'eglise de Kin hoa fu, mais elles ont esté reprises par les gentils deux Jours aprés leur depart. Le R. Pere fabri avec deux Peres Jesuites est employé a faire les cartes des Provinces du cu chuen, yun nan, Kuey cheu et hu Kuang. Le Pere fabri a tasché de faire rendre nostre eglise de chin tu fu et celle de mr Apiani qui est a chung King fu, mais il n'y a pas pû reussir. Le fu iuendit qu'il ne rendra pas la nostre a moins qu'il ne vienne un ordre de la cour. Les mandarins ont donné a un chrestien cinquante taels sur l'eglise de chung King, le quel les a mangé. Je ne scay pas ce qu'il aura fait a yun nan pour avoir la nostre. J'ay apris il y a long temps, que depuis nostre despart on avoit mis un maistre d'escolle." AMEP, 410:92.

425 Chapter 2.

all of these churches were allegedly seized again soon after the mapmakers departed, indicating that local officials did not fear reprimand by higher officials for disrespecting the imperial envoys' wishes. Thus, it appears local officials felt obliged to respect the mapmakers' wishes in their presence, but were ultimately accountable only to their own local hierarchy.

The third team of mapmakers, including Fabre-Bonjour, an Augustinian, was less successful in (temporarily) recovering the churches of Chengdu and Chongqing. Whereas negotiations concerning the churches of Guangxi, Quzhou, and Jinhua presumably occurred on the prefectural or district level, as Le Blanc suggests in describing the retaking of these churches immediately following the surveyors' departure from those cities, in Sichuan the surveying team negotiated directly with the governor of the province (*fuyuan* 撫院). This governor, Nian Gengyao, had become a protégée of the emperor and apparently insisted on receiving an order directly from the court before returning the two churches in his province. This meant that the European mapmakers in the third team of surveyors were unsuccessful in their negotiations and could not sufficiently impress the governor with their status as imperial envoys. Instead, a Christian in Chongqing, where no European missionaries were then present, received fifty taels in compensation for that city's church, cash he spent on food.[426] A doctor, Le Blanc mentions in another letter, had bought the church in Chengdu after missionaries had—under duress—handed over the property documents to local authorities. This doctor allegedly made repairs to the property, grossly estimated at five hundred taels, which Le Blanc was prepared to reimburse to facilitate the church's restitution.[427]

Nearly two years later, in a letter written on August 5th, 1716, Le Blanc transmitted updated information on the activities of this third team of mapmakers, who had since continued mapping the provinces of Yunnan and Guizhou:

> The Jesuit fathers, who had gone to the province of Yunnan to draw the map, in returning via Huguang have said to Mr. [Müllener][428] that the viceroy had taken the grounds of our church in Yunnan to enlarge

426 Ludovico Appiani was imprisoned in Guangzhou, while Johannes Müllener (1673–1742) hid in the mountains between Sichuan and Huguang. Both had been missionaries in Chongqing. von Collani (2009), 99–110.

427 AMEP, 410:242.

428 Presumably Johannes Müllener. In late 1713, when this team of mapmakers was in Sichuan, Müllener had asked Fridelli and Fabre-Bonjour's help to prevent hostile acts against Christians. However, Fridelli and Fabre-bonjour were unable to assist him. Baldini (2012), 279 (n. 137).

> the parade grounds adjacent to it, and that he had demolished everything I had built there.[429]

Le Blanc had obtained this information from the mapmakers as they returned to the capital from Yunnan via Guizhou and Huguang provinces. This short excerpt reconfirms the active support of the missionary-mapmakers for some of the more remote communities of Christians. They apparently worked in Yunnan to determine what had happened to the church in Kunming, which Le Blanc had built during his tenure as vicar apostolic of the province.

Besides confirming du Halde's claim that the mapmakers were able to regain churches confiscated from Christian communities throughout the provinces, these reports by Le Blanc aid reconstruction of local contacts made as a result of the surveys. They inform us of the power relations embedded in encounters with provincial and subordinate officials and allow us to identify some of the towns and villages where interactions took place. Moreover, Le Blanc mentions a number of details that contribute to a better understanding of these contacts. For example, he informs the reader about the fate of church buildings after local authorities had claimed them. In Guangxi and Kunming, authorities apparently housed, respectively, entertainers (*comediens*) and a schoolmaster (*maistre d'escolle*) in former church buildings, whereas they later apparently had the Kunming church demolished to enlarge the adjacent parade grounds. In Chengdu, on the other hand, a doctor purchased the church. Le Blanc also mentions the transaction costs related to church buildings in the provinces. As compensation for lost property, authorities paid the Christians of Chongqing fifty taels. Similar information is recorded in another source, the abovementioned *Lettres édifiantes et curieuses*. In a letter written from Nanyang 南陽 in Henan province, Jesuit missionary Jean Domenge (1666–1735) mentions Jesuit mapmakers buying property for 1600 *francs*:[430]

> It was Fathers Régis and de Mailla who bought the church where I am, when they were sent by the Emperor to draw the map of this province: it cost them 1600 francs.[431]

429 "Les peres Jesuites qui estoient aller dans la province de yun nan pour en faire la carte retournant par le hu Kuang, ont dit a mr. moneler que le viceroy avoit pris le terre de nostre eglise de yun nan, pour aggrandir la place d'armes qui estoit voisine et qu'il a fait desmolir tout ce que j'y avois fait bastir. " AMEP, 410:242.

430 Colloquial name for the official French currency at the time, the *livre tournois*.

431 "Ce furent les Peres Regis & de Mailla qui acheterent l'église où je suis, lorsqu'ils furent envoyés par l'Empereur pour faire la carte de cette province: elle leur coûta seize cens francs. " du Halde et al. (1702–76), 14:429–430.

Thus, Le Blanc and Domenge confirm that European mapmakers actively worked to support Christian communities they encountered in the provinces. They found a basis for providing this support in their status as imperial envoys, undoubtedly evoking respect and a certain standing in the eyes of administrators, a privilege local missionaries or Chinese Christians could not generally procure and compelling some of these authorities temporarily to restitute church buildings. This allows us to trace out and evaluate a number of contacts including the missionary-mapmakers, Qing officials of different levels, and Chinese Christians, letting us fill in some of the blanks regarding the many factors that shaped and supported the mapping project as a whole, including power dynamics and proselytizing efforts. On the other hand, the involvement of Jesuits, Dominicans, MEP missionaries, and an Augustinian calls for more careful evaluation in light of the Chinese rites controversy. After all, nearly all Jesuit missionaries favored tolerance of Chinese ancestral rites, whereas most MEP, Dominican, Augustinian, and Franciscan missionaries generally disallowed Chinese Christians such practices.

The Impact of the Chinese Rites Controversy

When the rites controversy came to a head following papal envoy de Tournon's audience at court, the emperor decided to expell most Dominican and MEP missionaries active in Jiangnan and Fujian to Guangzhou or Macao. He made this decision after they had refused to subscribe to "the teachings of Matteo Ricci" including acceptance of traditional ancestral rites. Only missionaries who subscribed to Ricci's methods received the *piao* (the special license allowing them to stay). Pillaging of certain churches and stigmatization of Chinese Christians in the Jiangnan area followed the expulsion of Dominican, MEP, and other missionaries.[432] Given these tensions, both sides, largely split between Jesuits and non-Jesuits, developed their own arguments and scrutinized the behavior and actions of the other side. This played out both locally and in Europe, as detailed reports home fueled in their turn even more heated discussions.

Le Blanc, an important actor during the rites controversy and author of most of the excerpts quoted above, had collaborated with apostolic vicar of Fujian Maigrot, who issued a mandate for his vicariate in 1693 banning use of the

432 Cams (2013).

terms "*Tian*" 天 and "*Shangdi*" 上帝 in reference to the Christian God. Both men were MEP missionaries and actively sought information to help make their case against the position most Jesuit missionaries held. Seen in this light, the narrative put forward by Le Blanc is surprisingly positive in its evaluation of the (mostly Jesuit) missionary-mapmakers, even more so because the controversy had reached its climax around the time of the mapping project, preventing many missionaries from returning to their stations. Antoine Guigne was another MEP missionary and staunch opponent of the Jesuit approach regarding the Chinese rites. Whereas Le Blanc advanced a positive image of the missionary-mapmakers' activities during the surveys, Guigne chose to link these activities explicitly to the rites controversy.[433] In a report sent from Guangzhou to the MEP superior in Paris on November 15th, 1714, he writes:

> Father de Mailla, who is drawing the map of Zhejiang and Fujian, where the Christians of the Dominicans are, writes that these Christians admitted to him that they had renounced the tablets of the dead so as to obey the Dominican fathers, but that they secretly erected new ones as they destroyed the old tablets in front of these Fathers… It is possible that some Christians of the Dominican fathers who are without a pastor, seeing Father de Mailla coming to them with the following of an ambassador and in a position to assist them, knowing that he is opposed to the doctrine and practice of the Dominicans regarding the tablets, talked in such a manner so as to pay court to this ambassador. Apparently, they were incited to do so by the servants of this Father, who have not failed to say that those who ban the tablets are know-nothings, muddle-heads and destroyers of the Christian religion, and that, if the Emperor knew that there were Christians who obey them on this point, he would severely punish them. That's what the servants of the Jesuits say throughout China.[434]

433 Guigne was an MEP missionary who reached Asia in 1706. After one year of missionary work, he received an order to return to Macao. Cams (2013), 53–54. He traveled to Beijing to plead his cause, but the emperor sent him to Guangzhou, where he remained for the next decade or so.

434 "Le P. de Maillard, qui a fait la carte de tche Kiang et de Fou Kien ou sont les chretiens des dominicains, ecrit que ces chretiens luy ont avoué qu'ils avoient renoncé aux tablettes des morts pour obeir aux Peres dominicains, mais qu'a mesure qu'ils detruisoient les vielles tablettes devant ces Peres, ils en erigiroient des nouvelles en cachette… Il se peut faire que quelques chretiens des PP. Dominicains qui sont sans pasteur, voyant venir chez eux le P. de maillard avec un grand train d'ambassadeur, et en etat de les assister, sachant qu'il est opposé à la doctrine et à la pratique des dominicains touchant les tablettes, auront parlé

It is remarkable that Guigne mentions the missionary-mapmakers' servants, or retinue. Although we know that each team of mapmakers included a number of personal servants, in the service of particular team members, the sources rarely discuss them and their identities remain obscure. In addition, when discussing de Mailla's servants, Guigne depicts them as collaborators with the Jesuit mapmakers, to an extent acting as interlocutors between de Mailla and the local Christians. If we can take Guigne's report seriously, these servants were well aware of the discussions surrounding the rites controversy. Unfortunately, no other sources confirm these elements.

In any case, Guigne establishes an explicit link between the mapmakers' special status as imperial envoys and the tensions between the Jesuits and the Dominicans in the context of the rites controversy. He refers to the missionary-mapmakers as possessing the capacity and retinue of an ambassador, which, Guigne believes, caused local Christians to play up to the wishes of de Mailla, at least verbally. This explicit reference to de Mailla's status confirms Le Blanc's reports concerning the power relations European mapmakers incited through their presence. In many cases, as discussed above, this allowed them to mediate between local officials and communities of Christians, most concretely achieving the restitution of church buildings. Later in Guigne's letter, he reiterates this point with yet another example:

> [I indicated that] the Emperor sent Fathers to all provinces for drawing their maps and that these Fathers had the capacity and the following of an ambassador… Father de Tartre, in passing the city of Shaozhou, where we have a church, obtained from the three most senior mandarins of the city an order to prevent the gentiles from causing disorder in the church. Since this order was given, a guardian quietly stays in our house. Father Labbe,[435] a French Jesuit, had me hoping that Father de Tartre would come to see me and assist me with what depends on him. However, he [de Tartre] wanted to do nothing for me and Christians have told me that he showed no goodwill. Yet I think that he would have come to see me and that he would have talked to the mandarins in my favor, had he

de la sorte pour faire leur cour a cet ambassadeur. Apparemment ils auront [estez] porté à cela par les serviteurs de ce P. qui n'auront pas manqué de dire que ceux qui defendent les tablettes ce sont des ignorans, des etourdis et des destructeurs de la religion chretienne, que si l'Empereur scavoit qu'il y eut des chretiens qui leur obeissent en ce point, il les feroit rigouresement punir. C'est ce que disent par toute la Chine les serviteurs des Jesuites." AMEP, 410:164.

435 Joseph Labbe (1677–1745).

> followed his inclination. But his confreres will never forget my refusal to allow Father Porquet,[436] declared excommunicate, to read Mass in one of our churches.[437]

Guigne states that de Tartre was able to obtain an order from the "three most senior mandarins" of Shaozhou, thereby protecting the MEP church in that city. He goes on to complain that the same missionary chose not to meet with him personally or to speak to the senior mandarins of Guangzhou in his favor, which, at least for Guigne, bore testimony to the Jesuits' malice due to specific events surrounding the rites controversy. As a French Jesuit, Labbe enjoyed easy access to de Tartre. Guigne, who clearly also enjoyed close relations with Labbe, therefore sought his help in getting de Tartre to obtain a similar order regarding the Guangzhou church. This reveals how the supposed dichotomy between the Jesuits and missionaries of other Catholic orders in the Chinese provinces is not always helpful in studying contacts and alliances, particularly on a local level. Rather, taken together, both Le Blanc and Guigne touch on the complexities and contingencies inherent to personal encounters, even when more centralizing forces or discourses tend to simplify and streamline actors into different factions.

Nevertheless, the rites controversy remained an important factor. Although Guigne's letter still allows us to trace local contacts, he wrote that letter specifically to argue against the position most Jesuits held. This alone renders it less reliable for one wishing accurately to gauge the motives behind local contacts among actors in Guangdong province. In establishing a clear link with the rites controversy, the excerpt thus raises a number of relevant questions: to what extent were missionary-mapmakers actively involved, or how did they find themselves implicated, in discussions surrounding the rites controversy? For

436 Louis Porquet (1671–1752).

437 "[Je marquoit que] l'empereur avoit envoyé des Peres par toutes les Provinces pour en faire la carte et que ces Peres avoient la qualité et le train d'ambassadeur… Le P. de Tartre passant par la ville de Xao Tcheu, ou nous avons une Eglise, obtint des trois premiers mandarins de la ville un ordre pour empecher les gentils de faire du desorde dans l'Eglise. Depuis cet ordre donne, le guardien reste tranquillement dans notre maison. Le P. Labbe J. françois m'avoit fait esperer que le P. de Tartre viendroit me voir et m'assisteroit en ce qu'il dependroit de lui. Cependant il n'a rien voulu faire pour moy et des chretiens m'ont dit qu'il n'a nulle bonne volonté. Je crois cependant qu'il seroit venu me voir et qu'il auroit parlé aux mandarins en ma faveur, s'il eut suivi son inclination; mais ses confreres n'oublieront jamais le refus que je fis au P. Porquet denoncé excommunié de dire la messe dans une de nos Eglises." AMEP, 410:164.

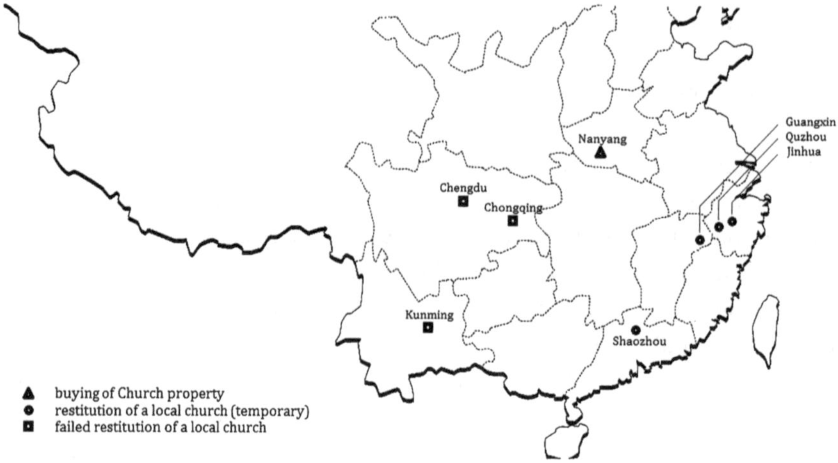

FIGURE 28 *Sites where the missionary-mapmakers purchased and recovered churches (1712–1716).*

example, did they use their status as imperial envoys to exert pressures on local Christians to subscribe to the Jesuit position, as Guigne suggests? Such questions exceed this study's scope.

Conclusion

This second intermission has explored the missionary activities of the European mapmakers during the land surveys, from which we can distill a number of relevant insights. First, the writings of Le Blanc and Guigne allow us to state with certainty that missionary-mapmakers used their capacities as imperial envoys to exert influence locally and in an attempt to further the interests of local Christian communities, as du Halde and others suggested. Second, given the factional tensions among the different Catholic orders, we must consider whether they did so selectively, as claimed by Guigne, as well as whether the missionary-mapmakers employed their status as imperial envoys for other purposes linked to tensions with their MEP and Dominican colleagues. Although these questions remain open, this discussion further contextualizes the land surveys and illustrates how some of the tensions discussed in the first intermission also had repercussions for the land surveys, even if they played out in differing ways.

CHAPTER 3

The Afterlife of Maps: Circulation, Adaptation, and Negotiation

諭內閣學士蔣廷錫、皇輿全覽圖、朕費三十餘年心力、始得告成。山脈水道、俱與禹貢相合。爾將此全圖、並分省之圖、與九卿細看。倘有不合之處、九卿有知者、即便指出。看過後面奏。[438]

康熙58年二月12日

Imperial order for grand secretary Jiang Tingxi. The *Overview Maps of Imperial Territories*, for which We have spent more than thirty years of mental effort, has finally been announced completed. Mountain ranges and river courses are all consistent with the *Tribute of Yu*. Together with the nine ministers, you will carefully look at this general map, and also at the provincial maps. Should there be inconsistencies, and there is someone among the nine ministers who know it, they should immediately point it out. After having looked at them, [you will] report back in person.

April 1st, 1719

∵

Immediately following the land surveys, the Imperial Workshops began producing different editions of the resulting atlas, the *Overview Maps of Imperial*

438 *Kangxi chao shilu, juan* 282. For a translation of the entire passage, see Fu (1966), 127–128. Jiang Tingxi (1669–1732) was one of the four earliest grand secretaries and would continue to serve under the Yongzheng emperor. Hummel (1943), 142–143. "More than thirty years" confirms that in retrospect the Kangxi emperor saw the preliminary work (Chapter 1) and the more intense period of land surveying (Chapter 2) as one and the same project. The *Tribute of Yu* (*Yugong* 禹貢) is the oldest known book on geography in East Asia and appears in the *Book of Documents* (*Shujing* 書經), one of the Five Classics of Chinese literature. This reference also confirms the emperor's attention to ancient precedents in establishing new Qing cartographic practice (Chapters 1 & 2).

© KONINKLIJKE BRILL NV, LEIDEN, 2017 | DOI 10.1163/9789004345362_007

Territories or *Huangyu quanlan tu* 皇輿全覽圖.[439] In later decades, these would form the basis for further adaptations and reduced maps in imperially commissioned compilations. At the same time, missionaries involved in the project sent a vast corpus of maps and descriptions back to Europe, igniting a period of intense broader exchanges relating to cartography. Much of the material eventually concentrated in Paris and inspired plans for a European incorporation of the Qing atlas. The young French cartographer Jean-Baptiste Bourguignon d'Anville (1697–1782) undertook the commission, basing most of his adaptations directly on maps the missionaries had sent. After finishing these adaptations, d'Anville also produced four overview maps, based only loosely on the Qing atlas, following careful negotiation of cartographic data among Beijing, Paris, and Saint Petersburg. This triangular exchange of cartographic material in the 1720s and early 1730s also informed a revision of the Qing atlas under the Yongzheng emperor and shaped the mapping of Central Asia in Saint Petersburg.

In 1735, finally, more than fifteen years after the first maps had reached Paris, d'Anville's forty-one regional and general maps of the Chinese provinces, Tartary, Tibet, and Korea appeared as part of an expensive four-volume description of continental East Asia entitled *Description géographique, historique, chronologique, politique et physique de l'empire de la Chine et de la Tartarie chinoise*.[440] Immediately after publication, d'Anville's collection of maps became the focus of much debate and enjoyed such wide circulation that still other cartographers would themselves later carefully rework these European adaptations from the Qing atlas. As a result of this long process of circulation, adaptation, and negotiation of cartographic data, both the Qing atlas and its European incorporation remained the authoritative cartographic works on continental East Asia until well into the nineteenth century.

With its main focus on the circulation of the Qing atlas and the process of its adaptation for the European reading public, this chapter discusses how maps gradually transferred between individuals, reworked and eventually adapted to fit new audiences' visual and cultural expectations. Ultimately, this led to an eventful afterlife for these maps on both ends of the Eurasian continent. I initiate my analysis with an examination of the different printed editions and manifestations of the Qing atlas. Next, I conduct a detailed analysis of its

439 I could not ascertain the Manchu name for the atlas.

440 du Halde (1735). The name "Tartary" referred at the time to almost the entire area lying between the Chinese provinces and Muscovy, much of it inhabited by nomadic peoples. It equates approximately with present-day northeastern China (or Manchuria), Mongolia, Xinjiang, and parts of Siberia and Central Asia.

European incorporation, discussing d'Anville's work as well as the first steps in the publication process. Finally, I will explore how certain gaps in cartographic information stimulated a process of negotiating indirectly obtained data among Beijing, Paris, and Saint Petersburg, on the one hand allowing revision of the Qing atlas while on the other facilitating the completion of d'Anville's four general maps. In all, this chapter aims to retrace the many connections that guided the circulation of these maps across Eurasia, in itself the result of the circulation of cartographic practices over previous decades, thereby illustrating their wide impact.

3.1 The Printed Life of the *Overview Maps of Imperial Territories*

None of the maps printed in the Imperial Workshops immediately following the land surveys were available for public sale before the twentieth century. Instead, at least for the remainder of the Kangxi emperor's reign, they remained at court for perusal by the emperor and senior court officials. Only significantly reduced versions eventually made it into large imperial compilations, treating other matters alongside geography, that circulated widely.

Walter Fuchs, in a comprehensive study on the subject, has argued for the production of different versions of the Qing atlas for consultation at the palace. At the time of writing, Fuchs had access to cartographic material kept at several institutions in Beijing. Based on a detailed analysis, he distinguishes among three different printed editions of the Qing atlas during the Kangxi reign. The first edition, he argues, printed in 1717 using woodblocks, contains twenty-eight separate and regional maps of the Chinese provinces, different parts of Tartary and Tibet, and Korea. The second, printed from copperplates by the Italian missionary Matteo Ripa and completed in 1719, consists of forty-one individual sheets combining to form one large map and incorporating updated depictions of the Tibetan highlands. A third edition, again from woodblocks, produced in 1721 as a revision of the first, includes seven updated regional maps, which also depict parts of today's Xinjiang province, in the place of the Tibetan maps of 1717.[441]

Although I largely follow Fuchs' distinction between the two main versions of the Kangxi-era Qing atlas, the woodblock atlas, and the copperplate multi-sheet map, I argue that the copperplate version was subject to the same revision process as the woodblock atlas during the years immediately following the completion of the land surveys. The 1717–1718 survey of the Tibetan

441 Fuchs (1943), 1:16 & 60.

highlands (Chapter 2) had improved cartographic knowledge of the region and prompted adjustments to both the woodblock and the copperplate atlases to accommodate the new data. My argument depends on a close comparison of extant copperplate sheets and a reinterpretation of Fuchs' own scholarship, in which he claimed to have identified a "draft copy" of the copperplate atlas.[442] I will show that the dynamics informing the differences between the first and second editions of both the woodblock and the copperplate versions were remarkably similar. In so doing, I discuss all printed editions resulting from the land surveys during the Kangxi reign as well as their later adaptations under subsequent emperors, laying out a foundation for further exploration of the afterlife of these maps.[443]

3.1.1 *The Woodblock Editions*

The carving and printing for the two woodblock editions of the Qing atlas, also known as the *Overview Maps of Imperial Territories per Province* or *Huangyu quanlan fensheng tu* 皇輿全覽分省圖, took place at the Imperial Workshops. This occurred in the *xiushuchu* 修書處, the court's printery, located just north of the palace's *Wuyingdian*. It seems the woodblock atlas formed a set with the *Imperially Commissioned Overview of Imperial Territories*, the large route book (produced simultaneously) discussed in Chapter 2. Both the maps and their textual companion, printed by and for the court, were never, it seems, available for sale as such before the fall of the Qing in 1911.

The first edition of the woodblock atlas (scale 1:2.000.000), carved and printed around 1717, contains twenty-eight separate and regional maps of varying sizes of:

- The Manchu homelands and outlying regions north of the Great Wall, including Shengjing 盛京 (present day Liaoning 遼寧), the Amur River basin, and the Ussuri River basin (5 maps);
- (Khalka) Mongol territories and outlying regions, including today's inner and outer Mongolia up to Lake Baikal and down to Hami (4 maps);
- The headwaters of the Yangzi, Yalong and Yellow rivers (3 maps);

442 Fuchs (1943), 1:25–27.

443 Several manuscript editions of the Qing atlas reside in the main libraries and archives of mainland China, but are not available for consultation at the time of writing. As a result, this chapter only considers printed editions. For a taste of the wealth of maps at The First Historical Archives of China 中國第一歷史檔案館, see Li (2012), 66–78.

- The Korean peninsula (1 map); and
- The fifteen Chinese provinces (15 maps).[444]

On maps of clearly delineated regions such as the Chinese provinces, Shengjing, Korea, and regions in the immediate vicinity of the Great Wall, those lands beyond the indicated provincial, state, and internal boundaries (such as the Great Wall) remain consistently left blank (figure 29). Together with the sequence of these maps within the atlas, these blank spaces suggest that the printed versions depended directly on the manuscript maps drawn in the field: particularly in the first edition woodblock atlas, the order of the different sets of maps as outlined above largely coincides with the order of the areas surveyed, just as those areas the maps depict or leave blank largely correspond to the geographic scope of the areas each team surveyed.[445]

The maps' empty spaces (beyond regional borders) contain the map titles as well as the names of bordering regions (only the latter in maps of the Chinese provinces (figure 37)). Degrees of latitude and longitude relative to the main meridian through Beijing appear beyond the gridlines on all four sides. The maps exaggerate the sizes of rivers because double lines depict their courses and mountain pictograms appear scattered rather randomly across the map in mountainous regions. These maps' precision, then, lies entirely in the placement of cities, towns, and, to some extent, the national, provincial, and internal boundaries within the gridlines. Squares indicate the prefectural, departmental, and district seats while small circles signify smaller towns. In areas lacking natural boundaries such as large rivers, dotted lines delineate provincial territories. The woodblock maps use only Chinese, although Manchu toponymy and terminology form the basis for Chinese-language renditions of place names and other information for areas outside the Chinese provinces to the north and west of the Great Wall.[446]

Compared to the first edition of the woodblock atlas, the revised edition first printed around 1721 is identical apart from the sheets covering Tibet and the northeastern part of today's Xinjiang (all other maps remain unchanged).[447]

444 Fuchs (1943), 1:14–18.

445 With the exceptions of Korea, for which teams only surveyed the main road connecting the capital with Qing territories; and the headwaters of the Yangzi, Yalong, and Yellow rivers, located in Kokonor, a region not surveyed until 1717–1718 (Chapter 2). These maps do not match the accuracy found on other sheets and therefore must borrow from earlier maps.

446 Stary (1997), 192.

447 Volume 2 of Fuchs (1943) includes facsimile versions of all 1717 and 1721 maps. I have not been able to consult a complete first edition of the woodblock atlas.

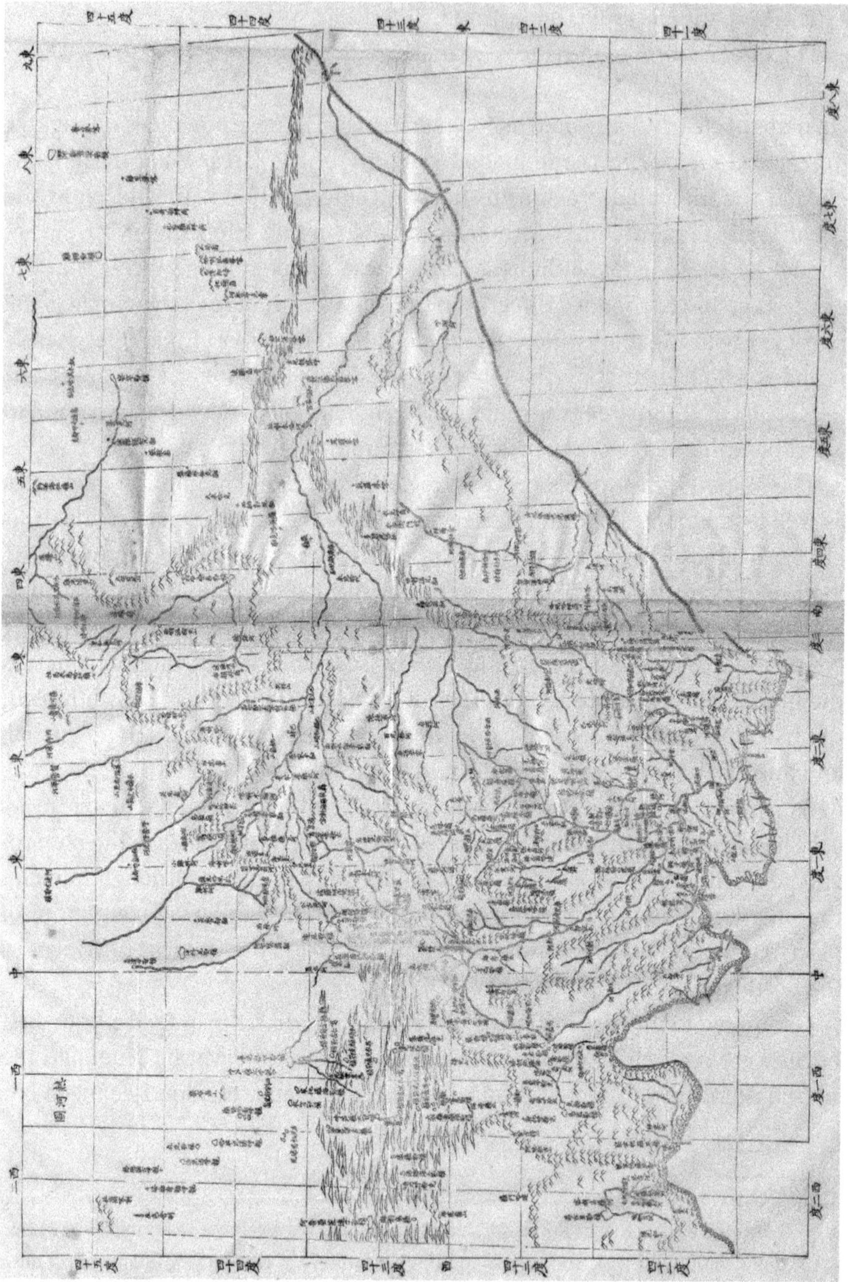

FIGURE 29 *Map of the region of Rehe, taken from a woodblock atlas. Territories in Shengjing (bottom-right) and within the Great Wall (bottom) are blank. All place names are in Chinese (c. 31 × 54 cm).*
ROYAL LIBRARY OF BELGIUM, LP VB 11.283 E (2), F.6.

TABLE 8 *The two woodblock editions compared.*

First woodblock edition sheets (1717)	Revised woodblock edition sheets (1721)
Heyuan tu 河源圖 (Upper Yellow River)	*Heyuan tu* 河源圖 (Upper Yellow River)
[no title] (Yalong river) [no title] (Upper Yangtze/Mekong/Salween)	*Jinsha langcang deng jiangyuan tu* 金沙瀾滄等江源圖 (Upper Yangtze/Mekong)
Hami tu 哈密圖 (Hami)	*Hami Gasi tu* 哈密噶思圖 (Hami-Gas) *Zawang A'erbutan tu* 雜旺阿爾布灘圖 (Tsewang Rabdan)
(*not included*)	*Lazang tu* 拉藏圖 (Central Tibet) *Yalu Zangbu tu* 雅魯藏布江圖 (Brahmaputra) *Gangdisi Alin tu* 岡底斯阿林圖 (Kailash)

According to Fuchs, the survey of the Tibetan highlands in 1717–1718 enabled the drawing of seven new maps covering the Tibetan regions depicted in the 1717 first-edition woodblock atlas' four maps plus additional territories in Tibet proper (table 8).[448] The carving for these sheets, however, only occurred in 1721 following a full revision of the Chinese renditions of place names.[449] The Imperial Workshops then printed the revised woodblock atlas simultaneously with its textual companion, the *Imperially Commissioned Overview of Imperial Territories*.[450]

One opportunity for improving Fuch's findings, which date back to the 1940s, lies in close study of these different editions' extant sheets. At present, unfortunately, maps in mainland China are difficult to consult at best, so that here I can discuss only a limited number of maps mentioned in

448 Fuchs (1943), 1:16.

449 Fuchs (1943), 1:43–44. Fuchs quotes a passage from the *Kangxi chao shilu*, which announces a revision of Tibetan place names. *Kangxi chao shilu, juan* 290.

450 Chen (1990), 107.

catalogues.[451] In Europe, on the other hand, the sheer quantity and dispersal of extant sheets makes consulting them all quite impracticable. Even though I have not been able personally to consult many of the copies mentioned in this chapter, I still intend to give a comprehensive overview of extant Kangxi-era sheets from the Qing atlas.[452]

I have been unable to locate a first-edition woodblock atlas, although at least one set reached Paris before 1720 and Fuchs himself allegedly owned one.[453] On the other hand, of the very rare revised edition of 1721, I have found only one complete copy, in the British Library.[454] Other extant copies are incomplete or adaptations: a nearly complete copy resides at the Royal Library of Belgium, given to the Jesuit College at Lille by François Noël (1651–1729);[455] the Library of Congress possesses a carefully executed manuscript version of fourteen Chinese provinces and the map of Shengjing, clearly based on the woodblock atlas;[456] and in the National Library of China in Beijing another incomplete copy of the revised edition survives.[457] The National Library of France, finally, harbors thirteen sheets with a blue silk border, all revised-edition maps of regions outside the Great Wall. The title of each map transcribed into Russian appears on the back, with place names transcribed in French on the front (figure 30).[458] The lore surrounding it claims it was a gift to Russian ambassador Lev Vasil'evich Izmailov (1685–1738) the same year of its production, but it perhaps more likely reached Saint Petersburg via correspondence at a later date.[459] Decades later, the atlas reached Paris either with Joseph-Nicolas

451 At the time of writing, maps kept in The First Historical Archives of China were unavailable for consulation.

452 An annex to this chapter contains a schematic overview of extant atlas sheets.

453 Fuchs (1943), 1:15. In this case, a 1717 atlas may still reside at the Peking University Library with the remainder of Fuch's collection, but I could not confirm this. For the atlas that reached Paris, see Gaubil (1970), 216.

454 British Library (hereafter BL), Maps C.11.D.15. A later reprint of these maps (c. 1850) is at BL OC 15270.E.4; Li (1996), 165–166.

455 Koninklijke Bibliotheek België (hereafter KBB), LP VB 11.283 E(3); Dudink (2006), 85–87.

456 Li (2004), 13–14. Unclear provenance, perhaps later copies. Viewable in Lin (2013), 75–78; or online at http://www.loc.gov/item/2002626779/.

457 *Yutu yaolu* (1997), 38.

458 BNF, Cartes et plans, Ge CC 4461 (RES).

459 Fuchs (1935), 398; Isnard (1917), 117; Cordier (1966), 1:184. The extant documents from the Ismailov embassy only mention that Kangxi showed Ismailov a map which they viewed together, initiating a discussion of different European countries. Demidova (1978), 1:231. I am indebted to Gregory Afinogenov for this reference.

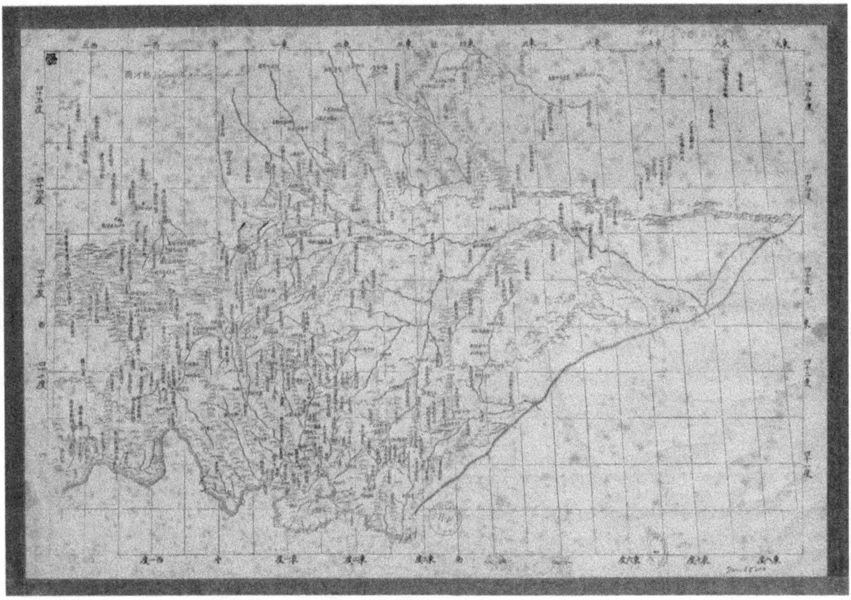

FIGURE 30.1 *The Rehe map from the revised woodblock atlas, with Russian and French transcriptions (c. 38 × 55 cm).*
BIBLIOTHÈQUE NATIONALE DE FRANCE, CARTES ET PLANS, GE CC 4461 (RES).

Delisle on his return to France in 1747 or following negotiations on the transfer of maps between the two cities by French cartographer Jean-Baptiste Bourguignon d'Anville during the 1740s.[460]

3.1.2 *The Copperplate Editions*

A stylistically very different version of the Qing atlas was printed from copperplates, a process introduced by the Italian missionary Matteo Ripa. Wang Qianjin has shown that the projection used for this version and, it seems, for the woodblock version as well, is that commonly known as Sanson-Flamsteed, probably introduced by either Joachim Bouvet or Pierre Jartoux in the late 1690s.[461] This sinusoidal projection allows measurement of true north–south

460 Dulac (2002), 265.

461 Wang (1990); Wang (1991); Wang (1995). Bouvet carried a map of Asia drawn on the basis of a Sanson-Flamsteed projection (Chapter 1) with him, whereas we know Jartoux to have occupied himself with the projection and the integration of the draft maps immediately after the land surveys (Chapter 2).

FIGURE 30.2 *The Rehe map from the revised woodblock atlas, with Russian and French transcriptions (c. 38 × 55 cm).*
BIBLIOTHÈQUE NATIONALE DE FRANCE, CARTES ET PLANS, GE CC 4461 (RES).

distances or latitudinal distances directly on the map where the parallels intersect with the main meridian.[462] Thus, another major practical advantage of this projection is that large maps divide readily into rectangular sheets easily reassembled after printing. Most extant copies of the copperplate atlas indeed consist of eight horizontal scrolls further divided into forty-one individual sheets (printed from forty copperplates) of about 40 × 66.5 cm (scale 1:4.000.000). Some sources mention more sheets, but include those lacking cartographic content. The sheets combine to form one large map of c. 3.2 × 4.7 meters (figure 31). One striking feature of the copperplate atlas is that place names for the fifteen Chinese provinces appear in Chinese, whereas those of areas outside the Great Wall and in Tibet and Korea are in Manchu.[463] This feature, when considered alongside the fact that the copperplate atlas is in fact a multi-sheet map rather than an atlas, suggests the Imperial Workshops first

462 Meskens (1998).

463 Place names for the Korean peninsula are in fact Manchu transcriptions of the Chinese pronunciations of Korean denominations. Stary (1997), 191.

produced it as a separate empire-wide map for the emperor's perusal, unlike the woodblock atlas that presumably formed a set with the *Imperially Commissioned Overview of Imperial Territories*.[464] In other words, the copperplate atlas was a Manchu representation of Qing space for the emperor and perhaps the imperial clan generally, while the woodblock atlas and its textual companion catered more towards the court's scholarly officials.

Another feature, also to be found in the woodblock atlas but far more noticeable in the copperplate one, is the empty spaces in eastern Taiwan and in Guizhou province. They appear entirely blank in print, but extant sheets in European libraries and archives often carry explanations added later.[465] The following note appears in an empty space on one of these sheets:

> Land called Miao tsu [sic]. Although it stands in the middle of the Chinese Empire, it is not subject to it. It has its own Prince. The people of this land are much stronger than those of China. They have different customs and language. [This land] is quite mountainous. The same I say of those two other lands lateral to it.[466]

Such manuscript comments on extant copperplate sheets in European institutions come mostly from the hand of Matteo Ripa, often credited as the copperplate atlas' sole author. Ripa's edited diaries, however, give a different perspective on his involvement in their production:

> The Emperor had long desired to have someone in his service who could engrave the geographic map mentioned above. He accordingly inquired whether… I understood anything besides music, mathematics, and painting… I said that I knew something of optics, and also the theory, although not the practice, of the art of engraving on copper with *aquafortis*. His Majesty was highly pleased to hear that, [and] although I had not

464 It remains unclear whether or not the copperplate version or multi-sheet map in fact appeared bound together in a book after printing. Hence, I choose to use the term "atlas" to refer to both the woodblock and the copperplate versions.

465 The extant sheets in Naples and London (see below) contain such additions. The sheets kept in Brussels carry notes in Latin, most likely by the hand of François Noël.

466 "Miao tzu" stands for the Miao people or *Miaozu* 苗族. "Terra chjamata Miao tsu. Benche stia in mezzo l'Imperio di Cina, non stà però soggeta all'Imperio della Cina. Ha Principe a parte. La di lei gente è assai più forte della gente di Cina. Ha costume, e favela diversa. è assai montuosa. L'istesse dico delle due terre à quella laterali." Stary (1997), 191.

> practiced the art of engraving, I was ready to attempt it, and he immediately ordered that I should begin to engrave. In the shortest time possible, I, with a point, traced a landscape upon a plate coated over with lampblack, as a preparation for the *aquafortis*; and I had scarcely done this when the Emperor desired to see it. As subjects on plates thus prepared present a very handsome appearance, the Emperor was delighted with it, and commanded his Chinese painters to draw a landscape, in order that I might afterwards engrave it. As soon as it was done it was shown, together with the original, to his Majesty, who expressed considerable delight and surprise at finding the copy so perfectly similar to the original without this being impaired; for this was the first time that he had seen an engraving on copper, the Chinese making theirs by fixing the drawing on a block of wood and cutting them both at once with a chisel… he ordered me to engrave and arrange together in the same manner the great map of the empire, which I afterwards executed in forty-four plates, as may be seen in the hall of our college.[467]

Ripa's unedited diary dates the imperial order that he engrave the maps onto copperplate to May 22nd, 1714, shortly after he had showcased his engraving skills with the thirty-six views of the imperial retreat at Rehe, but years before the surveys' completion in 1718.[468] What we do not know, however, is for precisely how long or when Ripa worked on his engravings, even though this date suggests his involvement in atlas production from early on. More unanswered questions exist with regard to Ripa's exact role in the production process. Based on a linguistic analysis of place names north of the Great Wall, Giovanni Stary has argued that Ripa could not have had much knowledge, if any, of the Manchu language, but must have relied instead on the collaboration of two clerks (*baitangga*) from the Imperial Workshops.[469] When we put

467 Ripa refers to the Collegio dei Cinesi di Napoli, which he founded after his return to Europe. Quote taken from an English-language narrative based on Ripa's writings. Prandi (1855), 66 & 88. English translation from an Italian original. Ripa (1832), 408–409 & 463–464. Ripa's actual diary corresponds to this text but gives the date of the imperial order. Ripa (1991–6), 2:136. Note that Ripa here mentions forty-four plates, presumably including some of the empty plates that make up the square (figure 31).

468 Ripa (1991–6), 2:136. The thirty-six views of Chengde, which Ripa refers to as "landscapes" in the quote above, have received extensive study. Gray (1960); Li & Wang (2006); Strassberg (2016).

469 Stary speaks of a "total ignorance of the Manchu language." Stary (1997), 188 & 191. For another article on the same subject, see Stary (2006). *Baitangga*, which I translate as "clerks," were helpers at the Imperial Workshops (Chapter 2). Albanese (2006), 57.

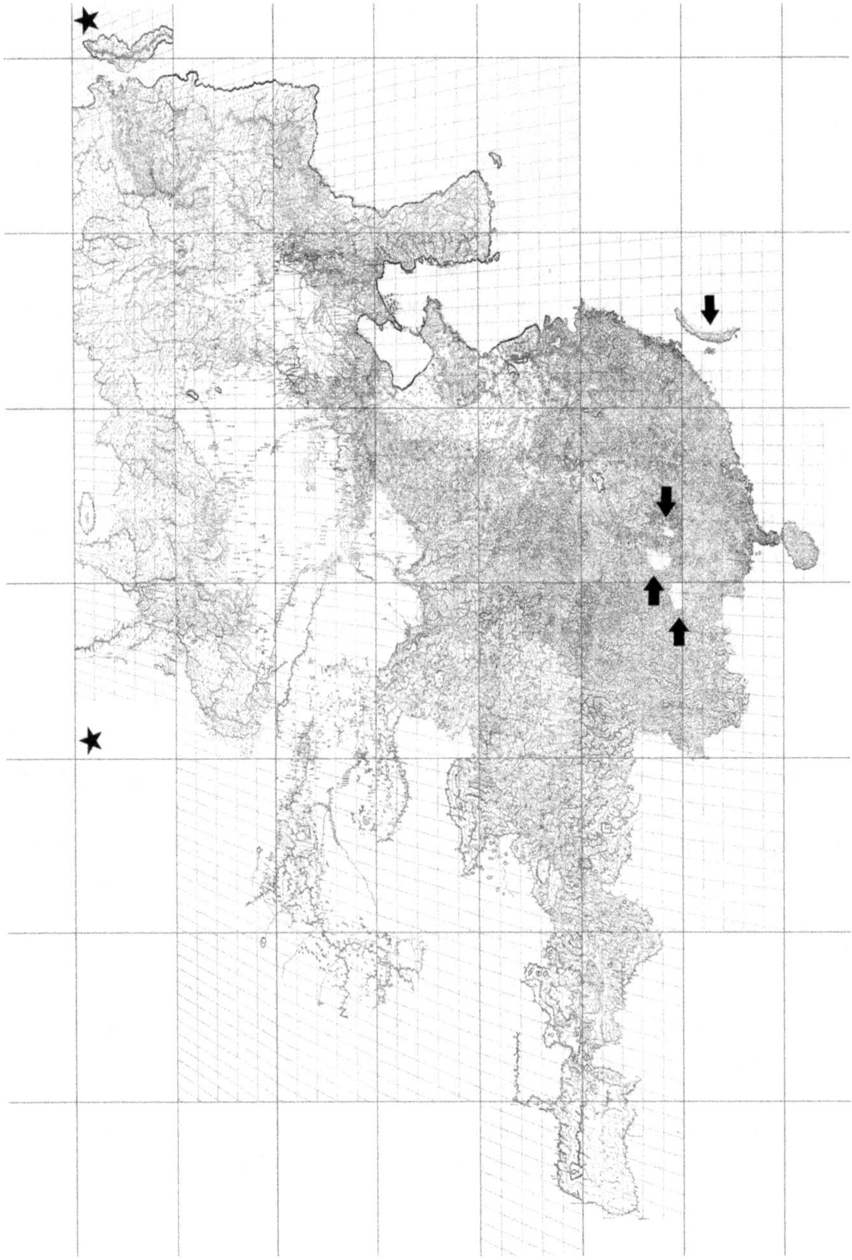

FIGURE 31 *All forty-one sheets from the 1719 revised edition of the copperplate atlas, printed from forty copperplates and forming one large map (c. 3.2 × 4.7m). The two sheets marked with an asterisk were engraved using the same copperplate. Arrows mark the empty spaces.*
DIGITALLY ASSEMBLED BY LIEVEKE OP TEN BERG.

Stary's claim next to the excerpt quoted above, in which Ripa claims to have merely engraved works others had already produced, it seems safe to assume he merely oversaw the technical execution of the copperplate printing process, since he had introduced the technique to the Qing court.

The technique of copperplate printing consists of engraving a design onto plates of polished copper subsequently coated with wax (Ripa claims to have used lamp-black instead). The actual engraving involves a sharp needle-like tool called a burin. The engravers likely executed coastlines, gridlines, and rivers first, followed by vegetation and depth elements such as shadowing along coastlines. Toponyms must have come last, since they constitute the most challenging aspect of copper engraving: the plate is a mirror image, requiring the engraving of all text in reverse. After all the engraving is done, the engraver dips the plate in nitric acid (which Ripa calls "*aquafortis*"), a process that permanently etches all the exposed areas of the copper plate while leaving unmarked those still coated with wax. Ink wiped over the cleaned plate will then adhere only to the etched lines. The printer then finally applies damp paper under pressure, producing a positive image of the plate. Evidence suggests that Ripa experimented with several local types of acid before perfecting his technique.

Whereas I have so far mentioned only the 1719 edition of the copperplate atlas, several elements suggest the existence of two different editions. Fuchs argues for the production of a draft version in 1717, listing important differences among the copies he was able to consult in China.[470] These discrepancies also appear in different extant copies in Europe. The Italian Geographic Society, the Royal Library of Belgium, and the British Library all possess copperplate sheets containing clear differences from printings of the 1719 copperplates.[471] Most important, these sheets display a rather sketchily executed depiction of Tibetan areas, for example in depicting areas in today's western Sichuan, whereas the 1719 edition contains more precise depictions and even includes routes and important temples (figures 32 & 33). The same is true for areas west of the Great Wall around Hami.

Detailed comparison confirms that these sheets were part of an earlier edition of the atlas lacking the updated data from the Tibetan survey of 1717–1718. This comparison of extant sheets could provide the basis for an attempt at

470 Fuchs (1943), 25–27.

471 Società Geografica Italiana (hereafter SGI), Cina 257. See also Cerreti (2001), 62–63. KBB, LP VB 11.283 E (2). See Dudink (2006), 89. BL, Maps K Top 116 15-15a-15b. I compared these sheets to the reproductions of the 1719 copperplate edition in Wang & Liu (2007), vol. 1. See also Li (1997), 127.

reconstructing the 1717 copperplate atlas (figure 34), although further consultation of extant sheets could help to improve it. The wall map kept at the Instituto Universitario Orientale in Naples, mentioned in the quote above and donated by Ripa himself, matches this reconstruction almost perfectly.[472] The existence of this 1717 edition copperplate atlas also corresponds with the evolution of the woodblock atlas, including, as discussed above, entirely revised Tibetan maps after 1717. The fact that Ripa claims involvement in atlas production from 1714 onwards further supports the possibility that the same revision process shaped the different editions of the woodblock and copperplate atlases. Based on these insights, we can construct an improved timeline for atlas production at the Qing court during the late Kangxi years (table 9). The table partly corresponds with the timeline proposed by Fuchs, illustrating that the production processes of both the woodblock and the copperplate atlases played out in parallel.

The annex inserted at the end of this chapter summarizes the extant copperplate sheets forming the basis for my analysis. Regarding those I have been unable to consult personally, I accordingly cannot at present determine whether they are first- or second-edition copperplate sheets. In mainland China, two different sets of unknown quantities of copperplate sheets reside at the Dalian Library, while twenty-eight copperplate sheets are at the Peking University Library, and nine sheets are in the National Library of China. The latter correspond to the two upper rows of the copperplate version and carry a number of Latin transcriptions.[473] Reproductions in a number of catalogues show notes in Italian, such as "*pozza*" (well) and "*loco di posta*" (postal station), on different sheets.[474] The autograph appears to resemble strongly that of Matteo Ripa. On the other hand, several European libraries also preserve incomplete copies of the copperplate atlas: the National Library of France owns twelve sheets, constituting the three upper rows depicting regions north of the Great Wall; the Royal Library of Belgium has an incomplete copy of the first copperplate edition; the Italian Geographic Society harbors two sets mixing both editions;[475] in Saint Petersburg, finally, three incomplete sets of sheets

472 Compared to the reproduction included in Petech (1953).

473 Dudink (2006), 90; *Yutu yaolu* (1997), 38; Huang (2010), 358–361; *Beijing daxue tushuguan cang guji shanben shumu* (1999), 115.

474 Huang (2010), 358–359 (bottom-middle).

475 KBB, LP VB 11.283 E(2); SGI, Cina 170 & 257. See also Dudink (2006), 87–92; Cerreti (2001), 62–63; Li (1996), 163. I am indebted to Ad Dudink for his excellent bibliographical references to extant copies of the copperplate atlas. The Brussels copy includes Latin notes in pencil.

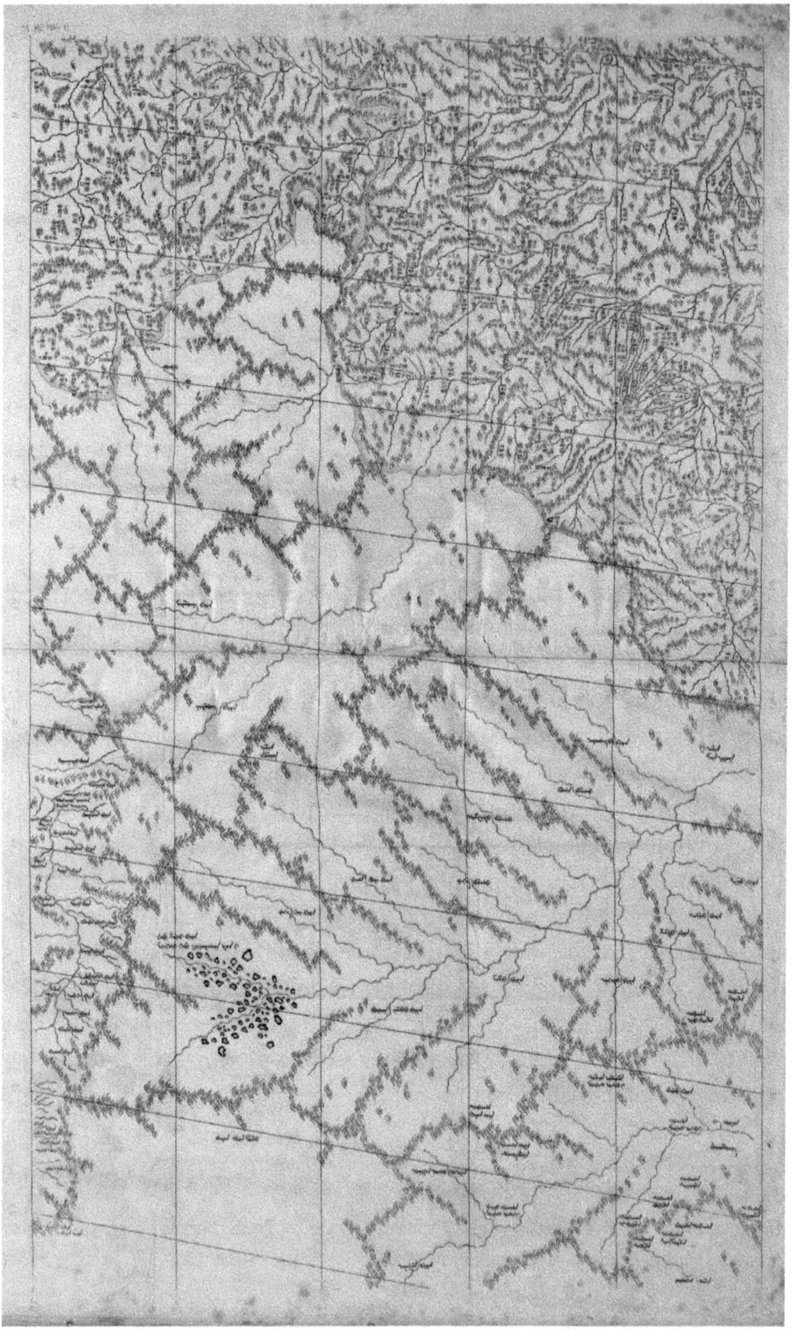

FIGURE 32 *First edition copperplate atlas, sheet 4 of row 5 (c. 40 × 66.5 cm). Territories in Sichuan (right) include place names in Chinese, while those in the Tibetan highlands (left) are in Manchu and contain large blank spaces.*
ROYAL LIBRARY OF BELGIUM, LP VB 11.283 E (2), FOL. 14.

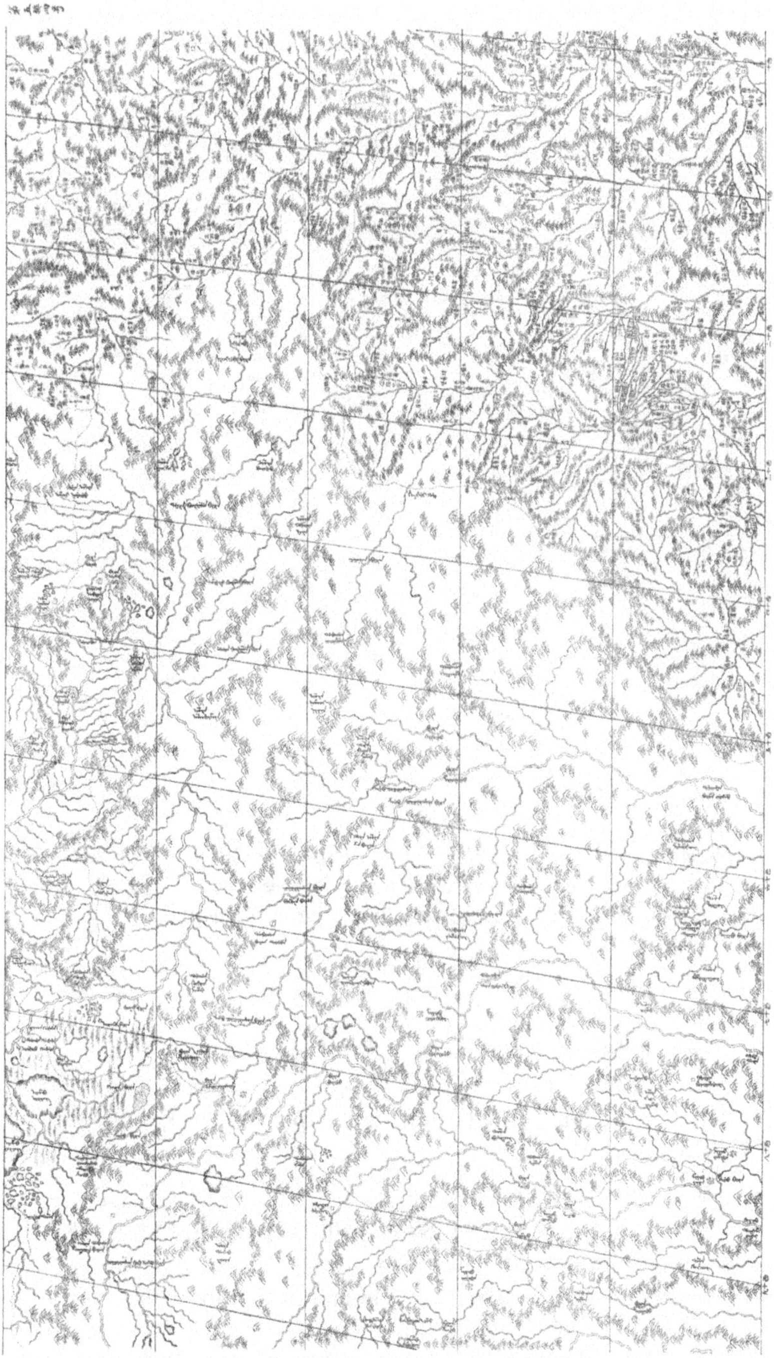

FIGURE 33 *Revised copperplate atlas, sheet 4 of row 5 (c. 40 × 66.5 cm). Tibetan areas include more detail compared to the corresponding first-edition sheet in figure 32. Dotted lines indicate routes.*
TAKEN FROM WANG & LIU (2007), VOL. 1.

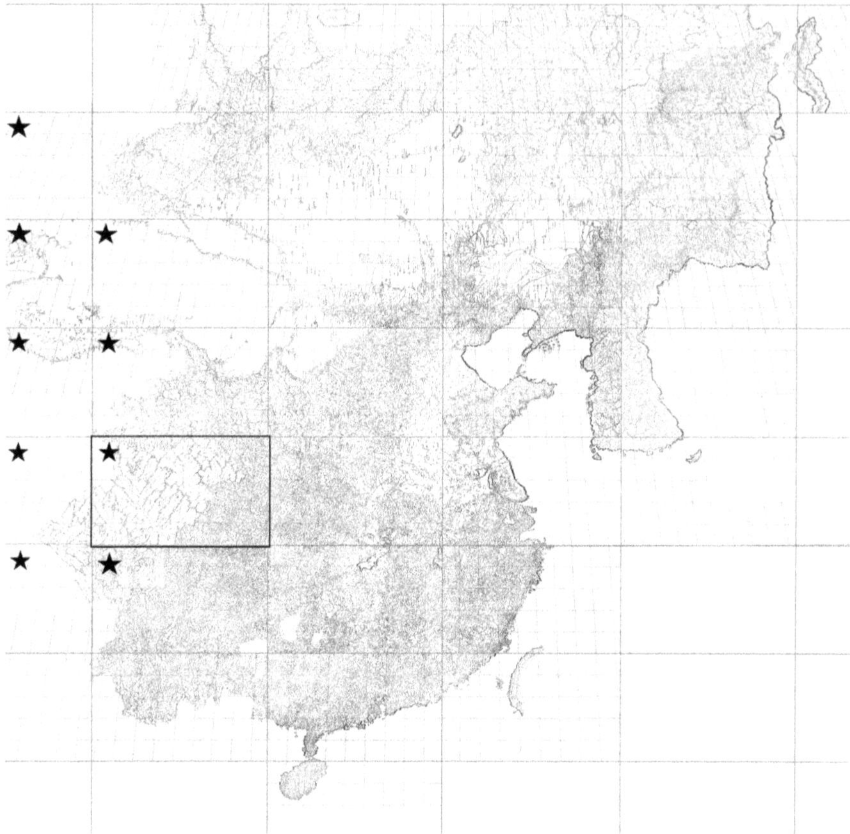

FIGURE 34 *Attempted reconstruction of the 1717 first edition copperplate atlas, with the Sichuan sheet (figure 32) inserted and informed by extant sheets. Sheets indicated with an asterisk were replaced in the revised edition. Compare to figure 31.*

TABLE 9 *Timetable of map production during the late Kangxi years.*

1717	– Return to Beijing of the last team of mapmakers including missionaries
	– A team not including missionaries goes to survey Tibet
	– Completion of **the first woodblock and copperplate editions**
1718	– Return of the team sent to survey Tibet
	– Completion of the revised maps in manuscript form[476]

476 Fuchs (1943), 1:35–36 & 60.

1719 – Completion of the **revised copperplate edition**
– Presentation of the revised copperplate edition to the emperor[477]
1720 – Thorough revision of Chinese renderings of Tibetan place names[478]
1721 – Completion of the **revised woodblock edition** and route book companion

---

covering areas north of the Great Wall are at the Institute of Oriental Studies of the Russian Academy of Sciences.[479] According to Giovanni Stary (I have not been able to verify this), Ripa personally handed one of these sets over to an Italian member of the Izmailov mission as a gift to the Russian Tsar.[480]

European institutions own at least four other copies brought back by Matteo Ripa himself, who returned to Europe aboard an English ship in 1724. These copies contain personal annotations and borders often traced in colored pencil. One such, a mixed copy on three large scrolls that Ripa donated to King George I (r. 1714–1727) as he passed through London, is at the British Library.[481] One of these scrolls forms part of a 1717 edition, thus including the first-edition Sichuan sheet (figure 32). The other two scrolls simply add parts of the revised edition to the 1717 sheets. All scrolls include notes in Ripa's own hand. Ripa also separately drew the coastlines as depicted on the copperplate atlas on four large sheets.[482] Second, a copy resides in Bologna alongside explanatory documents regarding the several experiments conducted before Ripa's perfecting of the printing process, identifying the Bologna copy as an early specimen Ripa kept for himself.[483] Third, as mentioned in the excerpt quoted above, the Instituto Universitario Orientale in Naples (formerly the Collegio dei Cinesi di

477 In the quote cited at the beginning of this chapter, the emperor asks grand secretary Jiang Tingxi to check the atlas for mistakes. A memorial to the Qianlong emperor credits Jiang Tingxi with responsibility for the production of the woodblock atlas. AGS, 177892-001.

478 Fuchs (1943), 1:43–44.

479 Described in a catalogue by Tatjana Pang. Pang (2001), 168–170.

480 Stary (2001).

481 BL, Maps K Top 116 15-15a-15b; Gray (1960), 40; Li (1996), 163; Albanese (2006), 64.

482 BL, Maps K Top 116 18.2 TAB.

483 This perhaps explains further inconsistencies in the sheets extant in Europe. Dudink (2006), 89–90. "tentò egli l'Impresa, e doppo varj esperimenti fatti prima coll'aqua forte, e poi col Bolino, le riuscì finalmente di intagliarla col Bolino, ed indi impressa al modo Europeo presentarla all'Imperadore, ritenendo per se pochissimi Esemplari, dei quali uno è il presente." Albanese (2006), 62.

Napoli) harbors a first-edition copy donated by Ripa, the college's founder.[484] Ripa himself transliterated place names, traced provincial borders in color, and wrote explanatory notes where he felt necessary. Finally, a copy in forty sheets donated to Holy Roman Emperor Charles VI (r. 1711–1740) on August 21st, 1726 in an effort to elicit financial support for the College in Naples resides in Vienna.[485]

Last, the original copperplates for the revised copperplate atlas came to light in Shenyang in the 1920s, enabling new prints published in 1929 and 2007.[486] The recent facsimile publication appeared in a set including both the Yongzheng and Qianlong editions discussed below and has attracted scholarly attention to the revised copperplate atlas. In contrast, scholars seldom consider the woodblock editions.

3.1.3 *Imperially Commissioned Compilations and Later Renditions*

The woodblock and copperplate versions of the Qing atlas described above, each of which appeared in two different editions, were not the only maps directly linked to the land surveys of the late Kangxi reign. A number of later compilations included strongly reduced maps, no less based on data from the surveys, often alongside textual descriptions of (parts of) Qing territories. The earliest version, dating from 1726, the *Comprehensive Collection of Books and Pictures of the Past and Present* or *Gujin tushu jicheng* 古今圖書集成,[487] includes maps of the different provinces and regions, strongly reduced and without gridlines, as well as 216 prefectural maps printed from woodblocks. Separate publications of maps included in the *Comprehensive Collection* occurred, it seems, throughout the eighteenth and nineteenth centuries, with titles such as *Maps from the Inner Palace* (*Neifu yu ditu* 内府輿地圖), *Maps of Qing Territories* (*Da Qing yu ditu* 大清輿地圖) or *Maps of the Unified Qing* (*Qing yitong yutu* 清一統輿圖).[488] These works sometimes include all

484 Petech (1953). The map carries the title *Yuzhi da Qing yitong quantu* 御制大清一統全圖, which does not appear in contemporary Qing sources. The handwriting of these characters is rather bad, suggesting Ripa wrote them himself. See also Li (1996), 163; Fatica (2006), 206–208; Albanese (2006), 62.

485 Albanese (2006), 66.

486 Wang & Liu (2007). The 1929 print, which I have not consulted personally, carries the title *Man-han hebi Qing neifu yitong yudi mitu* 滿漢合璧清内府一統輿地秘圖.

487 Fuchs (1943), 1: 48–56 & 60. Note that Jiang Tingxi, one of the grand secretaries and mentioned in the quote at the beginning of this chapter, was responsible for compiling these volumes.

488 The National Library of China (*Zhongguo guojia tushuguan* 中國國家圖書館, hereafter NLC) holds two copies, carrying the titles of *Neifu yu ditu* 内府輿地圖 and *Huangyu*

prefectural, provincial, and regional maps from the *Comprehensive Collection* or only contain the provincial and regional maps, sometimes with added grid-lines but never with degrees of latitude and longitude. We should also note the adjustment of the maps in these later publications reflecting administrative reforms implemented during the Yongzheng and Qianlong years.

Manuscript adaptations are also extant. The National Library of China, for example, harbors a set of twenty-eight manuscript maps, including one rare general map of Tibet and one schematic map of Mongolian banner areas, the *Map of Princely Territories beyond the Passes* or *Kouwai zhuwang tu* 口外諸王圖, neither of which exists in a known printed version.[489] The only other maps from this set that I have found reproduced in catalogues are those of Shandong and Tibet.[490] On the other hand, a highly similar set survives in the Peking University Library within a much larger collection of maps and recently enjoyed publication.[491] Some of these maps seem to be similar to the woodblock atlas of the Kangxi period, whereas the *Map of Princely Territories beyond the Passes* certainly had its origin in a time preceding the land surveys. These observations presumably led the volume's editors to date them back to "1719 at the latest."[492] However, several elements suggest a (much) later date. The map of Yunnan, for example, includes the prefecture of Zhaotong 昭通, meaning the map must date from after 1728, when administrative reform ended this prefecture's subordination to Sichuan province. Thus, if the two sets at the National Library of China and the Peking University Library are indeed as similar as my limited comparison suggests (I have not been able to consult these maps), both must date from a later period, reflecting administrative reform under subsequent emperors, particularly in the southwest.

The Imperial Workshops produced more thorough revisions of the Qing atlas over subsequent decades and under different emperors. Most important of these were the different Yongzheng and Qianlong-era renditions of the Qing atlas. Today we know the Yongzheng-era atlas, printed from woodblocks, as

quanlan tu fensheng tu 皇輿全覽圖分省圖. They contain only the 15 Chinese provinces, the map of Shengjing, and, only in the latter, a general map. Neither copy includes grid lines. *Yutu yaolu* (1997), 39 [0390 & 0391]. Another version of the *Neifu yu ditu* dating back to 1876 resides at BNF, Ge FF 14550 (RES); Li (1996), 168–169.

489 *Yutu yaolu* (1997), 38 [0384]. For more on the unique *Kouwai zhuwang tu*, see Ha-si-ba-gen & Cheng (2010).

490 Yan (1998), 186–187.

491 As part of Yao (2008), 1–65. This set only contains 18 maps. The Tibet map, included in the NLC set, is one of those missing here.

492 Yao (2008), 1. This volume reproduces all maps.

the *Yongzheng Map in Ten Rows* or *Yongzheng shipai tu* 雍正十排圖, but contemporary documents referred to it as *Complete Maps of Imperial Territories* or *Huangyu quantu* 皇輿全圖. It builds entirely on the style of the Kangxi copperplate atlas in including place names in Chinese and Manchu and in forming one large map by combining sheets, one hundred in total. The map includes areas far beyond the scope of the Kangxi-era copperplate atlas, essentially adding the entire Russian Empire from Kamchatka to Riga.[493] Even though apparently drawn on the basis of the grid system, with each square representing about 200 *li* on the ground, the Yongzheng edition also indicates latitudes on the left of each row and longitudes on the bottom of every sheet. This renders readings of latitude and longitude incorrect, suggesting the court's chief concern must have been the reading of distances. Even so, much of the data came directly from the Kangxi-era copperplate atlas, while European cartographies, route books, and accounts of imperial envoys supplied the raw data regarding areas not included in that atlas. Other differences with the Kangxi-era editions of the Qing atlas include an increase in the number of place names, the incorporation of administrative reforms in the Chinese provinces, and the depiction of historically significant journeys using dotted lines.[494]

Several Jesuit missionaries at court during the 1720s consulted on the revision process. For example, in 1725 the Yongzheng court asked Régis and Fridelli, two Kangxi-era mapmakers, to make a map of the lands between Qing territories and the Caspian Sea. Later, in the first months of 1727, André Pereira (1689–1743), de Mailla, and Gaubil compiled a depiction of Russian territories from Kamchatka all the way to Saint Petersburg and prepared it as a separate map.[495] In essence, these maps seem to have covered most of the lands the Yongzheng edition added to the scope of the Kangxi atlases (figure 36), although none of the missionaries appear to have been directly involved in the material production of the final sheets.[496] The Yongzheng revision finally reached print in late 1727 or early 1728 but, like the Kangxi-era atlases, was never available for sale.[497] It did, however, circulate more widely within Qing China than its predecessors: in the first half of 1728 the emperor personally

493 Reproduced in Wang & Liu (2007), vol. 2.

494 One example is the 1698 journey of Qing envoys to Khalka lands after the defeat of Galdan (Chapter 1).

495 Gaubil (1970), 182–183.

496 Gaubil (1970), 171–175; Pfister (1932), 1:532.

497 Yan (1998), 190–191.

sent copies to most provincial governors.[498] Yu Fushun has claimed that yet another Yongzheng version printed in 1729 followed further administrative reform in the Chinese provinces (I could not verify this independently).[499] Although a few Yongzheng editions have turned up in mainland archives and libraries, it is rare and no sheets appear to be extant in Europe, confirming the lack of missionary involvement in the material map's production as a whole and the Yongzheng court's general suspicion of them.[500] What is certain is that the production of the Yongzheng edition was inextricable from the multidirectional exchange of cartographic knowledge among Beijing, Paris, and Saint Petersburg in the aftermath of the circulation of the Kangxi-era atlas, as discussed in the third part of this chapter.

In the 1750s, the Qianlong emperor called for yet another new edition of the Qing atlas, including depictions of lands as far afield as Asia Minor, the Arabian Peninsula, Persia, and the Indian Subcontinent (figure 36). New surveys conducted, for the first time, to the west of Hami, greatly improved the depiction of those areas, reflecting Qianlong's control over them following his defeat of the Zunghars and incorporation of their lands, Altai Mountains down to Kashgar, into the Qing Empire. He asked several Jesuits to assist in the mapmaking efforts for these areas, although it seems they encountered more suspicion than under Kangxi, sensitive information kept carefully from their hands. A team of mapmakers, including Portuguese Jesuits Felix da Rocha (1713–1781) and José d'Espinha (1722–1788), two lamas, and two officials from the Directorate of Astronomy, He Guozong 何國宗 (d. 1766) and Mingghatu 明安圖, left Beijing for Hami on March 21st, 1756. They prepared a map of Qing-controlled northern Xinjiang as far west as the Ili River before returning to Beijing. In 1759, the same two Jesuits plus Ferdinand Augustin Haller von Hallerstein (1703–1774), another Jesuit, again embedded within a team of several Qing officials including He Guozong and Mingghatu, traveled to Xinjiang for a second round

498 A search in the online database *Yishi cang Kangxi Yongzheng chao man han wen zhupi zouzhe huibian mulu* 一史館藏康熙雍正朝滿漢文硃批奏摺匯編目錄 shows memorials by governors and governors-general acknowledging the arrival of the *Complete Map of Imperial Territories* and mentioning the atlas' adaptation from the Kangxi-era atlas. E.g. Henan: YZ 6/III/25—Shandong: 6/III/27—Zhejiang: 6/IV/6—Shaanxi & Sichuan: 6/IV/15—Yunnan: 6/IV/26—Jiangnan: 6/V/10—Guizhou: 6/XI/6.

499 Yu (1983), 75.

500 The 2007 reproduction in Wang & Liu (2007) relies on a copy kept at the Library of the Chinese Academy of Sciences. Other versions have turned up in The First Historical Archives of China. Yan (1998), 190–191.

of land surveying, this time mapping the more southern Qing-controlled regions around Kashgar. In both surveys, the mapmakers apparently followed practices dating back to the Kangxi period, determining the positions of major cities and major routes while filling in the blanks using existing maps and geographic information provided by local informants.

The different editions of multi-sheet maps produced under the Qianlong emperor using the Yongzheng-era (and thus the Kangxi-era) maps as a basis incorporated new data from the two Xinjiang surveys. The depictions of parts of Asia Minor, Persia, the Arabian Peninsula, and the Indian Subcontinent depended entirely on European cartographies and second-hand data on the routes connecting these lands to Xinjiang and Tibet.[501] Extant Qianlong editions of the Qing atlas listed in catalogues suggest it used both the Kangxi copperplate atlas with its Sanson-Flamsteed projection and the Yongzheng woodblock edition with its grid-based system as prototypes. First, there are indications of updating of the Yongzheng woodblock edition in ten rows around 1757, including place names in Manchu for regions outside the Great Wall.[502] If this is correct, it is possible the Imperial Workshops produced an update of the Yongzheng edition after the first team of surveyors had returned to the capital but before the second team went to survey southern Xinjiang in 1759.

Other Qianlong-era editions, recording all-new data from both Xinjiang surveys, reside in Beijing, London, and Rome. Three closely related versions of this edition are extant, all most likely produced with the assistance of Jesuit missionary Michel Benoist (1715–1774). In a letter dating back to 1773, Benoist mentions that he had engraved the same map at three different scales, with 1, 2 and 2.5 inches, respectively, between the lines of latitude.[503] Matthew Mosca, who reconstructed parts of the production process of the Qianlong multi-sheet maps based on Qing documents, also distinguishes among three versions, albeit

501 Millward (1999).

502 Extant copies seem to be in Beijing and Saint Petersburg, but I could not consult them personally. *Yutu yaolu* (1997), 35 [0393 & 0394]; Yao (1998), 200–201; Huang (2010), 365–366. The copy in Saint Petersburg is incomplete. Pang (2001), 170–171. Most catalogues date these copies back to 1757–1759.

503 Li (1996), 179–181. "On their [the Jesuits'] return from this distant expedition Khien-lung caused new maps, of different dimensions, to be made of the whole empire and the contiguous countries; they were on scales of 1 inch, 2 inches and 2.5 inches for each degree of latitude respectively. The superintendence of this work was assigned to Father Benedict [F. Benoist]." *Journal asiatique*, Vol. 9, 325, as quoted in Baddeley (2006), cxcii.

of 1.8, 2 and 2.5 inches, respectively.[504] The largest of these, the best-known of all three extant editions, commonly called the *Qianlong Map in Thirteen Rows* or *Qianlong shisan pai tu* 乾隆十三排圖 and most certainly linked with Benoist, was engraved on 104 copperplates using the Sanson-Flamsteed projection also used for the Kangxi copperplate editions. All place names are in Chinese only, but a poem by the emperor appears on the first sheet in both Chinese and Manchu.[505] A woodblock version of the same size that must have been the basis for this map consists of 103 sheets (lacking that bearing the emperor's poem) resides in the Vatican Library.[506] A second edition, identical in scope and content to the 104 copperplate edition but reduced to 78 sheets, is in the

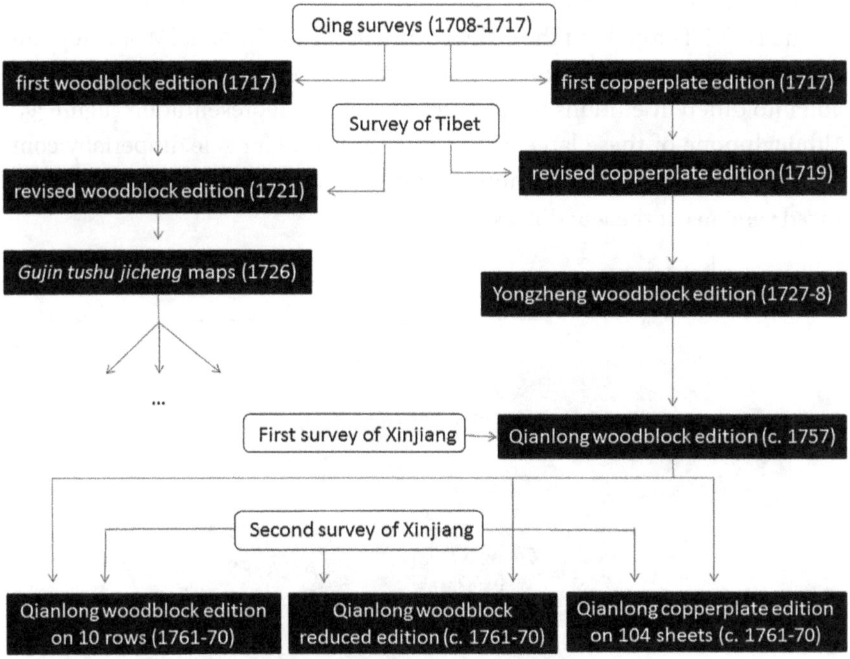

FIGURE 35 *Basic genealogy of most of the Qing editions mentioned above in relation to the land surveys.*

504 Mosca (2013), 112–113.

505 *Yutu yaolu* (1997), 40 [0395]. BL, Maps TAB 1b. A 1932 reprint was published by the Imperial Museum of Beiping. Reprint copies are in Taipei's Palace Museum at 021049-021156; and in Beijing, London, and Washington. Li (1996), 178–179; Li (2004), 14; *Yutu yaolu* (1997), 40 [0396]. This is also the atlas reproduced in Wang & Liu (2007), vol. 3.

506 Li (1996), 175–178.

National Library of China.[507] The third, also known as the *Qianlong Map from the Inner Palace in Squares* or *Qianlong fangge neifu yutu* 乾隆方格內府輿圖, grid-based in the style of the Yongzheng edition and printed on ten rolls, includes Manchu place names for regions outside of the Chinese provinces.[508] Further study of extant copies will hopefully specify the production process of these different versions and determine whether or not they correspond with the findings of Mosca.

This overview of printed editions, all of which directly incorporate data gathered during the 1708–1718 land surveys, has confirmed the initial production of two different versions of the Qing atlas at court: the first carved from woodblocks, the second engraved on copperplates. Each version appeared in two editions, the second including a revision for the Tibetan areas and the region around Hami. Throughout the eighteenth century, the Imperial Workshops under the Yongzheng and Qianlong emperors again revised the copperplate version into different editions using various modes of representation (figure 35). Although none of these large atlases were available for sale, imperially commissioned compilations and more popular publications included strongly reduced versions of the woodblock maps.

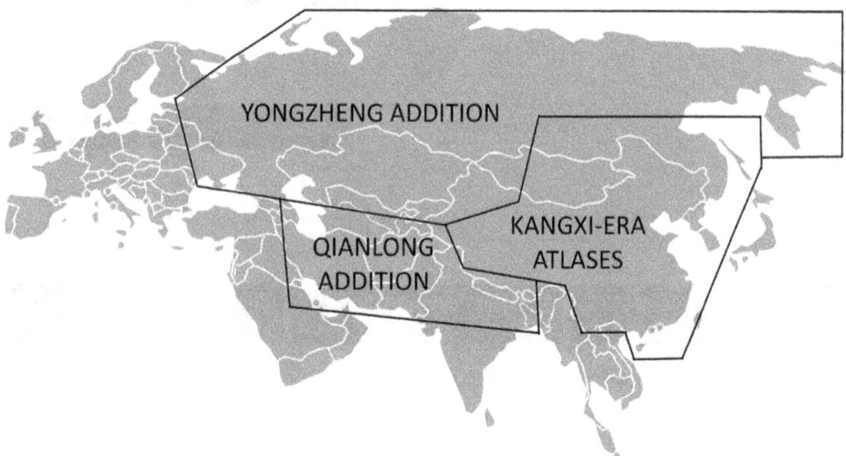

FIGURE 36 *The gradual increase in geographic scope from the Kangxi-era atlas to the Yongzheng and Qianlong editions.*

507 *Yutu yaolu* (1997), 40 [0397].

508 BL [IOR] X/3265. Li (1996), 179–181.

3.2 The European Incorporation of a Qing Atlas

The year 1735 saw the Parisian publication, more than a decade after the Imperial Workshops printed the Kangxi-era editions of the Qing atlas, of forty-one maps of the Chinese provinces, Tartary, Tibet, and Korea within the four large volumes of the *Description géographique, historique, chronologique, politique et physique de l'empire de la Chine et de la Tartarie chinoise*. In the preface, Jean-Baptiste du Halde, editor of this encyclopedic work on continental East Asia, confirms that most of these maps were adaptations of Chinese originals produced at the behest of the Qing emperor with assistance from European missionaries. Jean-Baptiste Bourguignon d'Anville, who also composed four new general maps of these regions, had executed the adaptations. Despite much scholarly attention to d'Anville's maps in past decades, we yet know little about the general context of their production. How did the original maps reach Europe? Why did more than fifteen years pass before their publication there, and to what extent were the Qing originals the basis for the published maps? In this section I will answer these questions by closely comparing d'Anville's maps to the different editions of the Qing atlas and reconstructing the production and publication processes behind different sets of d'Anville's regional maps.[509]

I begin by tracing the initial transmissions of the Qing atlas to Europe, where controversy immediately engulfed plans for a European adaptation. Next, a brief biography of d'Anville kicks off the story behind the actual publication process, reconstructing the lines of communication from the Beijing missionaries to d'Anville via du Halde. In a last step, I analyze the French cartographer's adaptations in detail, identifying the specific sets of maps—sent from Beijing to Paris at different times—that d'Anville used. This leads to the conclusion that d'Anville employed different versions of the Qing atlas in producing his celebrated maps of continental East Asia, all eventually inserted into du Halde's monumental *Description de la Chine*.[510]

3.2.1 *Early Transmissions and Reception in Europe*

Not long after the presentation of the Qing atlas to the emperor, and certainly before 1720, Pierre Jartoux managed, perhaps secretly, to dispatch copies of the

509 This section of Chapter 3 is a reworked version of an article published in *Imago Mundi*. Cams (2013).

510 Scholars sometimes assume the revised woodblock atlas was the main basis for d'Anville's maps. Wilkinson (2000), 148; Landry-Deron (2002), 143; Dudink (2006), 92.

first woodblock atlas (or its manuscript revision) to his confreres in Paris. As discussed above, Jartoux was part of the team of editors who reworked the draft maps at the Imperial Workshops. He included personal notes and transliterated the most important place names directly onto the copies he sent to Europe, most likely aboard a French ship.[511] Once in France, the maps proceeded to the Jesuit College in Paris, to the chamber of du Halde, scriptor of the French Jesuits and editor of the multi-volume *Lettres édifiantes et curieuses*, making edited correspondence from Jesuit missionaries across the world available to the European reading public.[512]

Xavier Ehrenbert Fridelli, another of the original mapmakers, in 1719 initiated a second early (and traceable) transmission. He sent maps to fellow Jesuit Coenraad Janning (1650–1723) by way of Ostend, then one of the more reliable sea routes between East Asia and Europe. Living in Antwerp, Janning was well-connected, on the one hand to the Ostend Company via the Maelcamp family, and on the other to the European Jesuit community.[513] Fridelli therefore suggested that Janning forward the maps to François Noël, a missionary who had returned from Asia and fully capable of transliterating place names, who lived in Lille.[514] The copperplate atlas sheets now at the Royal Library of Belgium presumably constituted one part of this transmission, but only Noël's one-time ownership of them substantiates this claim.[515] One source even claims that João Francisco Cardoso forwarded another set of maps to Lisbon, but no others

511 Gaubil (1970), 216.

512 One letter in this series, written by de Mailla, describes an episode of the mapping project in detail, discussing mapmaking operations on the island of Taiwan. du Halde et al. (1702–76), 14:1–85.

513 In the years 1718–1728, Antwerp became a dispatch center for all kinds of material to and from Qing China, as the "Via Ostendana" had become one of the most reliable sea routes between Western Europe and East Asia. Golvers (2014). The Jesuit College at Antwerp and the Maelcamp family played a major role in the exchange.

514 Fridelli wrote: "Ich wird (geliebt es Gott) folgendes Jahr die Tartarische Land-Carten, so Eur Erhwürden sehnlich begehen, nach Europam überschicken, nachdem ich dieselbe bereits letztverwichenes Jahr [1719] mit einem Ostendischen Schiff dem Patri Janningo also, wie sie allhier zu Peking ist gedruckt worden, nach Niederlanden zugefertigt hab: allwo sie Pater Noël aus der Sinischen in die Europaeische Sprachen übderzetwen kann." Fuchs (1943), 1:42; Dudink (2006), 91.

515 These copperplate sheets once belonged to Noël and remained at the Jesuit College in Lille. They constitute first-edition sheets, depicting only the Chinese provinces. The Tartary sheets appear to be missing, perhaps because Noël sent them to du Halde (see below). Dudink (2006), 91.

confirm this.[516] Later on, well into the 1720s, further transmissions, discussed below, would follow.

Nearly a decade passed before du Halde took the first steps towards publishing the maps Jartoux had sent, provoking the impatience of French Jesuits in both Paris and Beijing. As early as 1722, the treasurer of the Jesuit missions to Asia in Paris, Louis-François Orry, wrote thus to Beijing missionary Antoine Gaubil:

> I have not been able to see [the maps] for myself. They think they must keep everything secret out of fear that [others] will deprive us of the glory of discovery, so we only communicate through our *Lettres édifiantes*. Yet, everyone leaves us be and nobody supports us. This is what one gains from always wanting to play one's own game.[517]

When Gaubil himself wrote to the head of the French mission about the delay, he explained du Halde had explicit orders not to publish the maps immediately:

> Some complaints have been voiced here about Father du Halde: first, about how he keeps the maps for so long without having them published… they are wrong, and I say this in plain terms, because in 1720 Father Jartoux wrote [to ask him] not to publish these maps until further notice, due to several difficulties… I saw these letters myself in Paris before my departure, in the room of Father du Halde.[518]

According to Gaubil, one key reason behind the delay was fear that the maps, once published in Europe, might serve other Europeans as gifts for the Qing emperor, undermining the French Jesuits' credibility in his eyes. If this was

516 de Moyriac de Mailla (1777–1785), 1:clxxii; Pfister (1932), 1:623.

517 "je n'ai pas pu les voir moi-même, on croit qu'il faut tout retenir serré de peur qu'on ne nous enlève la gloire de l'invention, ainsi que nous ne communiquerons rien que par nos lettres édifiantes. Cependant, tout le monde nous laisse et personne ne nous soutient. Voilà ce qu'on gagne à vouloir toujours faire bande à part." Quoted from a letter of 21 December 1722 and kept at the Bibliothèque de l'Observatoire (BO), B 1/10/7. The letter was also addressed to Jean-Baptiste Jacques (1688–1728), who reached Asia together with Gaubil.

518 "on a fait ici quelques plaintes du P. du Halde 1° de ce qu'il garde si longtemps les cartes sans les faire paraître… on a certainement tort, et je l'ai dit nettement, car en 1720 le P. Jartoux écrivit positivement de ne pas publier les cartes jusqu'à nouvel ordre à cause de plusieurs difficultés… je vis moy-même les lettres à Paris, avant de partir, dans la chambre du P. du Halde." Gaubil (1970), 216. The delay is partly covered in Landry-Deron (2002), 120–121.

truly an important motive behind postponing the publication, it seems more than likely the Jesuits transported the Qing atlas to Europe secretly, without the Kangxi court's knowledge. Gaubil further states that du Halde wrote to the head of the mission in 1723 asking for permission to publish the maps, either under pressure from his critics or because news of the Kangxi emperor's death in 1722 had reached him.

The Parisian Jesuits only presented the different Qing maps in their possession (with the missionaries' hardwon permission) to the French king in 1725.[519] Had du Halde not deferred the publication of the atlas when the Qing maps first arrived, Guillaume Delisle, *Premier géographe du Roy* and member of the *Académie royale des sciences*, would have been the obvious choice to edit them. After all, the first French missionaries to the Qing had received their training at the *Académie* and entertained frequent contacts with some of its members, as well as with Joseph-Nicolas Delisle, Guillaume's brother who later became an astronomer for the Russian Empire in Saint Petersburg. Guillaume himself also frequently corresponded with different missionaries to the Qing, most notably with Czech astronomer Karel Slavíček (1678–1735), who had composed a detailed map of Beijing in 1718.[520] The celebrated cartographer passed away, however, in 1726, two years before the agreement between du Halde and the promising young cartographer d'Anville. We do not know whether Guillaume Delisle ever got to see the maps sent from Beijing.

3.2.2 *Contracting Jean-Baptiste Bourguignon d'Anville*

Jean-Baptiste Bourguignon d'Anville (1697–1782) was born in Paris at the end of the seventeenth century. He supposedly composed his first map at the age of twelve, based on information he had distilled from a number of historical texts. Whether or not this is more than legend, it certainly characterized his later work, for d'Anville, a typical early-eighteenth-century cartographer, never left his office to conduct or lead land surveys, relying instead on others' observational and descriptive work in drawing his maps. Nevertheless, d'Anville apparently had an eye for trustworthy data, making a point of excluding information not confirmed by other sources. He never blindly copied from earlier maps, as many cartographers of his time commonly did, but rather scrutinized every text he could lay hands on to extract and compare geographic data.

Having published his first map at fifteen and notwithstanding his relatively humble background as a tailor's son, d'Anville received the title of *Géographe*

519 Gaubil (1970), 216–217 & 302.

520 For a general overview of the scientific activities of this missionary, see Liu (2002).

du Roy at the age of twenty-two in recognition of his service to the crown as geography tutor to the young Louis XV (r. 1715–1774).[521] At the same time, he found a patron in Louis d'Orléans (1702–1752), son of the Regent.[522] Through his connections at court, d'Anville soon attracted the attention of Parisian elites, including Claude Bertrand Taschereau de Linières (1658–1746), the king's Jesuit confessor, who later introduced d'Anville to du Halde. As a result of these contacts, the young cartographer won the contracts to compose maps for a number of travel accounts and compilations throughout the 1720s and early 1730s. Thus consolidating his reputation and income, d'Anville could afford not only to continue in cartography, a costly occupation, but also to begin making his own choices himself regarding his maps' subject matter. Even so, not all his work was equally successful. While he was able to rectify earlier mistakes concerning Italy's geography solely on the basis of literary data, which a detailed survey subsequently and surprisingly confirmed, contemporary scientific field practice squarely refuted his writings on the shape of the earth. In 1775, however, at the end of his life, having drawn nearly two hundred maps, his reputation resounded across Europe.

D'Anville's repeated expressions of interest in the Qing atlas, news of which had reached the Parisian elites, led de Linières to recommend him to du Halde for the Qing maps' adaptation.[523] This was one of d'Anville's first major assignments, an important step in his career that contributed a great deal to his later fame. They agreed the cartographer would reduce and redraw the maps of the Chinese provinces (fifteen maps), Tartary (twelve maps), Tibet (nine maps), and Korea (one map) sent from Beijing, as well as compile four new general maps, all for insertion as fold-out maps in du Halde's books. Later, the list of cartographic material for the *Description* expanded through the acquisition of the maps of thirty-eight Chinese cities (on seven sheets) and a chart of the travels of Vitus Jonassen Bering (1681–1741) through indirect contacts from Saint Petersburg.[524] D'Anville's contract included not selling any of his adaptations elsewhere, individually or atlas form, without du Halde's explicit permission.

521 AN, Maison du roi, O[1] 63:157.

522 Dacier (1802), 3–44. Du Halde later became the Jesuit confessor to Louis d'Orléans.

523 d'Anville (1776), 8.

524 For the city maps, see Destombes (1976), 85–97. D'Anville corresponded with Joseph-Nicolas Delisle about the discoveries of Bering, but the king of Poland sent the chart itself to du Halde (discussed below).

Three separate documents spell out the specifics of the agreement. The first contract, drawn up between du Halde and d'Anville in 1728, stipulates the cartographer's duties regarding the maps of the Chinese provinces and Tartary, his remuneration in cash, books, and maps, and du Halde's right to correct the printer's proofs. This first contract also addressed intellectual rights: "[I, du Halde] will be happy to oblige the said Sir [d'Anville] by referring to his work in the prospectus that I have to give to the public of my books, and in the books themselves," ensuring full recognition of the cartographer's contribution.[525] Following common practice, the cartographer was also responsible for the engravings, for which he contracted the workshop of Delahaye, whom d'Anville elsewhere describes as "the most able [engraver] known to me."[526] At du Halde's request, however, a Mr. Humblot (said to be familiar with the techniques of Chinese painting and who had also supplied fourteen illustrations of cultural and ethnographical interest for du Halde's *Description*) designed the cartouches.[527]

3.2.3 *Intercultural Adaptation: d'Anville's Regional Maps*

Within a year of signing the contract, d'Anville had finished reducing and redrawing the maps of the fifteen Chinese provinces.[528] He based these entirely on the corresponding maps from the first woodblock edition of the Qing atlas Jartoux had sent to Paris: both sets of maps match closely and leave the territories outside the provincial borders blank (figures 37 & 38).[529] D'Anville made copies of these maps on oilpaper, which he then used in producing his own versions.[530] The only substantial difference is that d'Anville's maps lack place

525 "[Moi, P. du Halde] me ferai un plaisir d'obliger ledit Sieur en faisant mention de son travail, dans le plan que je dois donner au public de mon ouvrage, et dans l'ouvrage même." All contracts between du Halde and d'Anville are available in Cordier (1905), 391–400. The originals reside at the Bibliothèque de l'Institut (hereafter BI), Ms 5401:245–247.

526 "l'attention que j'ai eue de ne confier la gravûre de mes Cartes qu'au plus habile qui me fût connu." d'Anville (1776), 46. For an overview of d'Anville's work, see Haguet (2010), 88–101.

527 For his designs, Mr. Humblot consulted Chinese paintings provided by du Halde. A certain du Velaer, who had spent time on the island of Hainan as a director of the French Compagnie des Indes, had given the paintings to du Halde. du Halde (1735), 1:xlix.

528 Cordier (1905), 394.

529 I compared d'Anville's adaptations in du Halde's *Description* with the 1721 woodblock maps in Brussels. KBB, LP VB 11.283 E (1–4). D'Anville's draft maps of the Chinese provinces are preserved at BNF, Cartes et plans, Ge D 10569-83.

530 Fuchs (1935), 398. I could not find d'Anville's copies of the Chinese provinces, but identified those for Korea and Tibet at BNF (discussed below).

names for some of the localities plotted, presumably because Jartoux had not transliterated all of them.[531]

At the same time, d'Anville certainly had access to one of the copperplate editions of the Qing atlas as well. Unlike his maps of the Chinese provinces, most of the regional maps of Tartary descend directly from the copperplate version (figures 39 & 40). The place names are clearly transliterations from Manchu, the language used only in the copperplate atlas, and eleven of the twelve maps of Tartary match precisely the corresponding copperplate sheets in terms of land coverage. The only exception is the map of the region near the lower Amur River, combining two copperplate sheets into one larger map, probably because of the large blank space on one of the copperplate sheets.[532] In turn, d'Anville's map of the region surrounding Hami reflects unambiguously the use of copperplate sheets from the first 1717 edition. This further bolsters, although not conclusively, the hypothesis that François Noël forwarded to du Halde the Tartary sheets Fridelli had sent to Europe by way of Ostend which d'Anville would use for his adaptations.[533]

Whereas d'Anville based his maps of the Chinese provinces and of Tartary on different versions of the Qing atlas, his map of Korea appears to combine the woodblock and the copperplate versions in one and the same map: the scope of the map is nearly identical to that in both woodblock editions, but it also depicts the lands beyond the border with Manchuria, as only the copperplate version does. Moreover, the place names are apparently transliterations from Manchu but with Chinese pronunciations, another indication that d'Anville used the copperplate atlas.[534] However, d'Anville did not combine the two versions of the Qing atlas himself. He used a map called the *Map on Oilpaper of the Kingdom of Korea Given to the Jesuits by the Chinese*, which he copied onto Parisian oilpaper and which now resides in the National Library of France (figure 41).[535] Its title refers to the fact that Jean-Baptiste Régis, one of the original mapmakers, received permission to consult maps and textual

531 The first-edition woodblock atlas sent to Paris is no longer extant, but according to Fuchs its maps covering the Chinese provinces are identical to those in the revised version. Fuchs (1943), 15–16.

532 See above (figure 31—asterisk).

533 See above. Fuchs (1943), 1:28; Dudink (2006), 92.

534 Stary (1997), 191. Only the copperplate editions of the atlas used Manchu.

535 BNF, Cartes et plans, Ge DD 2987 (7325B): "Carte huilée du royaume de Corée fournie aux jésuites par les chinois." The map consists of two pieces of paper glued together. Gaubil (1970), 205. I am indebted to the BNF's Département des Cartes et plans for their help in determining the provenance of the oilpaper.

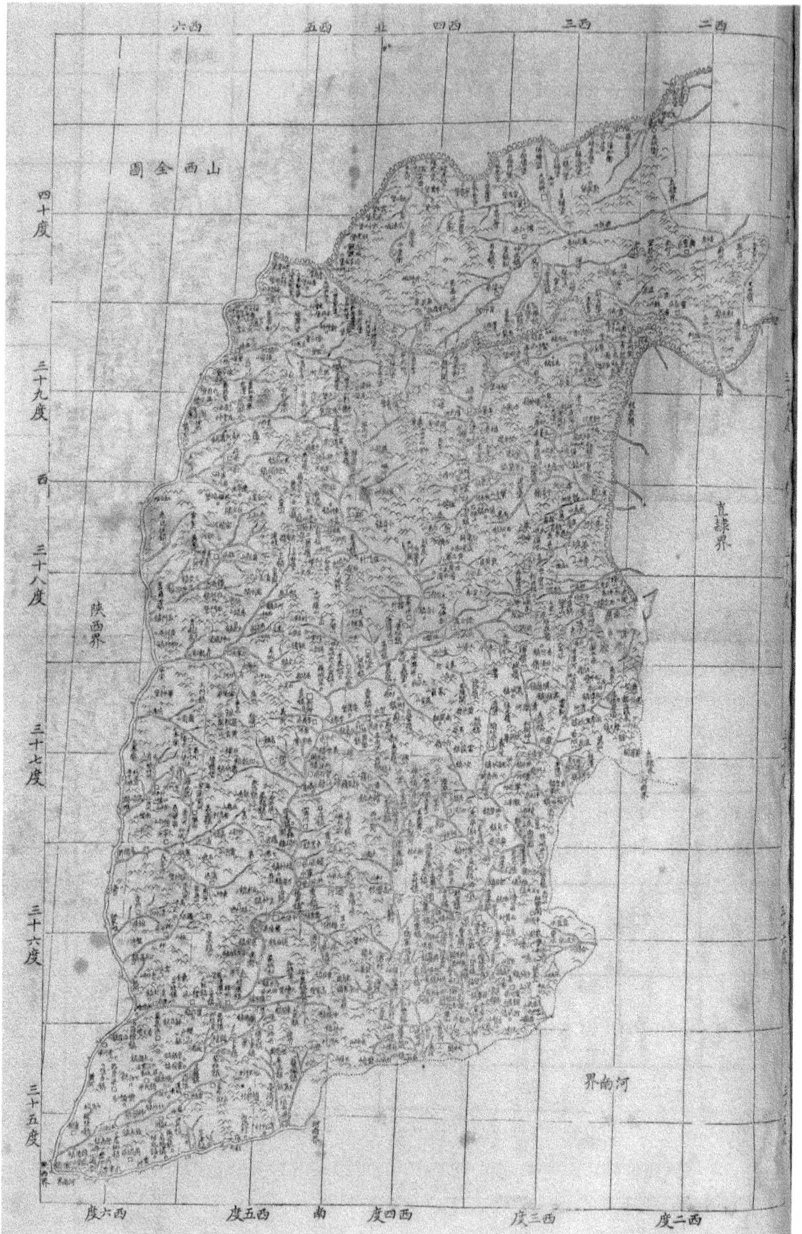

FIGURE 37 *The map of Shanxi province in the revised woodblock edition of the Qing atlas (c. 38 × 25 cm), identical to the Shanxi map in the first woodblock edition. Territories beyond the borders left blank.*
ROYAL LIBRARY OF BELGIUM, LP VB 11.283 E (3), F. 11.

FIGURE 38 *The map of Shanxi province as adapted by d'Anville (c. 35.5 × 25 cm). Not all plotted localities carry place names, presumably because Jartoux had not transliterated them.*
TAKEN FROM DU HALDE (1735), VOL. 1. MAURITS SABBEBIBLIOTHEEK (UNIVERSITY OF LEUVEN).

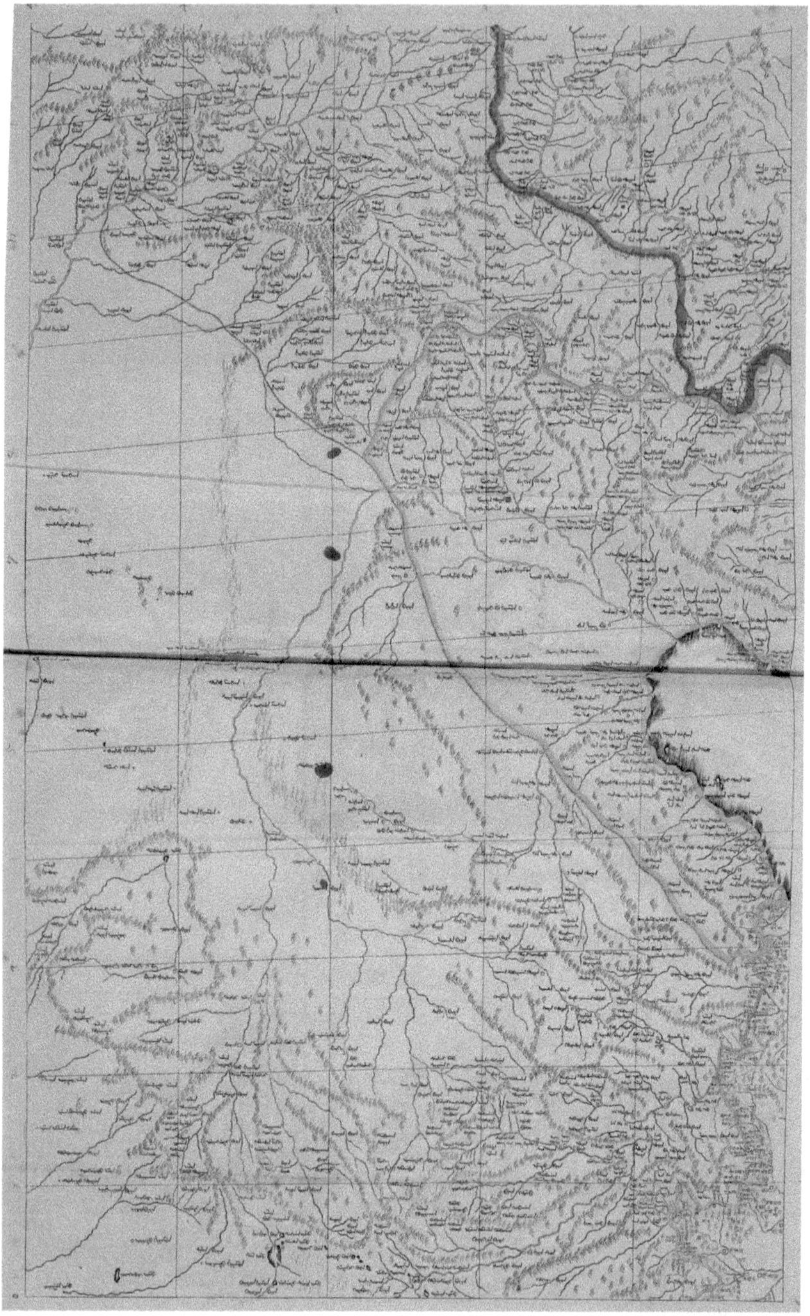

FIGURE 39 *The region of Shengjing in Tartary (today's Liaoning province) in the copperplate edition of the Qing atlas (c. 40 × 64 cm). Place names north of the Great Wall are in Manchu, those to its south in Chinese.*
ROYAL LIBRARY OF BELGIUM, LP VB 11.283 E (2), F. 10.

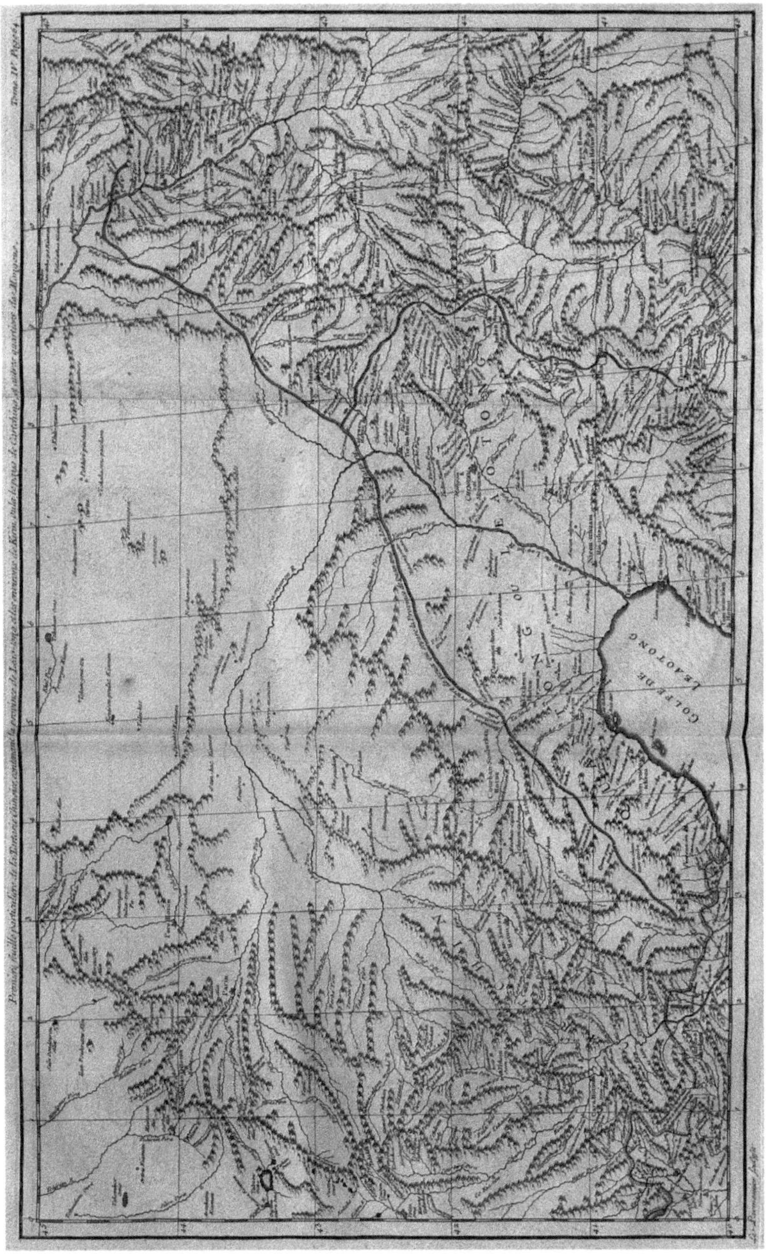

FIGURE 40 *The region of Shengjing as adapted by d'Anville (c. 29.5 × 48 cm). Place names north of the Great Wall are transliterations from Manchu.*
TAKEN FROM DU HALDE (1735), VOL. 4. MAURITS SABBEBIBLIOTHEEK (UNIVERSITY OF LEUVEN).

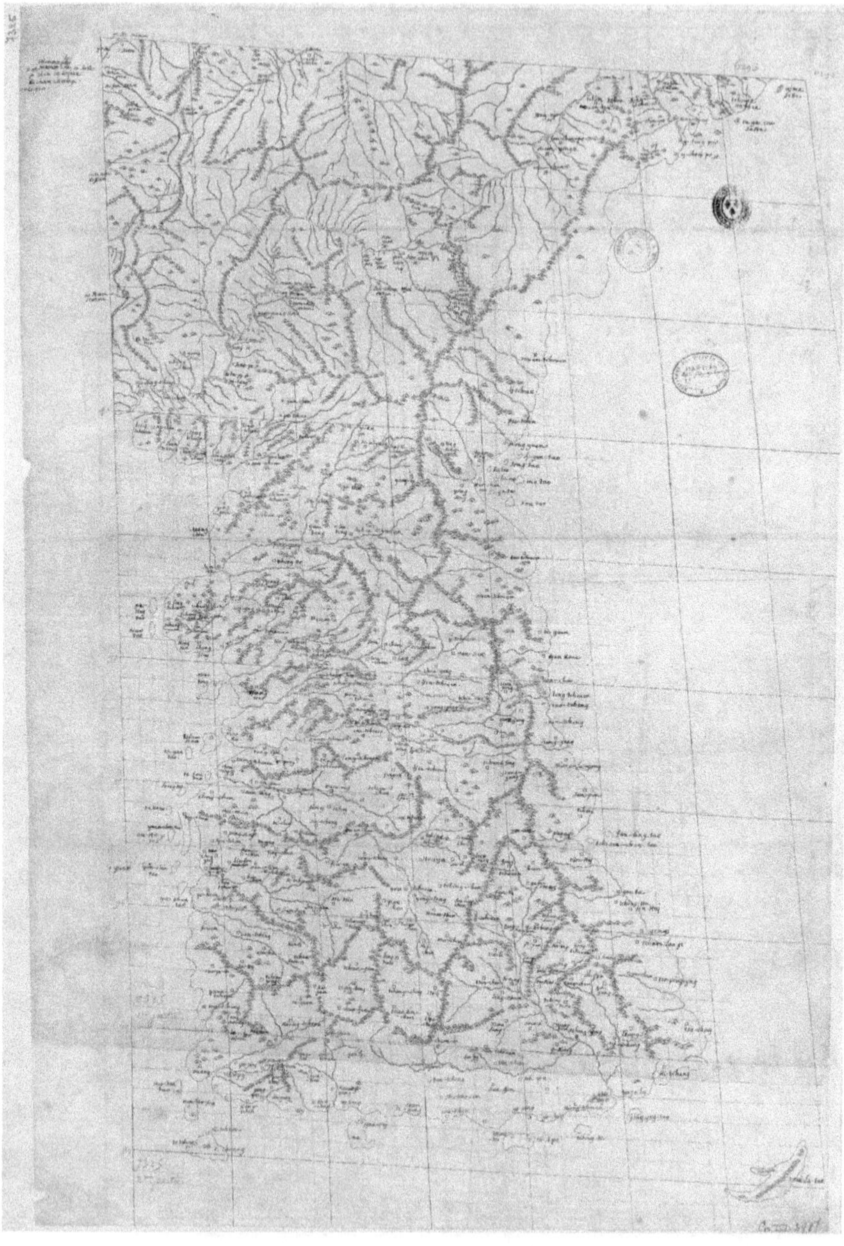

FIGURE 41 *Map of Korea on oilpaper used by d'Anville for his adaptation (c. 78.5 × 53 cm). The map combines different versions of the Qing atlas, constituting a copy of a map the Beijing missionaries sent to Paris. It includes a legend of Manchu toponyms (top-left).*
BIBLIOTHÈQUE NATIONALE DE FRANCE, CARTES ET PLANS, GE DD 2987 (7325 B).

descriptions of Korea at the imperial palace in the second half of the 1720s, copies of which he sent to Paris.[536] Régis appears to have had a keen interest in Korean geography, even discussing the matter with a Korean embassy to the Qing in 1720.[537] The map includes a small legend of Manchu toponyms, indicating copperplate sheets of Korea had been used in rendering place names. The original behind d'Anville's copy on oilpaper, composed by Régis, is also at the National Library of France. The two maps differ somewhat in the spelling of place names (figure 42).[538]

In 1729, before the completion of the adaptations of the maps of Tartary and Korea, d'Anville and du Halde signed a second contract increasing the cartographer's remuneration in exchange for also reworking the regional maps of Tibet and for compiling the *General Map of Tibet*.[539] Remarkably, d'Anville's set of regional maps of Tibet includes a second map of the region around Hami differing in scope and content from the Hami map among d'Anville's regional maps of Tartary. Comparing these two Hami maps shows that d'Anville must have based his maps of Tibet on the revised copperplate atlas, whereas first-edition copperplate sheets had served d'Anville in producing his regional maps of Tartary as discussed above. However, the Tibetan maps match neither of the copperplate editions precisely in scope, but seem to be cut-and-paste versions, with each map made up of two or more sheets. A closer look reveals substantial differences in detail, even from the revised copperplate atlas: areas left blank in the revised Qing atlas are not blank and the map places Lhasa at c. 29°30′ N/24°30′ W (*vis-à-vis* Beijing) rather than the c. 30°30′ N/25° W in the revised copperplate edition. Such discrepancies stem from the fact that Régis had sent a reworked map of Tibet to Paris in 1726 just as he had done for Korea.[540] After this scroll map arrived in Paris, d'Anville copied it onto the same type of Parisian oilpaper. It now resides in the National Library of France in three separate parts and, like the map of Korea, includes a legend and

536 In exchange for copying these maps, European missionaries provided the Yongzheng court with cartographic information from Europe (discussed in the third section of this chapter).

537 Lim (2012). Note that they also exchanged surveying instruments.

538 BNF, Cartes et plans, GE C-5323. The orthography of some of the Latin transcriptions differs. We can attribute this to the heated discussions on the orthography of East Asian place names at the time (discussed in the third section of this chapter).

539 The Tibetan maps also depict lands now part of western Sichuan and Xinjiang.

540 APF, Brotier 148:47-47v; Cordier (1915), 516–517; Gaubil (1970), 187; d'Anville, (1776), 14–17.

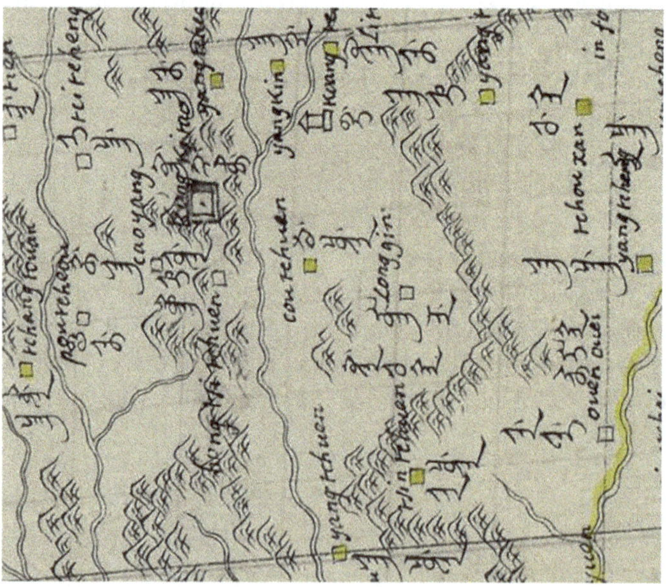

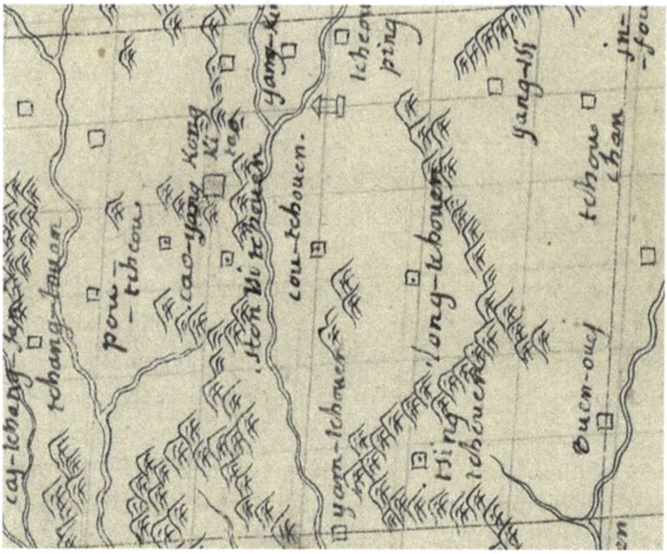

FIGURES 42.1 & 42.2 *The missionaries' map of Korea (left), from the copperplate atlas, and d'Anville's copy on oilpaper (right).*
BIBLIOTHÈQUE NATIONALE DE FRANCE, CARTES ET PLANS, GE C-5323 AND GE DD 2987 (7325B).

explanatory notes.[541] Content-wise, we can attribute the discrepancies between Régis' map and the copperplate atlas to Régis and his fellow Beijing missionary Gaubil's involvement in atlas production at the Yongzheng court in the 1720s and thus their access to improved data on Tibet.[542]

In sum, close analysis of d'Anville's regional maps reveals he employed both the woodblock and the copperplate versions of the Qing atlas, in their different editions, as bases for his adaptations (figure 43). In a response written several years later, attempting to refute criticism voiced against his work, d'Anville explicitly confirmed that his regional maps were true to the originals sent from Beijing:

> I am glad to announce, before anything else, that the regional maps of the provinces of China, of parts of Tartary, and even those of Tibet, all drawn to be included in the Description de la Chine by Reverend Father

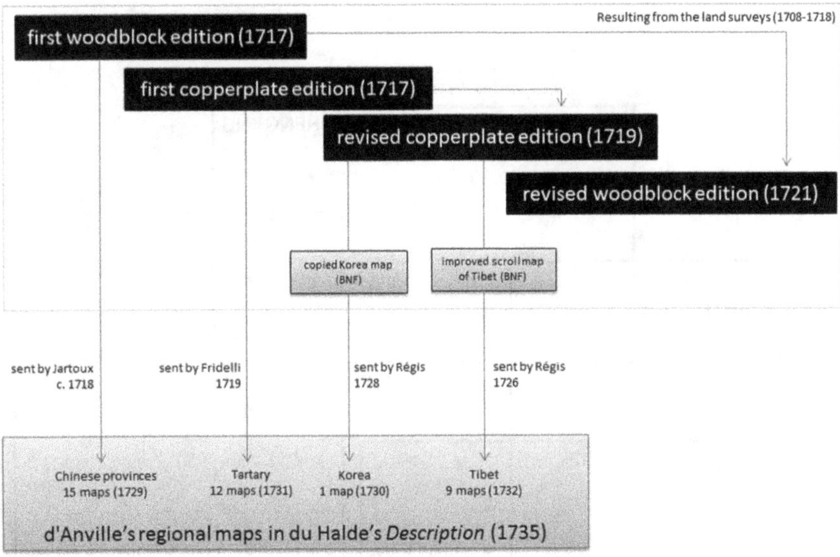

FIGURE 43 *D'Anville's use of different-edition Qing atlases for his regional maps.*

541 BNF, Cartes et plans, Ge DD 2987 (7348B) and Ge DD 2987 (7349B-50B). As described by d'Anville, this is a scroll map in three parts with routes across the region depicted in red. d'Anville (1776), 17. One part appears to be missing.

542 Gaubil later met with court officials, trained by Jartoux during the Kangxi-era surveys, who had made new maps of Tibet. This meeting took place during the Yongzheng court's preparations to revise the Qing atlas. The court also solicited Régis and Gaubil to provide cartographic data in preparation for the Yongzheng revision. Gaubil (1970), 182–183.

> du Halde S.J., are moderate and servile (if I may use this epithet) reductions of maps that were sent from China. It was thought necessary to retain even the style of the original design, so that this large and beautiful work could be communicated or presented to the public with the greatest faithfulness [to the originals].[543]

Another indication and direct consequence of d'Anville's close adherence to maps from different editions of the Qing atlas is that none of his regional maps includes the southernmost tip of Shengjing (the Liaodong Peninsula), which all Qing editions cover in one map or another. Instead, the French cartographer opted for the geographic scope of the copperplate version map for the region of Shengjing (figure 40), while using his oilpaper copy for the map of Korea (figure 41). As the copperplate and the woodblock versions of the original atlas determined, respectively, the geographic scopes of these two maps, d'Anville lost the southernmost tip of Shengjing between sheets (figure 44). In

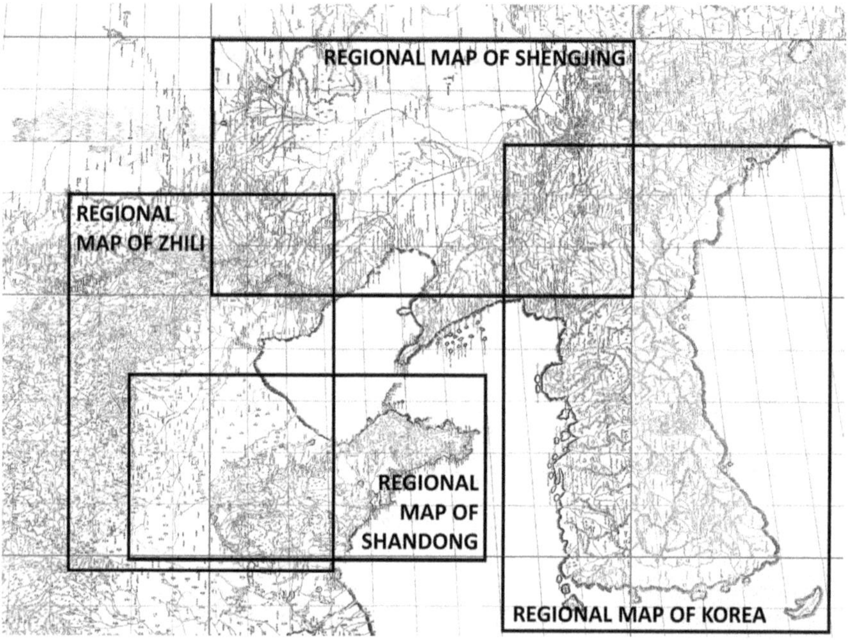

FIGURE 44 *The coverage of some of d'Anville's regional maps. None of these includes the southernmost tip of Shengjing (center).*

543 "Je suis bien aisé qu'on soit prévenu avant toute chose que les Cartes particulières des provinces de la Chine, les parties de la Tartarie, celles du Tibet même, selon que le tout a

spite of this lacuna, d'Anville completed all the regional maps du Halde had commissioned by 1733.

3.3 Beijing, Paris and Saint Petersburg: Negotiating the Gaps

With the thirty-seven regional maps of the Chinese provinces, Tartary, Korea, and Tibet completed, in 1733 d'Anville finally finished the reduction and redrawing of all regional maps du Halde had commissioned. Both men realized, however, that they could do more to increase the public's interest in the four-volume *Description* and its maps. They needed carefully to exploit their connection to Saint Petersburg, reaping new data on Siberia, Central Asia, and the overland routes to Beijing to include among the *Description*'s wealth of textual and cartographic material. On the other end of the Eurasian continent, at the court of the Yongzheng emperor, administrative reform and Qing expansion beyond Hami had created the need to revise the Kangxi-era atlas, and so the Yongzheng emperor asked the missionaries contribute their knowledge of regions beyond the scope of the Qing surveys. Here too, the missionaries' connection to Saint Petersburg proved of utmost importance. In exchange for providing information, the missionaries received permission to copy information from route books, travel accounts, and maps kept at the Qing court, some of which would of course eventually travel back to Europe. Last but not least, the expansion of the Russian Empire into Siberia meant that Saint Petersburg also desired more precise cartographic information of the lands along its southern borderlands. In the 1720s, Beijing, Paris, and Saint Petersburg thus came to rely on each other in their shared interest in filling the gaps in maps of Central and Northeast Asia.

This section argues that d'Anville's general maps and the Yongzheng-era atlas were products of an intercontinental exchange of cartographic knowledge via an extended web of personal connections connecting individuals in Beijing, Paris, and Saint Petersburg, amidst which stood the French Jesuits. First,

été dessiné pour entrer dans la Description de la Chine donée par le R.P. Du Halde Jésuite, sont une réduction modérée et servile (si je puis employer cette épithete) des Cartes qui ont été envoyées de la Chine. On a cru devoir conserver le goût même du dessein original, pour que ce grand et bel ouvrage fut communiqué ou remis au Public avec plus de fidélité." Archives départementales de l'Orne, SHAO 252J224. Images of the document published on 10 January 2012 on the *projet d'Anville*'s website by Lucile Haguet (http://danville.hypotheses.org/1271). Elsewhere, d'Anville mentions having copied the regional maps of Tartary and Tibet "without applying any changes." d'Anville (1776), 17.

I will discuss d'Anville's process in composing his general maps, for which the cartographer drew on a wide range of sources including material from Saint Petersburg. Next, I elaborate on the need, felt in Beijing and in Paris, to fill in the gaps in the Kangxi-era atlas and on its European incorporation, respectively. In the event, not surveying but a careful negotiation of indirect data from a wide variety of sources gathered in Paris, Beijing, and Saint Petersburg supplied this need. Lastly, I will conclude by elaborating on the reception of d'Anville's maps in both Europe and Asia, which polarized into a heated public debate and gave rise to further editing of d'Anville's maps.

3.3.1 *d'Anville's General Maps and the Paris-Saint Petersburg Connection*

Compared with his regional maps, the general maps du Halde had commissioned d'Anville to create *de novo* for the *Description* contain substantial differences because the maps sent from Beijing were not their exclusive basis. Even so, du Halde also provided a large share of the other information that d'Anville consulted in composing the general maps. As scriptor of the Jesuit community in Paris, du Halde had accumulated the writings of twenty-seven missionaries to the Qing, some of which contained useful geographic and astronomical data. For example, d'Anville used several accounts by Jean-François Gerbillon of his journeys north of the Great Wall in service to the emperor, discussed in Chapter 1, in creating his *General Map of Chinese Tartary*.[544] From such texts d'Anville distilled useful data, most notably concerning the routes depicted on the *General Map of China* and the *General Map of Chinese Tartary*. As several surviving manuscript drafts reveal, he first sketched these routes separately on smaller pieces of paper, only later incorporating them into his maps.[545] The regional maps depict none of these routes, confirming d'Anville's extra effort in producing his general maps. In addition, du Halde and his Parisian colleagues corresponded frequently with the French missionaries in Beijing. Gaubil's letters in particular testify to a rich exchange of ideas and information throughout the 1720s and early 1730s, often with regard to the geography of regions beyond the scope of the Kangxi-era Qing surveys.[546] French Jesuit

544 D'Anville himself as well as du Halde refer to d'Anville's reliance on Gerbillon. d'Anville (1776), 18–19; du Halde (1735), 1:xxvij & 4:87–422.

545 BNF, Cartes et plans, Ge D 10589-90. These depict routes from the imperial capital into Khalka lands.

546 Gaubil (1970). Gaubil's Parisian confreres, such as Étienne Souciet (1671–1744) and Nicolas Fréret (1688–1749), were also actively involved in the exchange.

missionaries in Beijing abstracted material from the route books and maps the Qing court allowed them to consult and interviewed envoys visiting the capital from distant regions to pay the emperor homage.[547] When the missionaries felt something was worth transmitting to Europe, they carefully translated the material and forwarded it to Paris. Around 1728, for example, Gaubil sent copies of revised copperplate edition sheets to Paris, along with notes for improvement.[548]

Even though du Halde and the missionaries in Beijing supplied such rich materials, d'Anville still actively sought other sources of information.[549] Du Halde claimed to have disliked the idea of using non-Jesuit material, but at the request of Gaubil and his colleagues in Beijing he permitted it.[550] Most important in this regard was d'Anville's correspondence with Joseph-Nicolas Delisle, head of the School of Astronomy attached to the Academy of Sciences in Saint Petersburg. With Delisle's aid, d'Anville got his hands on a chart of the latest discoveries of Bering, albeit indirectly: even though d'Anville corresponded with Delisle about Bering's discoveries, it was the king of Poland who sent the actual chart to du Halde so as to avoid endangering the position of Delisle, a Frenchman, at the Russian Academy of Sciences.[551] D'Anville first copied the chart onto oilpaper as he had done for the maps of Korea and Tibet.[552] The adapted chart eventually also made it into du Halde's *Description*, which some scholars have understood as the reason Delisle lost his post and returned to France in 1747.[553] D'Anville also used another non-Jesuit source, a Dutch manuscript chart of the coastlines of northern Japan depicting the journey of the ship Castricom through the waters surrounding Japan's northernmost islands around 1643. The Bering map and the Dutch maritime chart together inspired d'Anville to redraw the islands off the coast of Northeast Asia and to include

547 Gaubil (1970), 182–183; Cordier (1915), 518–527.

548 Gaubil (1970), 205–207. D'Anville appears to have made copies of geographic features from revised copperplate sheets for his own use, to which he added routes across the region, possibly in preparation of *The General Map of Chinese Tartary*. BNF, Cartes et plans, Ge DD 2987 (7308B-19B).

549 The cartographer referred to these in several of his justificatory texts, both published and unpublished. d'Anville (1776), 6–12; d'Anville (1737), 11.

550 d'Anville (1776), 15; du Halde (1735), 1:xlviii.

551 d'Anville (1737), 9. Note that another Delisle brother, Louis Delisle de la Croyère (1687–1741), also served in Russia and was a fellow-explorer with Bering.

552 BNF, Cartes et plans, Ge DD 2987 (7403, 1–4).

553 Isnard (1917), 18–19.

Japan in the *General Map of Chinese Tartary*.[554] After publication, d'Anville's decision to deviate so strongly from both the Qing atlas and, accordingly, from his own regional maps, in depicting this largely unexplored part of Northeast Asia would become a major point of criticism and caused inconsistencies among the maps included in du Halde's monumental work.

Eventually, all these materials, Jesuit and non-Jesuit, allowed d'Anville to complete the *General Map of China* (1730), the *General Map of Chinese Tartary* (1732), and the *General Map of Tibet* (1733). Immediately afterwards, in January 1734, du Halde and d'Anville signed a third and last contract stipulating publication rights and the final date of delivery. The only map that remained unfinished, the *Most General Map That Comprises China, Chinese Tartary and Tibet*, was to incorporate all of continental East Asia, including lands as far west as the Caspian Sea:

> I promise to deliver the General Map, which remains to be done, around the fifteenth of next April. If, by some accident or difficulty due to unexpected work, it should happen that this map is delayed by a fortnight, it will not annul anything that follows here below. I renounce the right or privilege of selling these maps separately for the amount of time the Father sees fit, and not to publish them separately from his printed work. The Reverend Father will not have any other person whosoever publish these, and reserves to me the right to separately publish the maps at a time he considers appropriate.[555]

This map was to be the masterpiece among all the maps du Halde had commissioned (figure 45). Again, d'Anville made full use of the Beijing missionaries' cartographic information, even though some of the territories depicted lay far beyond the hegemony of the Qing court. Around 1730, for example, an

---

554 BNF, Cartes et plans, Ge DD 2987 (7431, 1–2). D'Anville's draft, based on the chart, is BNF, Cartes et plans, Ge D 10592. Drafts of all of d'Anville's general maps are BNF, Cartes et plans, Ge D 10585-8.

555 "Je promets de donner le dessein de la Carte générale, qui reste à faire, autour du quinzième d'Avril prochain. Si par quelque accident ou difficulté de travail imprévue, il arrivoit que ce dessein fut retardé d'une quinzaine de jours, cela n'infirmera rien dans ce qui suit. Je me désisterai du privilège ou droit de la vente particulière des Caries, pour le tems qu'il plaira au Révérend Père de ne point débiter lesdites Cartes hors de son ouvrage imprimé: moyennant que le R. P. n'en fasse point débiter par quelque autre personne que ce soit, et m'en conserve le droit pour le tems qu'il jugera à propos d'ouvrir le débit particulier de ces Cartes." Cordier (1905), 395–396.

epistolary debate between d'Anville and Gaubil (via the Parisian Jesuits) arose regarding the position of Astrakhan, a Russian city on the western shore of the Caspian Sea.[556] In the course of this debate, Gaubil dispatched to Paris a manuscript map of the Caspian coastline remarkably similar in its depiction to the *Most General Map* (figures 45 & 46) and in accordance with the most up-to-date data.[557] The basis for this map was Gaubil's access to both European material, such as a 1723 map of the Caspian Sea by Guillaume Delisle (sent to him from Paris), as well as materials kept at the Qing court, such as the *Record of Sending of an Embassy to Remote Regions* (Chinese: *Yiyulu* 異域錄 / Manchu: *Lakcaha jecen de takūraha babe ejehe bithe*) by Tulišen 圖理琛 (1667–1741) discussed below.[558] Gaubil also maintained correspondence with both Joseph-Nicolas Delisle, with whom he discussed Guillaume Delisle's map of the Caspian Sea, which had reached Beijing in 1725, and Siegfried Bayer (1694–1738), another member of the Academy of Sciences in Saint Petersburg. Still another "Russian" contact of Gaubil was Lorents Lang (1690–1752), a Swedish-born official who travelled to Beijing in Russian service no fewer than six times between 1715 and 1739, and who likely also provided Gaubil with useful information.[559]

Even so, d'Anville was not content to rely exclusively on others in his quest for ever-better maps of the regions between Europe and the Qing. From his study, he turned his talents to scrutinizing the available literature for himself. The National Library of France holds a small sketch of the region of Zaysan, in present-day Kazakhstan, to take just one example.[560] The map clearly drew on information from the *Geographia nubiensis*, a Latin translation of an Arabic work of geography published in Paris more than one hundred years earlier.[561] In the event, d'Anville's maps did not include this bit of information, but du Halde did mention the book's usefulness in determining the longitude

556 Gaubil (1970), 96–97, 258 & 278–279; du Halde (1735), 4:465–468.

557 Comentale (1983). This map is now at BNF, Cartes et plans, Ge D 27249. Some place names are in yellow ink and seem to be later additions. D'Anville could have written these, since the proof copies of the *General Map of Tibet* contain similar additions. BNF, Cartes et plans, Ge D 10852. Around 1725, Gaubil also compiled a history of the Mongols, which he later sent to France. The manuscript map is likely an illustration of his work on Mongol history, all of which went to Paris in 1728.

558 Gaubil (1970), 116–117 & 182–183. Gaubil mentions that he received two maps by Guillaume Delisle, the *Carte des Pays voisins de la Mer Caspienne* (1723), and the *Mappemonde à l'usage du Roy* (1720). BNF, Cartes et plans, GE D-16764 & CPL GE DD-2987 (85 B).

559 Afinogenov (2016), 73.

560 BNF, Cartes et plans, Ge D 10591. The map is kept together with other maps of East Asia.

561 Sionita & Hesronita (1619), 213–214.

FIGURE 45 *D'Anville's* Most General Map *(1734), with clear similarities to Gaubil's manuscript map in the upper left-hand corner.*
BIBLIOTHÈQUE NATIONALE DE FRANCE, CARTES ET PLANS, GE DD 2987 (7144 B).

FIGURE 46 *Map of the lands between the Qing and the Caspian Sea, sent to Paris by Gaubil and used by d'Anville for his* Most General Map.
BIBLIOTHÈQUE NATIONALE DE FRANCE, CARTES ET PLANS, GE D 27249.

of Astrakhan.[562] Thus, d'Anville compared data from Saint Petersburg, Beijing, and the available literature and put these next to Gaubil's map before establishing the position of Astrakhan and other cities of Central Asia for his *Most General Map*. This serves to remind us that while d'Anville was dependent on an extensive network of individuals in Beijing, Paris, and Saint Petersburg, he himself also played a crucial role in actively soliciting and comparing data from a wide range of published and unpublished materials, although apparently only for his general maps.[563] In contrast, as argued above, his regional maps remained for the most part true to the Qing atlas, albeit employing various editions as well as, in the cases of Korea and Tibet, derived manuscript material the Beijing missionaries had provided.

With all maps finally completed, engraved, and delivered by May 1734, du Halde was able to publish his work (at great cost) in 1735.[564] In the years that followed, the *Description* would become the most renowned work on East Asia available to European readers. At the same time, a heated discussion over the Qing maps' content, discussed in this chapter's last part, soon displaced concerns that had been smoldering beneath the surface over and due to the roughly fifteen years' delay between the arrival of the first set of Qing maps in Paris and their European publication in the *Description*. First, I turn to a deeper analysis of the triangular exchange of cartographic material among Beijing, Paris and Saint Petersburg.

3.3.2 *The Saint Petersburg Connection to Beijing*

The connections among Paris, Beijing, and Saint Petersburg in the late 1720s and early 1730s via direct exchanges of cartographic data enabled d'Anville to make a substantial impact on European cartographic knowledge of East and Central Asia, particularly through his general maps and the adapted Bering chart. Yet, Saint Petersburg appears to have occupied a far more central place in the cartographic exchanges between Europe and the Qing than the individual Beijing-Paris and Paris-Saint Petersburg connections might suggest. It

562 du Halde (1735), 4:465–468.

563 The BNF's *Collection d'Anville* includes a revised woodblock edition of the Qing atlas. It has the Russian titles added in red ink, indicating acquisition through a Russian contact (figures 30.1 & 30.2). D'Anville likely obtained the atlas with the help of Delisle, who returned to Paris in 1747. Negotiations on the transfer of maps from Russia to France did in fact take place, after which d'Anville became a member of the Saint Petersburg Academy of Sciences. Dulac (2002), 2651.

564 Landry-Deron (2002), 37 (n.1). The high price of the original *Description* later became a subject of debate. Pinot (1932), 73.

is safe to say that missionaries in Beijing were also engaged in both direct and indirect exchanges of cartographic and geographic information with Saint Petersburg. These exchanges followed two different routes: the often interrupted overland channel of communication maintained by Russian embassies and trading caravans enabled direct correspondence, although it seems this route saw less frequent use until the 1730s, after the Treaty of Kiakhta between the Russian Empire and the Qing in 1727.[565] At the same time, the well-established but treacherous sea routes between Southern China and Western Europe provided another way to forward materials, mostly via the Parisian Jesuits.[566]

The material that best illustrates this connection between Beijing and Saint Petersburg is Gaubil's extensive correspondence. Soon after his arrival in Guangzhou, Gaubil wrote to Paris requesting maps from Saint Petersburg, a request his correspondence later repeats.[567] By the 1730s, Gaubil corresponded extensively with Delisle, Bayer, and Lang, all men in service to the Russian Empire. Gaubil indicated, for example, that he would send geographic information to Delisle via the Parisian Jesuits, a request Delisle voiced not long before setting off for Saint Petersburg. Gaubil even planned to send it in such a way that du Halde would not find out, perhaps because the Qing atlas was then still behind closed doors in du Halde's room.[568] Moreover, among the remainder of Delisle's own collection of cartographic material on Asia, now in the French National Archives, are several maps that could only have reached Delisle, directly or indirectly, through the agency of the Beijing missionaries.[569] These include maps depicting routes between mission stations in the Chinese provinces as well as a reduction of the map drawn by de Fontaney and Lecomte as discussed in Chapter 1. Another map constitutes a rough sketch of the upper reaches of the Ganges River as part of the Tibetan highlands, reduced from the revised copperplate version of the Qing atlas. Still other maps depict different sections of the Great Wall. Delisle perhaps personally carried these maps to Saint Petersburg in 1725, but given the secrecy with which du Halde handled sheets from the Qing atlas before 1728, it is also possible some of the Parisian

565 The Treaty of Kiakhta was a follow-up of the Treaty of Nerchinsk and the second border agreement between the Qing and the Russian Empire. Gaubil (1970), 373; BO, B/2/4, fol. 149. For an analysis of overland correspondence, see Afinogenov (2016).

566 Étienne Souciet, based in Paris, forwarded correspondence between Delisle and Gaubil in the late 1720s.

567 Gaubil (1970), 36.

568 Gaubil (1970), 96, 120, 187 & 205. Likely those kept at BNF, Cartes et Plans, GE BB 124(162,1-30RES).

569 BO, B/5/15; AN, Cartes et plans, 6JJ/74/A.

Jesuits later secretly forwarded Gaubil's copies to Delisle in Saint Petersburg. If they indeed shared parts of the Qing atlas with Delisle in Saint Petersburg while their adaptations remained unpublished in Paris, it must relate to the assistance Delisle provided Gaubil in determining the position of Astrakhan and other cities of Central Asia.[570] All this suggests that the cartographic exchanges among Beijing, Paris, and Saint Petersburg were more complex and multidirectional than their individual linkages suggest.

All information obtained from Delisle, Bayer, and Lang, strictly cartographic or otherwise, proved tremendously useful when the Yongzheng court requested that the Beijing missionaries draw maps in preparation for revising the Kangxi-era atlas. Preparations intensified in the first half of 1727, when the emperor ordered several European missionaries to contribute whatever knowledge they possessed about Central and Russian Northeast Asia. Kangxi's thirteenth son and the Yongzheng emperor's brother, prince Yinxiang 胤祥 (1686–1730), was in charge of the project and regularly invited the missionaries to his quarters, discussing cartography, exchanging new information, and ordering them to draw additional maps based on both European cartographies and sources provided by the Qing court. When discussing the Yongzheng edition in the first part of this chapter, I already briefly discussed the maps he asked the missionaries to produce. The first, commissioned in 1725 and drawn by Fridelli and Régis, covered the lands between Hami and the Caspian Sea, whereas the second, of "Siberia and Russia up to Saint Petersburg," was likely the direct basis for the maps of those areas the Yongzheng edition of the Qing atlas added to its Kangxi-era cartographic predecessor. Gaubil, de Mailla, and André Pereira spent twenty-five days preparing the map, which they later "engraved by order of the emperor."[571] When Gaubil speaks of engraving in this context, he likely refers directly to the Yongzheng edition of the Qing atlas, since the missionaries' map of Central Asia and Russia from Kamchatka to Saint Petersburg encompassed precisely those parts of Eurasia the Yongzheng edition added to the Kangxi-era atlas (figure 36). However, Gaubil and the other missionaries were unaware their map would become part of a larger multi-sheet map. Gaubil's correspondence suggests the Jesuits' place in the production of the Yongzheng edition atlas was of mere informants rather than the close involvement they had enjoyed under the Kangxi emperor.

570 Gaubil (1970), 96–97; Cordier (1915), 523–524.

571 I have not found any references to this map in catalogues. Gaubil (1970), 174–175; Bernard (1935), 466.

Not all the Jesuits involved in this exchange of cartographic knowledge with the Yongzheng court judged it identically. Gaubil writes that de Mailla, who had taken part in the Kangxi-era land surveys, expressed concerns over potential conflicts should visiting Russian embassies ever learn the Jesuits were sharing sensitive cartographic information from Saint Petersburg directly with the Qing court. After consultation among the Beijing Jesuits, they decided not to correct the position of the Russian cities of Tobolsk and Krasnoyarsk on the maps commissioned by prince Yinxiang.[572] Yet, despite such concerns about communicating sensitive information to the Qing court, Gaubil in particular apparently insisted on mutual exchanges with court officials rather than participating in a one-way transfer of data. In one of his letters, Gaubil recounts a specific instance of this mutual exchange:

> After the Emperor had given us the order to make a map from the north of Hami to Saint Petersburg towards the west and the northern seas towards the north, the Regulo one day ordered me to visit him and have a conversation with one of his main eunuchs. The latter told me that the prince wanted to be informed of the accounts I had [collected] for this map. I answered that I had rather exact and new accounts on the positions of Saint Petersburg, Moscow, and Archangelsk, as well as of the lands between those three cities; that I could not yet have ascertained the distance from Astrakhan to Moscow, or Tobolsk, or Hami; that I only had imperfect accounts on the kingdoms of Kazan and Siberia; and even that several of these accounts did not match. I added I was only very little informed on the lands of Tsewang Rapdan. The eunuch went to answer [to the prince] and the prince had me come to his presence... [He] ordered to have a route book in Manchu brought over and presented it to me, ordering me to read it in front of him... I got permission to take with me a good map of Tibet and of the sources of the Ganges. I found it to rather conform to the one Father Régis sent [to Paris] last year and I contented myself to draw from it that which was not on the map of Father Régis.[573]

572 Gaubil (1970), 183.

573 "Apres que l'empr nous eut donné ordre de faire une carte depuis le Nord de Hami jusqu'a Petersbourg vers l'ouest, et la mer du Nord vers le septentrion, le Regulo m'ordonna un jour de venir chez lui et de converser avec un de ses principaux ennuques. Celui-ci me dit que le prince vouloit etre au fait sur les memoires que j'avois sur cette carte. Je repondis que j'avois des memoires assez exacts et nouveaux sur les positions de Petersbourg, de Moscou et d'Archangel, aussi bien que des pays entre ces trois villes; que je n'avois encore pu etre bien seur de la distance d'Astrakan ou a Moscou, ou a Tobol, ou a Hami, que je

Gaubil's narrative gives us a good idea of how this flow of cartographic material at the Qing court manifested, although such exchanges did not continue for long. In 1731, Gaubil noted that, since the death of Yinxiang in 1730, he had been unable to engage in further cartographic exchanges with the court: "Today, I wouldn't know how to get copies of the maps [prince Yinxiang] had made, be it [maps drawn] by ourselves or by others. All of that is with the emperor and will perhaps never [again] be available."[574]

The missionaries thus carefully exploited the circulation of cartographic and geographic information, often via Paris, between Beijing and Saint Petersburg, which allowed them to assist the Yongzheng court in its atlas revision project while in turn gleaning cartographic data from its highest officials. This mutual exchange was possible only because the Yongzheng court needed access to the missionaries' cartographic knowledge of Central Asia and the vast Russian Empire for reasons of frontier management, and, perhaps, territorial ambitions. At the same time, Paris also gained quite directly from the circulation of these materials, since information the Yongzheng court gave to the Jesuits for their cooperation made its way back to Europe, providing d'Anville crucial aid in composing his general maps as discussed above. Lastly, in Saint Petersburg, Delisle worked to obtain precise information on Russia's southern and eastern borderlands for his Russian employers, on the basis of which he later composed an atlas of Russia, the *Atlas Russicus* (1745). He therefore also profited greatly from cartographic exchanges with the Beijing missionaries. Since neither the Qing court nor Saint Petersburg firmly controlled the Central Asian lands between their territories, detailed land surveys were inconceivable,

---

n'avois que des memoires imparfaits sur le royaume de Casan et la Siberie, et que meme plusieurs de ces memoires ne s'accordoient pas. J'ajoutay que je n'etois que tres peu au fait sur les pays du Tse vang raptan. L'eunuque alla rendre reponse, et le prince me fit venir en sa presence… [il] ordonna de faire venir un itineraire mantcheoux, il me le presenta en m'ordonnant de le lire devant lui. Je le fis. Il demanda si j'entendois ce que je lisois. Je lui dis que j'entendois un peu. C'est, dis-je, un itineraire pour les pays du Tse vang raptan… J'eus permission d'emporter une bonne carte du Tibet et des sources du Gange, je la trouvai assez conforme a celle que le P. Regis envoya l'an passe, et je me contentai d'en tirer ce qui n'etoit pas dans la carte du P. Regis." Gaubil (1970), 182–183. "Tse vang raptan" refers to Tsewang Rabdan (1643–1727), Galdan's nephew and successor, and Zunghar leader in today's northern Xinjiang. "Regulo" and "prince" here refer to prince Yinxiang, who led the Qing atlas revision project. The new information Gaubil obtained from the Manchu route book appears in part in Souciet (1729–32), 1:176–180.

574 "je ne saurois aujourdhuy avoir copie des Cartes qui'il a fait faire, soit par nous, soit par d'autres, tout cela et chez l'empr et n'en sortira peut etre jamais." Gaubil (1970), 289.

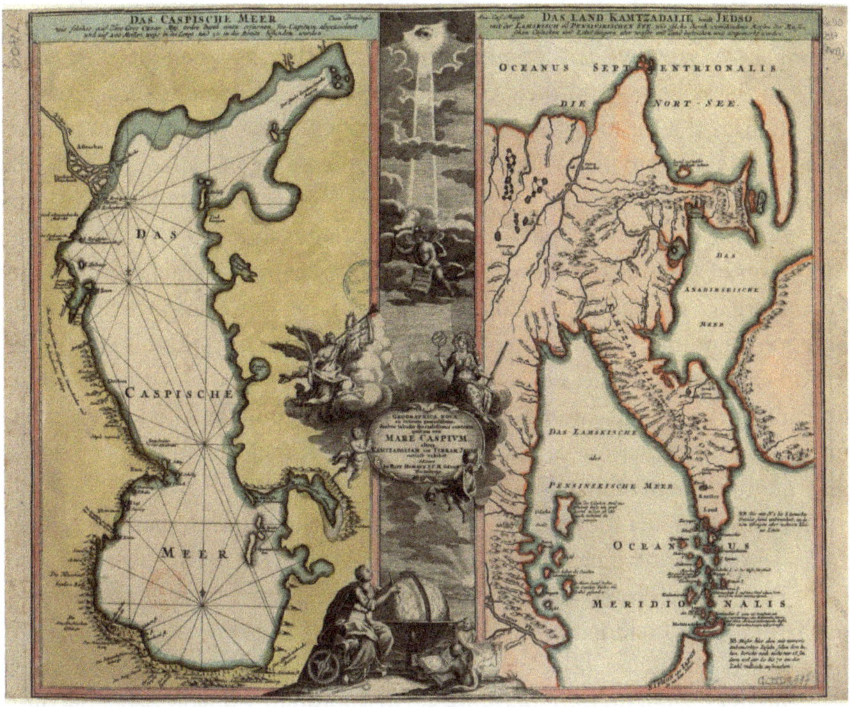

FIGURE 47 *Homann's map, featured in a discussion between Gaubil and the emperor's brother, entitled* Geographica nova ex oriente gratiosissima, duabus tabulis specialissimis contenta quarum una Mare Caspium altera Kamtzadaliam seu Terram Jedso.
TAKEN FROM HOMANN'S *ATLAS MAIOR CUM GENERALES* (1725). BIBLIOTHÈQUE NATIONALE DE FRANCE, CARTES ET PLANS, CPL GE DD-2987 (7409).

leaving mapmakers to distill useful information from travel accounts, route books, envoys' diaries, and other secondary materials. The circulation of such raw data among Beijing, Paris, and Saint Petersburg allowed all parties involved to produce better maps of these regions. Only a few, most notably Gaubil, du Halde, and Delisle, stood at the center of this Beijing-Paris-Saint Petersburg network of cartographic exchanges and strove to maintain the uninterrupted circulation of cartographic information.

At the same time, this rather informal network of personal connections was not the only framework for cartographic exchange among the three centers. Cartography also featured in direct diplomatic exchanges between Russian embassies and the Qing court without missionary mediation. For example, a Russian official offered the Yongzheng emperor an atlas produced by Johann

Baptist Homann (1664–1724), printed in Nuremberg,[575] suggesting the missionaries' fears of sensitive cartographic material traveling between the Qing court and the Russians beyond their control were far from unfounded. Homann's depiction of the Caspian Sea and the Isle of Jesso, two of the most contentious points in cartography at the time, later came up in a meeting between Yinxiang's collaborators and Gaubil (figure 47).[576] This makes it likely the Jesuits later took Homann's depictions of Russian lands into acount in compiling their map of Central Asian and Russian territories from Kamchatka to Saint Petersburg for the Yongzheng court. That could not, however, have been their only source of information, since Gaubil's correspondence with his colleagues in Paris and with Delisle, Bayer, and Lang undoubtedly provided more up-to-date information on Central Asia. Moreover, Gaubil built and had access to an extensive library on Central Asia from which he carefully distilled geographic information, much as d'Anville did in Paris.[577]

A map not in fact immediately related to the Qing atlas perhaps best illustrates the extent of such direct and indirect exchanges between Beijing and Saint Petersburg. Preserved at the French National Archives is a Latin version of the map also included in the *Book Recording the Sending of an Embassy to Remote Regions* (figure 48).[578] This book, published bilingually in Chinese and Manchu in 1723, relates the Qing embassy to the Torgut Mongols (a people then living just north of the Caspian Sea), a long journey through Russian-controlled

575 "Le Regulo nous a fait examiner en sa presence un atlas offert par les Russiens. C'est celuy de Jean Hotman [sic], imprimer depuis peu a Nuremberg. Le Regulo fut fort attentif a la figure nouvelle donnee a la mer Caspienne et au pays de Jesso apelle Campsatalia par les Russiens." Gaubil (1970), 174; Bernard (1935), 466.

576 Gaubil (1970), 174. "Jesso," "Jeso," "Ieso," "Jedso," or "Jeço" were different names for one or more islands the identification of which was the key subject of debate. Different eighteenth-century sources identify these islands as Hokkaido, Sakhalin, or Kamchatka. After the publication of d'Anville's *General Map of Chinese Tartary*, this debate would grow ever more intense, because d'Anville includes "Jedso" as Hokkaido and Ieso or Jesso as an extension of Kamchatka. E.g. Brucker (1881), 517. For d'Anville's contribution to this lively debate, see d'Anville (1737).

577 Golvers (2008); Gaubil (1970), 187.

578 AN, Cartes et plans, 6/JJ/74(13D). The map, entitled *Tabula Geographica Tartaria a Mari Orientali Usque ad Mare Caspium*, is carefully executed, apparently a perfect copy of the map included in Tulišen's journal. Rivers, mountains, and the number of place names all correspond perfectly. Most place names are not from the Manchu edition, however, but transliterated from Chinese toponyms, with the exception of well-known places, rivers, or lakes, such as the Caspian Sea or Lake Baikal, indicated in Latin. Short explanatory sentences on the map, but not included on the Manchu edition, appear rendered into Latin. This map was contemporary to the land surveys but not drawn on the basis of field measurements.

Siberia, led by Tulišen, between 1712 and 1715.[579] Combining personal anecdotes with a detailed description of peoples and geographic features between the Qing and the Caspian Sea, the journal is a rare account of an embassy to a region far beyond the Qing's sphere of influence. Once Gaubil got his hands on a Chinese copy of the text, he translated parts of the journal selectively, reflecting his intent to distill from it all geographically useful information, and sent it to Paris, where his translation appeared excerpted in a 1729 publication.[580] The Latin map preserved in Paris, however, circulated in a very different manner. First, one of the missionaries at court sent it to Rome. A short note attached to the map, written by Delisle, substantiates this:

> Map of the journey of a Mandarin named Tulišen, sent by the emperor of China to the land of the Turgut in 1718 [sic]. This map was composed by [Tulišen] himself and copied and translated by the Jesuit missionaries. They sent it to Rome, from where it was sent to General Keith in Saint Petersburg, who gave it to me.[581]

Composed on the road, published and translated into Chinese and Latin in Beijing, sent to Rome, copied, and forwarded to Saint Petersburg, this Manchu map experienced a journey that testifies to the wide, multidirectional and complex circulation of cartographic material from Beijing to Saint Petersburg.

3.3.3 *d'Anville's Maps: Reception and Further Adaptations*

After the publication of d'Anville's maps in du Halde's *Description* in 1735, his confreres in Beijing waited impatiently for the four large volumes' arrival. In their eyes, the books would help their mission secure more support by showing the European public how the French missionaries had improved European knowledge of continental East Asia. When the four volumes finally reached Beijing in the first half of 1737, the missionaries scrutinized them exactingly

579 The embassy planned to continue to Saint Petersburg at the Russians' invitation, which did not materialize. Tulišen's original account and map reached the emperor on April 30th, 1715. Fu (1966), 1:116–117 & 120.

580 Souciet (1729–32), 1:150. I am indebted to Wu Huiyi for these insights.

581 "Carte du route du mandarin tartare nommé Toulichen envoié en 1718 [sic] par l'empereur de la chine aux Pays de Tourgouts. Cette carte a été dressé par lui-même et copiée et traduite par les missionnaires Jesuites qui l'ont envoié en Rome, d'où elle a été renvoié à Petersbourg à m le General Keith qui m'en a fait present." AN, Cartes et plans, 6/JJ/74 (13D). 'General Keith' refers to James Francis Edward Keith (1696–1758), a Scottish soldier who was recruited into the Russian military by Peter the Great and eventually obtained the rank of General of Infantry.

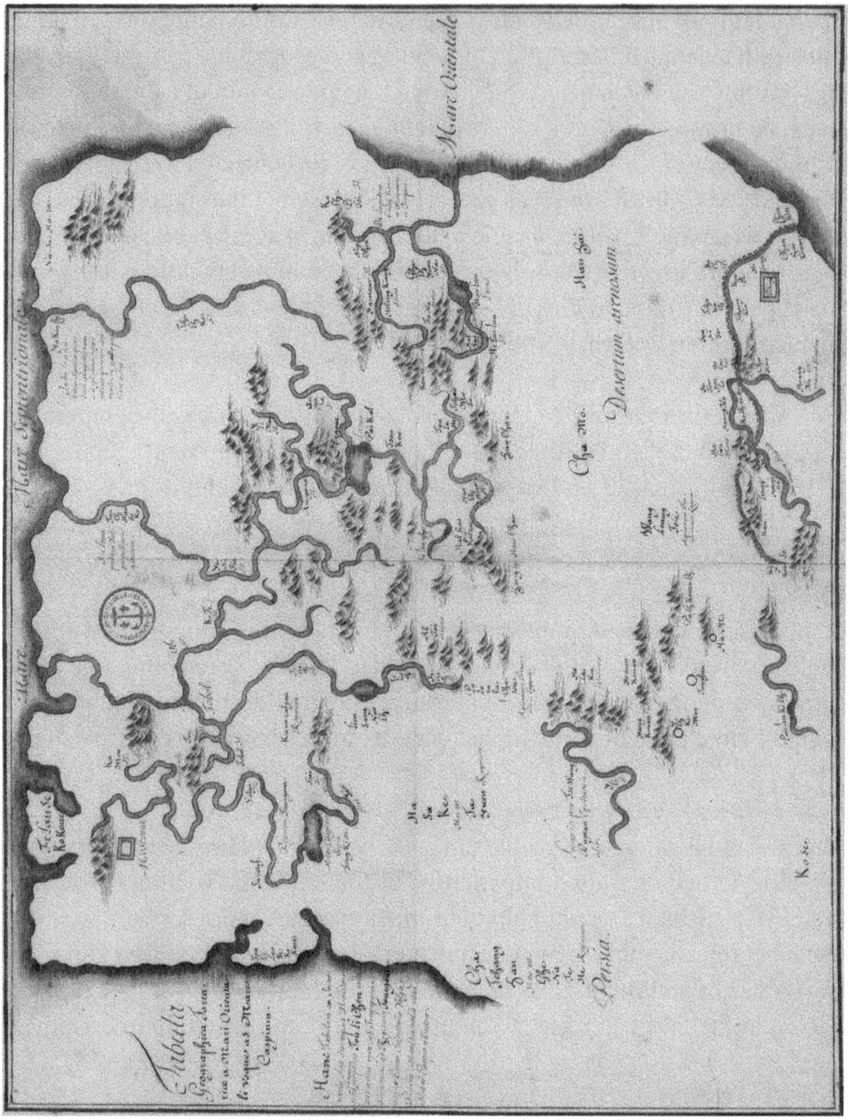

FIGURE 48 *Latin version of the map included in Tulišen's* Book Recording the Sending of an Embassy to Remote Regions.
ARCHIVES NATIONALES, CARTES ET PLANS, 6/JJ/74 (A13 D1).

because word had reached them of criticisms voiced in Europe about the lack of justification for some of the data included on d'Anville's maps.[582] Some, for example, questioned how many *li*, and thus *chi*, the Qing's most basic unit of

582 Gaubil (1970), 458 & 470–471.

length adjusted precisely for the purpose of consistent mapmaking, make up one degree of latitude.[583] Another complaint, even within the ranks of the Parisian Jesuits, was the high price of du Halde's four-volume *Description*.[584] More important, however, was the tumult over the depiction of coastlines on d'Anville's *General Map of Chinese Tartary*. On this map, d'Anville had decided to deviate from the depictions on the Qing maps and on his own regional maps, preferring data that had reached him from Saint Petersburg, which he judged superior.[585] In Beijing, de Mailla took the lead in criticizing the work of both du Halde and d'Anville. His primary complaints concerned the spelling of place names and information included on d'Aville's general maps.[586] The heated discussions ended only after d'Anville published a number of justifications, refuting much of the criticism.[587]

These criticisms seem scarcely, however, to have undermined the work's general popularity: 1736, hardly a year after the *Description* first appeared in Paris, saw the printing of a pirated edition in The Hague. The Dutch version lacked any of the maps because the publisher had no access to the copperplates, which remained in Paris.[588] Instead, a separate atlas with the title *Nouvel atlas de la Chine, de la Tartarie chinoise, et du Thibet*, containing maps copied directly from du Halde's book, slightly edited, then re-engraved, appeared in 1737.[589] It included all of d'Anville regional and general maps, but not the thirty-eight city maps inserted in du Halde's *Description*. As a supplement to this pirated atlas, the editors also added a description of the region surrounding Bukhara, in present-day Uzbekistan, authored by Philipp Johann von Strahlenberg (1676–1747), a Swedish officer whom the Russians had captured and sent to the city of Tobolsk, where he collected information on much of Central and Inner Asia between 1711 and 1722.[590] D'Anville had used von

583 Landry-Deron (2002), 147.

584 In 1735, Fréret wrote the following in a letter to Gaubil: "J'ai eté trompé dans mon attente. Vous verrez cette collection et vous jugerez si le public a tort d'etre mécontent de ce qu'on [sic] faire payer 200f d'un ouvrage en blanc dont on a cherché agrossir le volume pour en augmenter le prix et dans lequel il ne trouve que ce qu'il avoit déjà vu." BO, B/1/10.

585 See above. For d'Anville's response to criticism of the coastlines of Tartary, see d'Anville (1737).

586 d'Anville (1776), 18–30.

587 d'Anville (1737); d'Anville (1738); d'Anville (1761); d'Anville (1776).

588 du Halde (1736).

589 *Nouvel atlas de la Chine, de la Tartarie chinoise et du Thibet* (1737).

590 An implicit reference to Strahlenberg's work is in du Halde (1735), 4:465; Knüppel (2008). His account, including cartographic material, was later published in Stockholm. von Strahlenberg (1730). Peter Perdue has elaborated on von Strahlenberg's contribution to cartography. Perdue (1998), 281–282.

Strahlenberg's description in composing his *Most General Map*, but since du Halde had included little information about this region in the *Description* itself, the Dutch editors added it to the atlas as if to justify the separate pirated edition.[591] They also advanced further justifications for publishing a pirated edition of d'Anville's maps:

> After having decided to bring together, and as a fifth volume, all the geographic maps that belong to the *Description de l'Empire de la Chine & de la Tartarie chinoise* published by Father du Halde, we think it necessary to explain the reasons behind this arrangement. The main reason is the format of our edition [making it easier to consult the large maps]... Moreover, we intended to please the public, because by separating the maps from the *Description*, those who wish to skip the latter can separately consult the former, in whole or in part.[592]

The same foreword rather straightforwardly admits "significant differences between our maps and those of the Paris edition."[593] Indeed, although at first sight the pirated atlas' maps appear to be exact copies of the originals, closer examination reveals several important differences. The prints and paper are of slightly inferior quality and the mountain pictograms differ. More importantly, routes and other information found only on some of d'Anville's general maps appear on the pirated regional and provincial maps. The pirated maps of both Tartary and Tibet also have separate cartouches, which the originals lack. The Dutch edition of his maps appalled d'Anville, not only because of the simple fact of piracy and publication separate from the *Description* but due also to differences in quality (figure 49):

591 Du Halde states that this information was the work of d'Anville. du Halde (1735), 4:465–468.

592 "Avant jugé à propos d'assembler en corps & comme en un Cinquieme Volume toutes les Cartes Géographiques qui apartiennent à la *Description de l'Empire de la Chine & de la Tartarie chinoise*, publiée par le Père du Halde, nous avons cru devoir expliquer les raisons de cet arrangement. La principale est le format de notre Edition. D'ailleurs on a eu dessein de faire Plaisir au Public; parce qu'en séparant ainsi les Cartes de la Description, ceux qui voudront se passer de la derniere, pourront avoir séparement les prémieres, en tout ou en partie." Taken from the opening paragraph of the preface of the Dutch pirated atlas. *Nouvel atlas de la Chine, de la Tartarie chinoise et du Thibet* (1737), 3.

593 "la difference notable qu'il y a entre nos Cartes & celles de l'Edition de Paris." *Nouvel atlas de la Chine, de la Tartarie chinoise et du Thibet* (1737), 3.

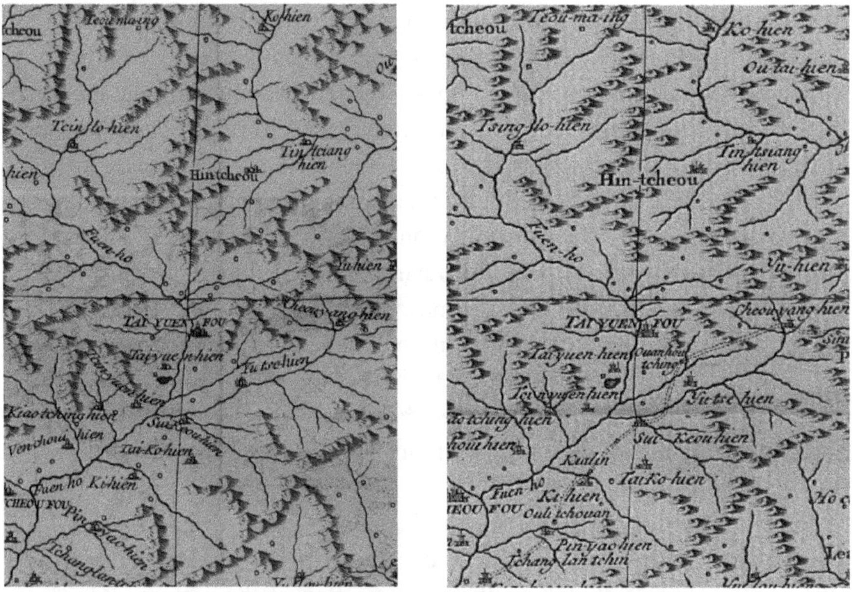

FIGURES 49.1 & 49.2 *Detail from d'Anville's Shanxi map (left) as taken from du Halde (1735) and detail from the Shanxi map in the pirated* Nouvel atlas *(1737) (right). Mountain pictograms differ and an additional route crosses the region.*
MAURITS SABBEBIBLIOTHEEK (UNIVERSITY OF LEUVEN).

> Almost all of these maps are not only terribly inferior to the original ones in terms of execution, but also very badly engraved in all respects... The mountains on the Dutch map are in very bad taste: they are little hills, detached and randomly scattered, without being joined together in order to express the natural; the typeface is puny and of equally bad taste... We even find some rivers that are not depicted on the original map as they are on the Dutch copy, and if we would take pains to scrutinize it and submit all copies to a careful examination, there is no doubt that we would find many more mistakes of this sort.[594]

---

594 "presque toutes ces Cartes non-seulement fort au-dessous des Cartes originales pour l'execution, mais fort mal gravées à tous égards... les montagnes sont d'un fort mauvais goût dans la Carte Hollandoise: ce sont des petites buttes toutes détachées & semées au hazard, sans aucun enchainement qui exprime le naturel; l'Ecriture est maigre, & d'un goût également mauvais... quelques Rivieres qui ne se comuniquent pas dans la Carte originale comme dans la copie Hollandoise, & si l'on se donnoit la peine d'y regarder de bien près, & de soumettre toutes ces copies à un sévère examen, il n'est pas douteux qu'on y trouveroit bien des fautes de cette espece." d'Anville (1738), 770–771.

Notwithstanding d'Anville's personal objections to the *Nouvel atlas de la Chine*, the pirated editions of du Halde's *Description* (1736) and d'Anville's maps (1737) were more popular among contemporary readers than the original Paris editions, chiefly because of their more reasonable price. Another factor contributing to the Dutch atlas' popularity was its larger format, which made its maps easier to consult than the foldout originals in the *Description*. Eventually, the success of the Dutch atlas led to its republication in Amsterdam almost fifty years later with the same pirated engravings,[595] and to the reprinting in Paris, too, of d'Anville's maps, for the first time in a separate atlas published in France in 1785, at the same time the final volume of de Mailla's *Histoire générale de la Chine* also appeared on the market.[596] This edition, printed from the original plates of d'Anville's maps, also included the thirty-eight city maps and fourteen illustrations of cultural and ethnographical interest from the *Description* that the Dutch atlas had omitted.

As for du Halde's texts, besides the French-language pirated edition from The Hague mentioned above, many translations appeared throughout Europe, only one of which included a substantial portion of the cartographic material. The English translation, gradually published by Edward Cave in London between 1738 and 1741, included maps John Green, an alias for Braddock Mead (c. 1688–1757), had reworked and improved somewhat.[597] The wide circulation and multiplication of editions of both du Halde's monumental work as a whole and of d'Anville's maps in particular ensured that, for the European public, the French cartographer's maps remained the authoritative work on East Asia until well into the nineteenth century. A quick glance at late-eighteenth and early-nineteenth-century maps of Asia makes clear that many cartographers borrowed directly from d'Anville's depictions of continental East Asia. Only the expansion of first-hand knowledge about the region through increased European imperialism and evolutions in cartographic technology eventually overtook the impact of d'Anville's maps on European cartographies of East Asia.

In this section, I have argued that both d'Anville's general maps and the Yongzheng-era atlas were products of a remarkable exchange of cartographic material across the Eurasian continent through an extensive network of people in Beijing, Paris, and Saint Petersburg. The French Jesuits stood at this

595 *Nouvel atlas de la Chine, de la Tartarie chinoise et du Thibet* (1785).

596 Grosier (1785). Another edition of this atlas defies precise dating: *Atlas général de la Chine, de la Tartarie chinoise, et du Thibet. Pour servir aux différentes Descriptions et Histoires de cet Empire* (1790?).

597 For more on Braddock Mead's editing of d'Anville's maps, see Foss (1985), 361–376.

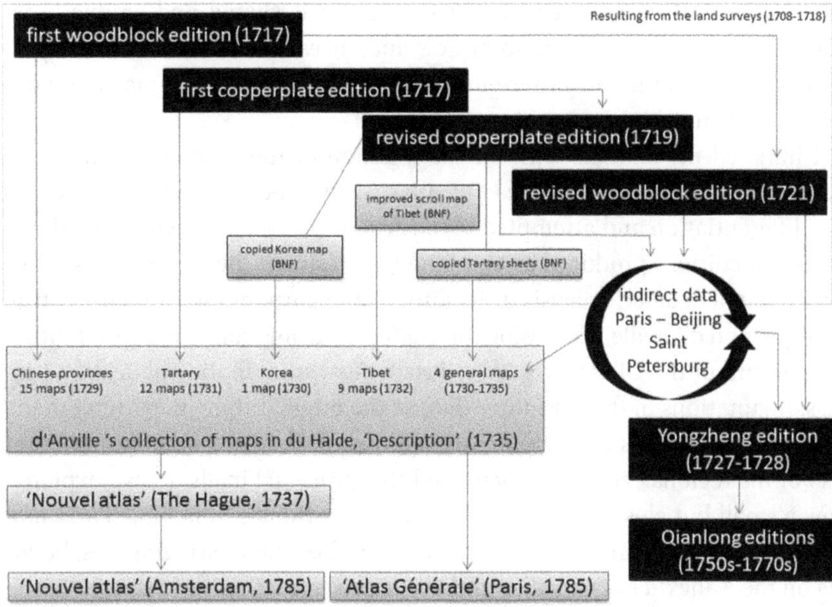

FIGURE 50 *A genealogy of most cartographic material discussed in this chapter.*

network's center, transmitting invaluable information from one continent to another and back, creating conditions that allowed cartographers on both ends of Eurasia to fill in the cartographic gaps left by the Kangxi-era surveys. These gaps, expanses of land between the European powers and the Qing, had aroused mapmakers' interest and curiosity in part precisely because of their absence from Kangxi-era atlases and therefore also from their European adaptations, at a time when both Russia and the Qing were working to increase their influence over Central Asian lands. Careful comparison and negotiation of indirect data among Beijing, Paris, and Saint Petersburg, however, rather than the mounting of new surveys, allowed cartographers in Paris and Beijing to fill these gaps, compiling, respectively, the Yongzheng revision of the Qing atlas (1727–1728) and the general maps included in du Halde's *Description* (1730–1735) (figure 50).

Conclusion

This chapter's main focus has been the circulation of the Qing atlas produced as a direct outcome of the land surveys, including the process of its many adaptations and its wide impact on cartographic depictions of East Asia. It has reconstructed the processes by which individuals shared and gradually reworked maps, adapting them to fit new audiences' visual and cultural expectations, leading ultimately to eventful reincarnations for these maps on both ends of the Eurasian continent.

I initiated my analysis with an examination of the different printed editions and manifestations of the Qing atlas. Most important in this regard was the differentiation and attempted reconstruction of the different printed versions and editions produced as an immediate result of the land surveys. I also engaged in a detailed analysis of its European incorporation by Jean-Baptiste Bourguignon d'Anville, discussing the gradual transmission to Europe of different editions of the Qing maps, as well as the first steps in the publication process. Adaptations in format rather than in the original Qing maps' underlying data, as in d'Anville's use of different editions of the Qing atlas for different sets of his regional maps, characterized this process. Finally, I have explored how a small but closely connected group of individuals in Beijing, Paris, and Saint Petersburg sought to widen the scope of these new cartographies, building on the Kangxi-era Qing atlas and in the process gathering and negotiating a wide range of indirect data. This led to an unprecedented circulation of cartographies in the 1720s and 1730s, propelled by the need, felt, albeit for differing reasons, in all three capitals, to fill in the blanks on the map of continental East Asia.

Thus, from the first version of the Qing atlas to its later editions under subsequent emperors and its intercultural appropriation and proliferation in Europe, this chapter has reconstructed the afterlife of a family of maps. It has traced the many connections that guided their circulation around the globe and back, igniting chain reactions in the production of cartographic knowledge on East and Central Asia. These efforts ultimately resulted in an extensive genealogy of maps, within which many unclear connections and questions of provenance remain. These genealogies stand for the perhaps-obvious fact that mapmaking does not occur only across a blank piece of paper. New maps inevitably feed off older maps, embedded, as it were, in long chains of individual maps each representing particular states of cartographic knowledge, limited by their own unique production contexts. As such, part of their substance is inherited, produced in dialogue with preceding maps that provide the foundation for the new, as the new will in turn for the next generation of maps. Naturally, links between new and preceding maps can be stronger or weaker. For example, the links between the maps discussed in this chapter are relatively strong, even, as I have argued, between d'Anville's maps and their Qing atlas originals. On the other hand, links with maps beyond the scope of this study may be much weaker, such as the Qing atlas' borrowing of data, stylistic elements, and methods of material production from previous maps and works of geography produced by Ming or Qing scholars, or d'Anville's adherence to a style of mapmaking influenced to no small degree by contemporary cartographic practices in Europe. However, whether the links between individual maps or atlases are weak or strong, the deep continuity binding them is both obvious and intuitive, if at times squarely ignored by historians of cartography.

Such continuity illustrates that we might best think of the production of new cartographies as a process rather than a moment in time: it takes place in the gaps between the different stops represented by the material maps and atlases. Reconstructing the story of this particular family of maps not only leads to a better understanding of the processes informing map production but also speaks to the particular generalities of knowledge production, that it occurs gradually, non-linearly, and likely subject to contingencies and complexities of numerous kinds. In other words, this chapter argues that the production of new cartographies is indistinguishable from the circulation of cartographic data, and therefore, at least in early modern times, that the production of cartographic knowledge was highly subject to, although not limited by, the circulation of material maps across political and cultural boundaries. This chapter is then both a reminder and a prime example of how "growth of scientific knowledge is intimately bound up with geographic movement."[598]

598 Livingstone (2003), 177.

Annex: Extant Kangxi-era Sheets (Printed)

BL	British Library (London)*
BNF	Bibliothèque nationale de France (Paris)*
BUB	Biblioteca Universitaria di Bologna
DL	Dalian Library
IOS	Institute of Oriental Studies (Saint Petersburg)
IUO	Insituto Universitario Orientale (Naples)
KBB	Koninklijke Bibliotheek Brussel (Brussels)*
NLC	National Library of China (Beijing)
ONB	Oesterreichische Nationalbibliothek (Vienna)
PUL	Peking University Library (Beijing)**
SGI	Società geografica Italiana (Rome)*

* Personally consulted

** At the time of writing, I had not been able to establish precisely the Kangxi-era sheets kept at the First Historical Archives in Beijing.

Beijing (*Yutu yaolu*, 1997 & *Beijing daxue tushuguan cang guji shanben shumu*, 1999)
Woodblock print (13 sheets): NLC
Copperplate print (9 sheets): NLC
Copperplate print (28 sheets): PUL, NC3070/2787
Bologna (Albanese, 1999)
Copperplate print (41 sheets): BUB
Brussels (Dudink, 2006)
Woodblock print (revised edition—30 sheets): KBB, LP VB 11.283 E (3).
Copperplate print (first edition—14 sheets): KBB, LP VB 11.283 E (2).
Dalian (Dudink, 2006)
Copperplate print (two upper rows): DL
Copperplate print (third row): DL
London (Li, 1996; Simon, 1977)
Copperplate print (mixed—sheets pasted on 3 scrolls): BL, Map, K. 116. 15, 15a, 15b
Woodblock print (full revised edition): BL, Maps C.11.D.15
Naples (Li, 1996)
Copperplate print (unknown—sheets pasted as one wall map): IUO
Paris
Woodblock print (revised edition—13 sheets): BNF, Cartes et plans, Ge CC 4461 (RES).
Copperplate print (unknown—12 sheets): BNF, Cartes et plans, Ge DD 2987 (7296B–7307B).

Rome (Li, 1996)
Copperplate print (mixed—5 scrolls + 1 sheet): SGI, Cina 257
Copperplate print (mixed—1 scroll + 5 sheets): SGI, Cina 170
Saint Petersburg (Pang, 2001)
Copperplate print (unknown—18 sheets): IOS, C68 mss.
Copperplate print (unknown—2 sheets): IOS, C69 mss.
Copperplate print (unknown—7 sheets): IOS, C70 mss.
Vienna (Li, 1996)
Copperplate print (unknown—40 sheets): ONB, KI 116 120

CONCLUSION

Unlocking Dichotomies: Revisiting Cross-Cultural Circulation

The trajectories of surveying instruments and maps, distilled from close readings of a wide variety of sources, have guided this study through the chains of events surrounding the build-up, practical execution, and aftermath of a large-scale mapping project. As a premise, I opted to start not from the many discussions that have surrounded the mapping project and its resulting atlas, nor to subscribe wholeheartedly to any existing conceptual framework for studying intercultural contact,[599] attempting to give voice to the many individuals involved in the circulation of instruments and maps. This has led to a rather descriptive study and to a perspective strictly bound by the circulation of material objects, even though I have attempted to explore briefly the missionary aspect through two intermissions. Finally, in this conclusion, I hope to take part in some of the broader discussions in which the Qing mapping project has featured. In general, I believe this study's contribution to the existing literature is summarizable as a challenge to four conceptual dichotomies that pervade the literature: (1) traditional versus scientific practice; (2) global versus local networks; (3) the circulation versus the production of knowledge; and (4) China versus Europe. To be sure, recent scholarship has already called some of these dichotomies into question, which partly inspired me to confront all four at once.[600] As a direct result of a focus on the circulation of instruments and maps, I argue that a reevaluation of all four dichotomies provides a unique perspective on the circulation of the sciences between Europe and continental East Asia in the seventeenth and eighteenth centuries.

599 As discussed in Standaert (2002).

600 Jami's work calls for a reevaluation of the dichotomy between the production and the circulation of knowledge. Jami (2012); Hostetler has argued for a reevaluation of the dichotomy between "traditional" and "scientific" practices. Hostetler (2009); and both Jami and Hostetler also show the difficulties involved in the "China" versus "Europe" dichotomy. See also Hart (2013).

© KONINKLIJKE BRILL NV, LEIDEN, 2017 | DOI 10.1163/9789004345362_008

On Qing Imperial Cartography: Traditional vs. Scientific Practice

With its focus on scientific practice through manipulation of instruments and the role of the frontier in the development of new Qing cartographic practice, Chapter 1 has called into question the idea that a new cartographic practice simply and fully displaced another preexisting cartographic practice. Many scholars who have written on cartography as practiced in continental East Asia, including Walter Fuchs, Joseph Needham, Theodore Foss, Cordell Yee, and Richard Smith, tend to accept the idea that the Qing mapping project constituted an essentially Jesuit or European project, embraced by the Kangxi emperor but only temporarily practiced alongside a more traditional "Chinese" cartography.[601] Over the last fifteen years, scholars have tried to reframe this idea to allow more credit to the Qing side, for example Laura Hostetler arguing that the mapping project and resulting (copperplate) atlas were part of a Eurasian shift toward early modernity, or Benjamin Elman interpreting an inward turn of cartography during the early Qing as a reaction to cartographic representations introduced by the Jesuits, as well as to increased contacts with the wider world. Regardless of these efforts, however, the assumption that a new type of cartographic practice, be it "Jesuit"/"European," "scientific," "specialist," or "early modern," replaced, or existed separately from, a preexisting cartographic practice, whether we called it "Chinese," "traditional," or "pre-modern," remains firmly in place.[602]

In contrast, I challenge the assumption that the Kangxi, Yongzheng, and Qianlong cartographic projects were intermezzos executed by Jesuit missionaries who simply imported European techniques of land surveying, in, using the words of Cordell Yee, a straightforward "Manchu adoption of European techniques." In Chapter 1, I argued that the late seventeenth century saw the gradual exploration, adoption, and, finally, full appropriation of specific techniques of land surveying as practiced by the Parisian *Académie royale des sciences* into a preexisting Qing cartographic practice in an effort to improve existing cartographies.[603] The Qing court selected only certain elements of newly

601 Smith (2013); Yee (1994), esp. 177–185; Foss (1994); Elman (1981); Needham (1979); Fuchs (1943).

602 Han (2014); Bai & Li (2014); Whiteman (2013); Hostetler (2009); Elman (2007), esp. 38, 42 & 45; Hostetler (2007); Perdue (2005), 449–453; Hostetler (2001).

603 Recent scholarship confirms, in the case of the Qing, rather effective appropriation into existing frameworks of practices introduced by individuals and goods arriving from afar. Jami (2012); Puente-Ballesteros (2011).

introduced techniques of land surveying to incorporate within its existing practices, mostly the instrumental aspects of practices developed in Paris, while rejecting other essential features, such as the accurate measurement of base lines for triangulation or the observation of Jupiter's satellites, and retaining and refining important components of inherited Qing practice, not least the measurement of road distances with ropes. The entire process took shape over decades and led eventually to the establishment of a new imperial cartographic practice. Moreover, Catherine Jami's observation that the Kangxi emperor reverted to ancient precedents in support of new techniques was certainly true for cartographic practice as well, as evidenced by his attention to the authority of the *The Tribute of Yu* and the Zhou-era units of length contained therein.[604]

This integration of surveying techniques entails a remarkable spatial factor. Whereas many of the cross-cultural scientific exchanges in seventeenth- and eighteenth-century Eurasia took place within well-defined spatial boundaries, in this story the nature of land surveying and the context for the exploration of two divergent practices meant that much of the encounter and integration materialized *in via* throughout the northern frontier. Naturally, this spatial dynamism thoroughly impacted the encounter: the north-to-south routes taken by the surveyors through large swaths of scarcely populated lands in 1698, for example, allowed the early detection of an incompatibility between new techniques for determining latitude and then-current assumptions as to the length of one degree of latitude. The solution to this problem, redefining the length of one degree of latitude and re-standardizing the *chi*, directly facilitated the Qing court's establishment of a new, more consistent, cartographic practice. In other words, the evolution described in this book was in no small measure a product of the environment in which it took place: the Qing's vast northern frontier and the roads thither.

Even so, most scholars discussing the mapping project focus not on the techniques employed during the land surveys but instead on the modes of representation adopted in the resulting atlases. For many of them, the central question is how to comprehend the atlas within the whole of "Chinese" or "late imperial" cartography.[605] After all, from a modern perspective, it may seem problematic that the Qing court let newer and older representations simply exist alongside

604 The *Qinding huangyu quanlan*, for example, records road distances in the newly standardized *yingzaochi* 營造尺, as well as their equivalents in the old *zhouchi* 周尺. Chapter 2. For the emperor's attention to the *zhouchi*, see Chapter 2; Jami (2012), esp. 245–252 & 384.

605 Smith (2013); Yee (1994); Needham (1979).

each other, seemingly oblivious to their incompatibility. This observation lies at the basis of dismissing the Kangxi, Yongzheng, and Qianlong cartographic projects as "Jesuit intermezzos." Three new elements presented in this study highlight the need to consider the Qing imperial atlases on their own terms. The first is the strong link between the development of a new cartographic practice on the one hand and Qing conquest and frontier management on the other. Hostetler, Elman, and others have long since established that the Kangxi emperor embraced new techniques of mapmaking due to practical concerns, namely, more effective control of his vast lands and new frontiers, all the more urgent in the face of increased contacts with other powers. In other words, the unique characteristics of Qing rule directly motivated the creation of new practices and, therefore, new modes of representation, in a push to "define the extent of Qing imperial space."[606]

The second observation is that the Qing court produced two different versions of the atlas, each for a different audience. The woodblock atlas contains place names only in Chinese and features a textual companion, the *Imperially Commissioned Overview of Imperial Territories*, which drew on and amended the accumulated knowledge of several centuries of literati scholarship. This illustrates that the woodblock versions of the Qing atlas catered to inherited literati tastes in its presentation of geographic and cartographic data.[607] They also separately produced the copperplate atlas, including toponyms in Manchu for areas outside the Great Wall and designed to form one large map. Moreover, the very technique of copper engraving was new to the Qing court and apparently appealed to the emperor's aesthetic tastes, indicating the copperplate atlas was fundamentally a Manchu representation of Qing territories and territorial ambitions for the emperor and, by extension, the imperial clan.

The third observation is that different surveying teams surveyed the different areas represented on the maps and atlases according to different standards, or not at all, as in the case for the region around Lake Baikal. Whereas we can consider the depictions of the Chinese provinces reliable, the teams surveyed only certain routes in the north, leaving outside areas blank or filling them with information from other travelers. On the other hand, in Korea and Tibet, Qing teams carefully surveyed only the main roads connecting Qing-controlled territories to Seoul and Lhasa, depicting other areas on the basis of locally produced maps and second-hand data. In short, all three of these

606 Elliott (2000), 624.

607 As Cordell Yee noted, two tendencies characterized cartography in continental East Asia, mensurational *and* textualist. Yee (1994), esp. 109. Also note that it was the woodblock edition maps and not the copperplate ones that later inserted into large compilations. Chapter 3.

observations complicate efforts to understand the Qing imperial atlases solely within a tradition of "Chinese" or "Jesuit" cartography.

Thus, in observing how the Qing court appropriated only selected cartographic practices both in technical *and* in representational terms, we come to see the mapping project no longer as a complete or temporary departure from "traditional" practice in favor of a "scientific" one but rather as an attempt to marry selected new techniques of land surveying and modes of representation to existing practices. As a result, improved hybrid practices developed, constituting creative and carefully negotiated answers to the Qing court's contemporary needs and dynamics.

On the Role of the Individual: Global vs. Local Networks

When studying cross-cultural circulation in the seventeenth and eighteenth centuries, historians tend to evoke the concept of globalization, noting the first instances of profound interactions between local and global networks and dynamics.[608] The use of this global–local dichotomy in approaching cross-cultural circulation has generally led to a strong emphasis on those elements connecting the two levels, in this case the exceptionally mobile Jesuit missionaries.[609] As a consequence, in scholarship dealing with the Qing mapping project, the Jesuits often appear as maintaining the most central social connections, thereby defining the entirety of the mapping project.[610] This study has called into question the Jesuit missionaries' presumed singular centrality. Studying the peregrinations of instruments and maps has forced me to divide my attention among the many actors involved in their trajectories. In other words, the overwhelming emphasis in past scholarship on the role of the Jesuit missionaries and their social connections gives way in this study to a displacement of the Jesuit missionaries in appreciation for different clusters of equally central connections.

Chapter 2 has brought several of these clusters, most of which constitute hitherto unexplored aspects of the mapping project, into the discussion. I have shown how, during the land surveys, a wide variety of individuals took part in the mapping project, from the emperor to the European missionaries and

608 Alternatively, some use the platitudinous term "glocalization" to describe this perceived shift.

609 E.g. Hostetler (2007).

610 Hence, several studies speak of "Jesuit cartography in China." E.g. Fuchs (1943); Foss (1988); Foss (1994); Yee (1994); Han (2014); Ribeiro (2014); Bai & Li (2014); etc.

their servants to Qing officials of all ranks and backgrounds to the runners in the service of local administrators. By zooming in on several Qing officials who were part of the mapmaking teams, I discovered the Imperial Workshops as one of the major centralizing forces during the land surveys, suggesting a major cluster of social connections in support of the project and connecting a number of forgotten individuals. These include the artisans who produced and adapted Qing surveying instruments, the workshop overseers representing the emperor during the land surveys, the senior workshop officials with direct lines of communication to the throne, the mathematicians who edited the draft maps, and the Hanlin scholars who compiled the textual companion to the resulting atlas. Other chapters and the intermissions have likewise drawn a wide range of individuals into the story of the mapping project, including, among many others, instrument builders in Paris, Chinese Christians in the provinces, and scientists in Saint Petersburg. Among the long list of individuals implicated in the mapping project, European and Asian, missionary and official, elite and commoner, Han and Manchu, many have remained obscured against the global–local dichotomy, even though the mapping project depended on all of them in more or less equal measure. Thus, by following the most basic connections between individuals on the basis of material circulation, this study has created a more inclusive, more level playing field with more space for the many individuals involved.

This approach has led to the telling of many stories, often in great detail. Along the way, I stumbled upon different institutions, at times historically visible and powerful, such as the Society of Jesus or the Qing imperial throne, at other times historically discreet but indispensable ones such as the Imperial Workshops. Yet, no matter how relevant these institutions may have been in steering historical outcomes, they too consisted of individuals who enjoyed contacts far beyond institutional limits. Instead of studying the impacts of certain prominent institutions such as the Society of Jesus or the Parisian *Académie royale des sciences*, therefore, this study has focused on the voices of the individuals involved in the mapping project and through them traced out the trajectories of instruments and maps connected to the Qing mapping project. Precisely by looking at the mapping project from this most basic level of individual connections, without imposing (too many) conceptual categories in advance, this study has brought many previously unexplored actors and dynamics to light. This highlights how distinguishing socially between the global and the local can hinder clear understanding of the actual processes of circulation, often characterized by far greater social complexity and by connections between individuals unbound by the local–global distinction.

On Instruments and Maps: The Circulation vs. the Production of Knowledge

The close attention paid to the trajectories of instruments and maps in this study has put major emphasis not only on their physical movements in space (guided by social connections) but also on the many adaptations they underwent in the process. In Chapter 3, linking the many printed editions of the resulting maps and atlases permitted construction of a genealogy of cartographic material. Based on these insights, I argued that the material maps capture the state of cartographic knowledge at a certain point in time as understood by or negotiated among the individuals involved in their production. Similar observations are possible with regard to the surveying instruments discussed in Chapters 1 and 2, forming their own (albeit somewhat looser) genealogy connecting prototypes built in Paris to their technological descendants produced in the Imperial Workshops. Thus, whereas the material production or physical circulation of such goods is relatively unproblematic, the circulation of the knowledge they embody is inseparable from the production of that knowledge: at any given point in time, individuals may or may not construct their own bodies of knowledge informed by whatever reaches them through their own connections and personal contacts. Much like texts, therefore, material goods can serve as the media of intellectual, technological, and cultural exchanges of knowledge, as snapshots of what circulated, in leaps and bounds, back and forth, through individuals and their connections.

Seen in this light, one of the crucial factors directly impacting circulation is the vulnerability of the individual. After all, if personal contacts and social connections enable circulation, then factors that shape or impact individuals can have profound consequences on that circulation. The two intermissions inserted in this book, although diverting from the trajectories of instruments and maps, fully engage this idea. Both pay careful attention to just a few of the complexities and contingencies impacting the mapping project, mostly centered on more missionary aspects. Moreover, Chapter 2 featured several episodes of unforeseen circumstances and frustrations among team members, presenting immediate repercussions for the land surveys' practical execution: one missionary returned to the capital without imperial permission; an expedition leader's personal servants' drew imperial punishment for causing disturbances in the relay system, etc. All of this illustrates the contingencies and complexities inherent to circulation and the inseparability of scientific practice from the individual. This may be a rather obvious point to make, but it is still worth emphasizing that complex and ever-changing individual connections, themselves shaped and molded by a wide range of factors, make knowledge circulation possible. The more important among these factors include power dynamics, often created through the involvement of powerful institutions, spatial dynamics, made relevant by physical, political, and cultural distances, and individual unpredictability.

On Interculturality: China vs. Europe

Throughout this study, it has become clear that the circulation of land surveying techniques and cartographic representations followed the footsteps of empire. In an age of ever-greater intercontinental mobility, a need emerged in multiple imperial centers for the tools necessary to explore and gather information regarding one's own territories *vis-à-vis* those of others. Exceptionally pragmatic emperors, kings, and tsars, guided by their personal motivations and interests, or those of their principal advisors and aides, seem to have instigated this process. This explains how someone like Cassini, the son of a Tuscan and educated at the Jesuit College of Genoa, found employment in service to French science, developing important new tools and methods the lure of which in turn helped Frenchmen to win places in the service of the Russian and Qing courts, and so on.[611] In fact, in the practical quest to build their empires, none of these rulers seem to have concerned themselves particularly with skilled personnel's homelands or places of birth, except perhaps when particular tensions raised questions as to allegiance. At the same time, this study challenges the idea that scientific encounters like these took place in-between "China" and "Europe," with the Jesuit missionaries as the only mobile elements connecting the two blocs. Although this abstraction has proven fruitful, in this case tracing the encounter requires that we step away from the dichotomy, displace the Jesuit missionaries, and bring several other actors, or groups of actors, into the equation. The Manchus are an obvious choice for matters related to Qing China, but perhaps equally crucial in my reconstruction of the mapping project were the Khalka, the Zunghar, and the Russians.

Therefore, instead of concerning itself primarily with "civilizations" or cultural blocs such as "China," "Europe," "the West" and "the East," let alone the somewhat uneasy anachronisms of nationalities and ethnicities, this study started by considering each and every individual involved in a circulation held together by factual exchanges in instruments and maps. This approach creates a space in which to take into account the plethora of personal views as to the

611 Also argued in Jami (2014) & Hart (2013), 261–263.

categories constituting their social world. For example, for the Jesuit missionary, one way to understand the social world they inhabited in Asia, and which is apparant in their writings, involved differentiating among Europeans of different countries and Catholic orders and between Chinese and Tartars, the latter further consisting of Eastern and Western Tartars, as Verbiest and Gerbillon often note. At the same time, from documents the Kangxi emperor left behind, it becomes clear that he often defined his social world in terms of Manchus, Han, Mongols, Tibetans, and Westerners, all of them his subjects, albeit each administrated in a different way. A commoner in one of the Chinese provinces, who would perhaps have paid more attention to differences in appearance and conduct among his fellow locals, Han from further afield, and foreigners, the latter certainly including the Manchus (at least during the early Qing), would have experienced the social world in yet another way. Within, across, and beyond these categories, many more personal categories naturally existed, made relevant or surfacing according to specific circumstances in which the individual in question found him- or herself.

With all this in mind, let me return to the literature on the Qing mapping project, since nearly all of it focuses upon the question of whether it constitutes a Chinese or a European project, even while carefully avoiding strong opinions and promoting the idea of shared authorship. This study has effectively bypassed questions of cultural authorship by engaging fully the complexities involved in the events surrounding the mapping project. I have shown how objects and individuals circulated across geographical, political, and cultural boundaries, thereby profoundly shaping both objects and individuals; at the same time, I have shown that we cannot easily define any object or individual in terms of these boundaries. This is due to the simple fact that those involved in the mapping project, as well as historians describing the events of their lives, experience and understand these boundaries differently. In this light, it seems arbitrary to adopt the categories of only one group of individuals, such as those of the Jesuit missionaries, whereas it seems highly problematic to impose one's own contemporary categories, even though one can never avoid this entirely. I argue that, since the resulting atlases constitute essentially hybrid or intercultural products, the question of whether they constitute Chinese or European products and the related question of authorship is redundant at best.

In all, my intention in challenging these four dichotomies is not to argue that the categories of which they consist do not exist or are imaginary. On the contrary, I recognize that these dichotomies have proved fruitful in studying circulation in history and that they engage personal realities that have shaped and continue to shape history. I have, however, attempted to add nuance to approaches that impose these dichotomies onto historical events in advance.

One way better to explain what this study contributes, in my view, is to point out that it engages the material world so as to create more space for the irreducible complexities that shaped the conceptual worlds and social realities of the individuals involved. I do this because, from the perspective of historical distance, material circulation is easier to grasp, so long as historical materials survive to suggest it took place; whereas the conceptual world is not accessible in the same manner, chiefly because different individuals approach and conceptualize the world using different (layers of) categories. Of course, this distinction between the material and the conceptual need not become yet another constraining dichotomy, but rather provides another of several ways forward in studying cross-cultural circulation in history, to regain a sense of the larger picture.

I have begun this book by noting that John Lewis Gaddis, in his book *The Landscape of History*, draws a parallel between cartography and history, stating that both "provide a way of reversing divisibility, of retrieving unity, of recapturing a sense of the whole, even though it can never be the whole."[612] In challenging four dichotomies that pervade the literature, and in trying to construct an alternative cartography of cross-cultural circulation in history, I have attempted to do just that.

612 Gaddis (2004), 32.

References and Bibliography

Consulted Archives and Libraries

AGS *Neige daku dang'an* 內閣大庫檔案 Archives of the Grand Secretariat (Taipei).
AN Archives nationales (Paris).
AMEP Archives des Missions étrangères de Paris (Paris).
APF Archives de la Province de France (Vanves).
ARSI Archivum Romanum Societatis Iesu (Rome).
BI Bibliothèque de l'Institut (Paris).
BL British Library (London).
BNF Bibliothèque nationale de France (Paris).
BO Bibliothèque de l'Observatoire (Paris).
HYL Harvard-Yenching Library (Cambridge, MA).
IF Institut de France (Paris).
KBB Koninklijke Bibliotheek van België (Brussels).
NCL *Zhongyang guojia tushuguan* 中央國家圖書館 National Central Library (Taipei).
NLC *Zhongguo guojia tushuguan* 中國國家圖書館 National Library of China (Beijing).
SGI Società Geografica Italiana (Rome).

Consulted Digital and Online Databases

Kangxi chao shilu 康熙朝實錄 (Accessible via zh.wikisource.org/zh/ 康熙朝實錄).
Siku quanshu 四庫全書 (Digital Heretage Publishing).
Yishiguan cang Kangxi Yongzheng chao manhan wen zhupi zouzhe huibian mulu 一史館藏康熙雍正朝滿漢文硃批奏摺彙編目錄 (Accessible via www.ihp.sinic.edu.tw).
Zhongguo fangzhi ku 中國方志庫 (Beijing Erudition Digital Research Center).

Bibliography

Acta Pekinensia: Western Historical Sources for the Kangxi Reign. Macao: Ricci Institute, 2013.

Afinogenov, Gregory D., *The Eye of the Tsar: Intelligence-Gathering and Geopolitics in Eighteenth-Century Eurasia*. Ph.D. dissertation Harvard Univ., 2016.

Albanese, Andreina, "La carta geografica di Matteo Ripa: Caratteristiche dell'esemplare della Bibliotheca Universitaria di Bologna." In Michele Fatica (ed.), *La missione cattolica in Cina tra I secoli XVIII–XIX: Matteo Ripa e il Collegio dei Cinesi*, Naples: Instituto Universitario Orientale, 1997, pp. 135–183.

Albanese, Andreina, "Matteo Ripa e la carta geografica dell'Imperio Cinese commissionata da Kangxi." In Michele Fatica (ed.), *Matteo Ripa e il Collegio die Cinesi di Napoli (1682–1869): Percorso documentario e iconografico*, Napoli: Università degli studi di Napoli, 2006, pp. 49–70.

An Shuangcheng 安雙成 (ed.) (trans.), *Qingchu Xiyang chuanjiaoshi manwen dang'an yiben* 清初西洋傳教士滿文檔案譯本. Dengzhou: Daxiang chubanshe, 2014.

Antonucci, Davor, "Some Notes from Western Sources on Galdan." *Etudes orientales* 25:1 (2008), pp. 29–51.

Antonucci, Davor, "Pereira's Trip to Tartary in 1685." In Saraiva, Luís (ed.), *History of Mathematical Sciences: Portugal and East Asia IV*, Singapore: World Scientific Publishing, 2012, pp. 115–134.

Aomen lishi ditu jingxuan 澳門歷史地圖精選. Beijing: Huawen chubanshe, 2000.

Atlas général de la Chine, de la Tartarie chinoise, et du Thibet. Pour servir aux différentes Descriptions et Histoires de cet Empire, Paris: Dezauche, 1790[?].

Baddeley, John F., *Russia, China, Mongolia....* 2 Vols., London: Macmillan and Co., 1919. Reprinted Mansfield Centre (CT): Martino Pub., 2006.

Bai Hongye 白鴻葉 & Li Xiaocong 李孝聰, *Kangxi chao "Huangyu quanlan tu"* 康熙朝《皇輿全覽圖》. Beijing: Guojia tushuguan chubanshe, 2014.

Baldini, Ugo, "Guillaume Bonjour (1670–1714): Chronologist, Linguist, and 'Casual' Scientist." In Luís Saraiva (ed.), *History of Mathematical Sciences: Portugal and East Asia IV*, Singapore: World Scientific Publishing, 2012, pp. 241–294.

Baqi manzhou shizu tongpu 八旗滿洲氏族通譜. Reprinted Shenyang: Liao Shen shushe, 1989.

Bartlett, Beatrice S., *Monarchs and Ministers: The Grand Council in Mid-Ch'ing China, 1723–1820*. Berkeley: Univ. of California Press, 1991.

Beijing daxue tushuguan cang guji shanben shumu 北京大學圖書館藏古籍善本書目. Beijing: Beijing daxue tushuguan, 1999.

Bello, David A., "To Go Where No Han Could Go for Long." *Modern China* 31.3 (2005), pp. 283–317.

Bernard, Henri, "Les etapes de la cartographie scientifique pour la Chine et les pays voisins: Depuis le XVIIe jusqu'à la fin du XVIIIe siècle." *Monumenta Serica* 1:2 (1935), pp. 428–477.

Bernard, Henri, "Note complémentaire sur l'Atlas de K'ang-hi." *Monumenta Serica* 11 (1946), pp. 191–200.

Bosmans, Henri, "L'œvre scientifique d'Antoine Thomas de Namur, S.J. (1644–1709)." *Annales de la société scientifique de Bruxelles* 44 (1924–5), pp. 169–207.

Bosmans, Henri, "L'œuvre scientifique d'Antoine Thomas de Namur, S.J.: Vice-Président effectif et Président intérimaire de l'Observatoire de Pékin (1644–1709)." *Annales de la société scientifique de Bruxelles* 46 (1926), pp. 154–181.

Bouvet, Joachim, *Portrait historique de l'empereur de la Chine,* Paris: Estienne Michallet, 1697.

Brucker, Joseph, "La mission de Chine de 1722 à 1735: quelques pages de l'histoire des missionnaires français à Péking au XVIIIe siècle, d'après des documents inédits." *Revue des Questions Historiques* 29 (1881), pp. 491–532.

Brucker, Joseph, "Communication sur l'exécution des cartes de la Chine par les missionaires du XVIIIe siècle d'apres des documents inédits." In *IVe Congrès International des Sciences Géographiques tenu à Paris en 1889: compte rendu publié par le Secrétariat général du Congrès Paris,* Vol. 1, Paris: Société d'éditions scientifiques, 1890–1, pp. 378–396.

Cams, Mario, "The Early Qing Geographical Surveys (1708–16) as a Case of Collaboration between the Jesuits and the Kangxi Court." *Sino-Western Cultural Relations Journal* 34 (2012), pp. 1–20.

Cams, Mario, "The Early Qing geographic Suveys (1708–1718) as seen through the 'Folded Memorials with Vermillion Endorsement of the Kangxi Reign'." In Rens Krijgsman (ed.), *Globalization and glocalization in China,* Leiden: Shilin, 2012, pp. 203–250.

Cams, Mario, "The Discovery of Chinese Rites Controversy Documents in a Branch of the Bibliothèque nationale de France." *Sino-Western Cultural Relations Journal* 35 (2013), pp. 48–56.

Cams, Mario, "The China Maps of Jean-Baptiste Bourguignon d'Anville: Origins and Supporting Networks." *Imago Mundi* 66:1 (2013), pp. 51–69.

Cams, Mario, "Restituting Church Buildings and Negotiating Church Factions: Missionary Mapmakers and the Making of Local Networks (1712–1716)." *Frontiers of History in China* 9:4 (2014), pp. 489–505.

Cao Hongjun 曹紅軍, "Kangxi 'Huangyu biao' de bianzhuan ji qi zai Suzhou de kanke guocheng kao 康熙《皇輿表》的編撰及其在蘇州的刊刻過程考." *Xin shiji tushuguan* 新世紀圖書館 (2007:3), pp. 53–55.

Cerreti, Claudio (ed.), *Carte di riso: genti paesaggi, colori dell'Estremo Oriente nelle collezione della Società geografica Italiana.* Roma: Società geografica Italiana, 2001.

Chaney, Théodore, *Vie du P. Romain Hinderer de la compagnie de Jésus, l'apotre du Sacré-Cœur dans l'Eglise de Chine au dix-huitième siècle 1668–1744.* Tournai: Decallonne-Liagre, 1889.

Chang, Michael G., *A Court on Horseback: Imperial Touring & the Construction of Qing Rule, 1680–1785.* Cambridge (MA): Harvard Univ. Asia Center, 2007.

Chang, Sen-dou, "Manuscript Maps in Late Imperial China." *The Canadian Cartographer* 11:1 (1974), pp. 1–14.

Chen Guodong 陳國棟, "Wuyingdian zongjianzao He-shi-hen: 'Liyi zhi zheng' shijian zhong de yi wei neiwufu renwu 武英殿總監造赫世亨:「禮儀之爭」事件中的一位內務府人物." In *Liang'an gugong di san jie xueshu yantaohui: shiqi, shiba shiji (1662–1722) zhongxi wenhua jiaoliu* 兩岸故宮第三屆學術研討會:十七、十八世紀 (1662–1722) 中西文化交流 3 (2011), pp. 257–290.

Chen Li 陳力, "Xinchu 'Qinding huangyu quanlan' ji 新出《欽定皇輿全覽》記." *Sichuan daxue xuebao* 四川大學學報 (1990:1), pp. 105–107.

Cheng Zhi 承志, "Manwen 'Wula deng chu difang tu' kao 滿文《烏喇等處地方圖》考." *National Palace Museum Research Quarterly* 26:4 (2009), pp. 147–236.

Cheng Zhi 承志, "Nibuchu tiaoyue jiebei tu de huanying: Manwen 'Heilongjiang liuyu tu' yanjiu 尼布楚條約界碑圖的幻影: 滿文《黑龍江流域圖》研究." *National Palace Museum Research Quarterly* 29:1 (2011), pp. 1–74.

Cheng Zhi 承志, "'Huangyu quanlan tu' dongbei dadi cehui kao: yi manwen dang'an wei zhongxin 《皇輿全覽圖》東北大地測繪考: 以滿文檔案為中心." Paper presented at the International Conference on Ming-Qing Studies (2015).

Chu Ping-Yi, "Trust, Instruments, and Cross-Cultural Scientific Exchanges: Chinese Debate over the Shape of the Earth, 1600–1800." *Science in Context* 3 (1999), pp. 385–411.

Chu Ping-Yi, "Remembering our Grand Tradition: The Historical Memory of the Scientific Exchanges between China and Europe, 1600–1800." *History of Science* 41:2 (2003), pp. 193–215.

Ch'u T'ung-Tsu, *Local Government in China under the Ch'ing*. Cambridge (MA): Harvard Univ. Press, 1988.

Comentale, Christophe, "Une carte inédite du père Gaubil, s.j.: Chine et Tartarie." In *Chine et Europe: évolution et particularités des rapports est-ouest du XVIe au XXe siècle*, Actes du IVe colloque internationale de sinologie, Centre de recherches interdisciplinaire de Chantilly, 1983, pp. 126–133.

Conley, Tom, "Ingénieurs du Roy, Ingénieur du Moy: Self and Space in Montaigne and Descartes." In David W. Sabean & Malina Stefanovska (eds.), *Space and Self in Early Modern European Cultures*, Toronto: Univ. of Toronto Press, 2012, pp. 258–277.

Cordier, Henri, "Situation de Ho-lin en Tartarie: Manuscrit Inédit du Père A. Gaubil S.J., publié avec une introduction et des notes." *T'oung Pao* 4.1 (1893), pp. 33–60.

Cordier, Henri, "Du Halde et d'Anville (cartes de Chine)." In *Recueil de mémoires orientaux: textes et traductions publiées par les professeurs de l'école spéciale des langues orientales vivantes à l'occasion du XVIe congrès international des orientalistes réunis à Alger*, Paris: Leroux, 1905.

Cordier, Henri, "Mélanges géographiques et historiques: Manuscrit inédit du Père A. Gaubil S.J." *T'oung Pao* 16.2 (1915), pp. 515–561.

Cordier, Henri, *Bibliotheca Sinica*. 2 Vols., Taipei: Ch'eng Wen, 1966.

Curtis, Emily B., *Glass Exchange between Europe and China, 1550–1800: Diplomatic, Mercantile and Technological Interactions*. Surrey: Ashgate Publishing Ltd., 2009.

Da Qing huidian shili 大清會典事例. Beijing: Jingshiguan shuju, 1899.

Dacier, Joseph-Bon, "Éloge de d'Anville." In Barbie du Bocage & Jean Denis, *Notice des ouvrages de M. D'Anville*, Paris: Fuchs, 1802, pp. 3–44.

Dai, Yingcong, *The Sichuan Frontier and Tibet: Imperial Strategy in the Early Qing*. Seattle: Univ. of Washington Press, 2010.

d'Anville, Jean-Baptiste Bourguignon, *Lettre de M. d'Anville, geographe ordinaire du roi, au R.P. Castel, jesuite. Au suject des Pays de Kamtchatka et de Jeço. Et réponse du R.P. Castel*. Paris: s.l., 1737.

d'Anville, Jean-Baptiste Bourguignon, "Réponse de M. d'Anville." In *Mémoires pour l'Histoire des Sciences et des Beaux-Arts*, 1738, pp. 768–773.

d'Anville, Jean-Baptiste Bourguignon, "Mémoire sur le Li, mesure itinéraire des Chinois." *Mémoires de l'Académie des Inscriptions et Belles-Lettres* 28 (1761), pp. 487–502.

d'Anville, Jean-Baptiste Bourguignon, *Mémoire de M. D'Anville, Premier Géographe du Roi, Des Académies Royales des Belles-Lettres, & des Sciences. Sur la Chine*. Beijing: s.l., 1776.

Daumas, Maurice, "Quelques fabricants d'instruments scientifiques anciens." *Revue d'histoire des sciences et de leurs applications* 3 (1950), pp. 364–370.

Daumas, Maurice, *Scientific Instruments of the Seventeenth & Eighteenth Centuries and their Makers*. London: Batsford, 1972.

de Moyriac de Mailla, Joseph A. M., *Histoire générale de la Chine, ou annales de cet empire, traduites du "Tong-kien-kang-mou"....* 13 Vols., Paris: Pierres et Clousier, 1777–85.

de Thomaz de Bossierre, Mme Yves, *Un belge Mandarin à la cour de Chine aux XVIIe et XVIIIe siècles: Antoine Thomas 1644–1709*. Paris: s.l., 1977.

de Thomaz de Bossierre, Mme Yves, *Jean-François Gerbillon, S.J. (1654–1707): Un des cinq mathématiciens envoyés en Chine par Louis XIV*. Leuven: Ferdinand Verbiest Foundation, 1994.

Dehergne, Joseph, "Les Chrétienités de Chine de la période Ming (1581–1650)." *Monumenta Serica* 16 (1957), pp. 1–136.

Dehergne, Joseph, *Répertoire des Jésuites de Chine de 1552 à 1800*. Rome: Institutum Historicum Societatis Iesu & Paris: Letouzey et Ané, 1973.

Demidova, Natalia F. (et al), "Stateinyi spisok poslannika L.V. Izmailova o posol'stve v Tsinskuiu imperiiu." In *Russko-kitaĭskie otnosheniia v XVIII veke. Materialy i dokumenty, vol. I*, Moscow: Nauka, 1978.

Destombes, Marcel, "Les originaux chinois des plans de ville publiés par J.-B. du Halde, s.j., en 1735." In *La mission française de pékin aux XVIIe et XVIIIe siècles*, Actes du colloque internationale de sinologie, Centre de recherches interdisciplinaire de Chantilly, 1976, pp. 85–98.

Dong Guangbi 董光璧 (ed.), *Zhongguo jinxiandai kexue jishu shi* 中國近現代科學技術史. Changsha: Hunan jiaoyu chubanshe, 1995.

du Bocage, Barbie & Jean Denis, *Notice des ouvrages de M. D'Anville*. Paris: Fuchs, 1802.

du Halde, Jean-Baptiste et al. (eds.), *Lettres édifiantes et curieuses écrites des missions étrangères par quelques Missionnaires de la Compagnie de Jesus*. 34 Vols., Paris: Le Clerc, 1702–76.

du Halde, Jean-Baptiste (ed.), *Description géographique, historique, chronologique, politique, et physique de l'empire de la Chine et de la Tartarie chinoise*. 4 Vols., Paris: Le Mercier, 1735.

du Halde, Jean-Baptiste, *Description géographique, historique, chronologique, politique, et physique de l'empire de la Chine et de la Tartarie chinoise*. 4 Vols., The Hague: Henri Scheurleer, 1736.

Du Shiran & Han Qi, "The Contribution of French Jesuits to Chinese Science in the Seventeenth and Eighteenth Centuries." *Impact of Science on Society* 167 (1992), pp. 265–275.

Dudink, Ad, *Chinese books and documents (pre-1900) in the Royal library of Belgium at Brussels*. Brussels: Archives et bibliothèques de Belgique, 2006.

Dulac, Georges, "Science et politique: les réseaux du dr. António Ribeiro Sanches (1699–1783)." *Cahiers du monde russe* 43:2 (2002), pp. 251–273.

Elliott, Mark C., "The Limits of Tartary: Manchuria in Imperial and National Geographies." *The Journal of Asiatic Studies* 59:3 (2000), pp. 603–646.

Elliott, Mark C., "The Manchu-Language Archives of the Qing Dynasty and the Origins of the Palace Memorial System." *Imperial China* 22:1 (2001), pp. 1–70.

Elliott, Mark C., *The Manchu Way: The Eight Banners and Ethnic Identity in Late Imperial China*. Stanford: Stanford Univ. Press, 2001.

Elliott, Mark C., "Frontier Stories: Periphery as Center in Qing History." *Frontiers of History in China* 9:3 (2014), pp. 336–360.

Elman, Benjamin A., "Geographical Research in the Ming-Ch'ing Period." *Monumenta Serica* 35 (1981–3), pp. 1–18.

Elman, Benjamin A, *On Their Own Terms: Science in China, 1550–1900*. Cambridge (MA): Harvard Univ. Press, 2005.

Elman, Benjamin A, "Ming-Qing Border Defense, the Inward Turn of Chinese Cartography, and Qing Expansion in Central Asia in the Eighteenth Century." In Diana Lary, *Chinese State at the Borders*, Seattle: Univ. of Washington Press, 2007, pp. 29–56.

Fairbank, John K. & Ssu-yu Teng, "On the Transmission of Ch'ing Documents." *Harvard Journal of Asiatic Studies* 4:1 (1939), pp. 12–46.

Fairbank, John K. & Ssu-yu Teng, "On the Types and Uses of Ch'ing Documents." *Harvard Journal of Asiatic Studies* 5:1 (1940), pp. 1–71.

Fan, Fa-ti, "Science in Cultural Borderlands: Methodological Reflections on the Study of Science, European Imperialism, and Cultural Encounter." *East Asian Science, Technology and Society: An International Journal* 1 (2007), pp. 213–231.

Fatica, Michele (ed.), *Matteo Ripa e il Collegio die Cinesi di Napoli (1682–1869): Percorso documentario e iconografico*. Napoli: Univ. degli studi di napoli l'orientale, 2006.

Fauque, Danièle, "Un instrument essentiel de l'expédition pour la mesure de la terre: le quart de cercle mobile." In Henri Lacombe & Pierre Costabel, *La figure de la terre du XVIIIe siècle à l'ère spatiale*, Paris: Gauthier-Villars, 1988, pp. 209–221.

Feng Kam-Wing 馮錦榮, "Qingchu Xiyang celiang xue de fazhan: yi Chen Ding (1650–1732) ji qi 'Gougu shu' wei li 清初西洋測量學的發展: 以陳訂 (1650–1732) 及其《句股術》為例." *International conference on East–west Studies: Tradition, Transformation and Innovation*, National Univ. of Singapore, 1998.

Feng Kam-Wing 馮錦榮, "Mapping the Universe: Two Planispheric Astrolabes in the Early Qing Court." In Chan Alan K.L. et al. (eds.), *Historical Perspectives on East Asian Science, Technology and Medicine*, Singapore: Singapore Univ. Press, 2002, pp. 448–464.

Feng Mingzhu 馮明珠, *Kangxi dadi yu taiyangwang Luyi shisi tezhan: Zhongfa yishu wenhua de jiaohui* 康熙大帝與太陽王路易十四特展: 中法藝術文化的交會. Taipei: National Palace Museum, 2011.

Ferguson, John C., "*Chinese Foot Measure*." *Monumenta Serica* 6 (1941), pp. 357–381.

Findlen, Paula (ed.), *Early Modern Things: Objects and their Histories, 1500–1800*. London: Routledge, 2013.

Foss, Theodore, *A Jesuit Encyclopedia for China: A Guide to Jean-Baptiste du Halde's "Description… de la Chine" (1735)*. Ph.D. dissertation Univ. of Chicago, 1979.

Foss, Theodore, "The editing of an atlas of China: a comparison of the work of J.-B. d'Anville and the improvements of John Green on the Jesuit/K'ang-hsi atlas." In *Imago et mensura mundi: Atti del IX Congresso di storia della cartographia*, Rome: Instituto della Enciclopedia Italiana, 1985, pp. 361–376.

Foss, Theodore, "A Western Interpretation of China: Jesuit Cartography." In Charles E. Ronan & Bonnie B.C. Oh (eds.), *East Meets West: The Jesuits in China 1582–1773*, Chicago: Loyola Univ. Press, 1988, pp. 209–251.

Foss, Theodore, "Jesuit Cartography: A Western Interpretation of China." *Review of Culture* 21:4 (1994), Macao: Instituto Cultural de Macao, pp. 133–156.

Frémontier-Murphy, Camille & Daniel Alcouffe (eds.), *Les instruments de mathématiques: XVIe–XVIIIe siècle*. Paris: Editions de la Réunion des musées nationaux, 2002.

Fu Lo-shu, *A documentary Chronicle of Sino-Western Relations (1644–1820)*. 2 Vols., Tucson: Univ. of Arizona Press, 1966.

Fuchs, Walter, "Ueber einige Landkarten mit Mandjurischer Beschriftung." *Manzhou xuebao* 滿洲學報 2 (1933), pp. 1–17.

Fuchs, Walter, "Materialien zur Kartographie der Mandjuzeit I." *Monumenta Serica* 1:2 (1935), pp. 386–427.

Fuchs, Walter, "Materialien zur Kartographie der Mandjuzeit II." *Monumenta Serica* 3 (1938), pp. 189–231.

Fuchs, Walter, *Der Jesuiten-atlas der Kanghsi-zeit: Seine Entstehungsgeschichte nebst Namensindices für die Karten der Mandjurei, Mongolei, Ostturkestan und Tibet.* 2 Vols., Beijing: Fu-jen Univ., 1943.

Gaddis, John L., *The Landscape of History*. Oxford: Oxford Univ. Press, 2004.

Gandolfo, Romolo, "Bhutan and Tibet in European Cartography (1597–1800)." In *The Spider and the Piglet: Proceedings of the First Seminar on Bhutan Studies*, Centre for Bhutan Studies Thimphu, 2004, pp. 90–136.

Gaubil, Antoine, (ed. Renée Simon), *Correspondance de Pékin 1722–1759*. Genève: Librarie Droz, 1970.

Golvers, Noël, "An Unnoticed Letter of F. Verbiest, S.J., on his Geodesic Operations in Tartary (1683/1684)." *Archives internationales d'histoire des sciences* 50 (2000), pp. 86–102.

Golvers, Noël, "Non-Chinese (Oriental and Western) Sources on Central Asia at the Jesuit Libraries of China (ca. 1650–1750)." In K. d'Hulster & J. Van Steenbergen, *Continuity and Change in the Realms of Islam, Orientalia Lovaniensia Analecta*, Leuven: Peeters, 2008, pp. 261–285.

Golvers, Noël, "Communication and exchange of knowledge between West/East (seventeenth/18 cent.): the 'routes,' illustrated on the case of the '*Via Ostendana*'." Paper presented at *The Materiality of Chinese-Western Relations, in the Ming-Qing periods: Methodological Approaches, Empirical Cases*, Leuven (May 26–27), 2014.

Golvers, Noël, "The Correspondence of Antoine Thomas, SJ (1644–1709) as a Source for the History of Science." *Ziran kexueshi yanjiu* 自然科學史研究 33:2 (2014), pp. 131–144.

Golvers, Noël, "Jesuit Correspondence from China: The Two 'Tartary letters' of Ferdinand Verbiest, SJ (1682 & 1683) and Their Oldest Printed Edition (Paris, 1684) as a Case Study." *Sino-Western Cultural Relations Journal* 38 (2016), pp. 43–55.

Golvers, Noël & Efthymios Nicolaidis, *Ferdinand Verbiest and Jesuit Science in seventeenth century China: An annotated edition and translation of the Constantinople manuscript (1676)*. Athens/Leuven: Institute for Neohellenic Research & Ferdinand Verbiest Institute, 2009.

Gongzhongdang Kangxi chao zouzhe 宮中檔康熙朝奏摺. 7 Vols., Taipei: Guili gugong buwuguan, 1977.

Gray, Basil, "Lord Burlington and Father Ripa's Chinese Engravings." *The British Museum Quarterly* 22:1–2 (1960), pp. 40–43.

Greenberg, John L., *The Problem of the Earth's Shape from Newton to Clairaut: The Rise of Mathematical Science in Eighteenth-Century Paris and the Fall of "Normal" Science.* Cambridge (UK): Cambridge Univ. Press, 1995.

Grosier, Jean-Baptiste, *Atlas général de la Chine pour servir à la Description générale de cet Empire*. Paris: Moutard, 1785.

Gunn, Geoffrey G., *First Globalization: The Eurasian Exchange, 1500–1800*. Lanham (MD): Rowman & Littlefield, 2003.

Guo Fuxiang 郭福祥, "Qing qianqi gongting Xiyang kexue yiqi de zhonglei he laiyuan 清前期宫廷西洋科學儀器的種類和來源." *Asian Studies* III (XIX), 2 (2015), pp.117–136.

Guy, Kent, *Qing Governors and Their Provinces: The Evolution of Territorial Administration in China, 1644–1796*. Seattle: Univ. of Washington Press, 2010.

Haguet, Lucile, "J.-B. d'Anville as Armchair Mapmaker: The Impact of Production Contexts on his Work." *Imago Mundi* 63:1 (2010), pp. 88–101.

Halsberghe, Nicole, *Xin zhi lingtai yi xiang zhi: Vertoog over de nieuwgebouwde instrumenten op het observatorium, Ferdinand Verbiest, Beijing, 1674*. 3 Vols., Unpublished Ph.D. dissertation KU Leuven, 1992.

Han Qi 韓琦, "Junzhu he buyi zhi jian: Li Guangdi zai Kangxi shidai de huodong ji qi dui kexue de yingxiang 君主和布衣之間: 李光地在康熙時代的活動及其對科學的影響." *Qinghua xuebao* 清華學報 26:4 (1996), pp. 421–445.

Han Qi 韓琦, "Sino-French Scientific Relations through the French Jesuits and the Académie Royale des Sciences in the Seventeenth and Eighteenth Centuries." In Stephen Uhally & Xiaoxin Wu (eds.), *China and Christianity: Burdened Past, Hopeful Future*, London/New York: M.E. Sharpe, 2001, pp. 137–147.

Han Qi 韓琦, "Cartography during the Times of the Kangxi Emperor: The Age and Background." In Roberto M. Ribeiro & John O'Malley, *Jesuit Mapmaking in China: D'Anville's Nouvelle Atlas de la Chine (1737)*, Early Modern Catholicism and the Visual Arts Series 11, Philadephia: St Joseph's Univ. Press, 2014, pp. 51–62.

Han Qi 韓琦, "Kangxi di zhi zhishu yu 'xixue zhongyuan' shuo xinlun: 'Yuzhi sanjiaoxing tuisuan falun' de chengshu ji qi beijing 康熙帝之治術與'西學中源'說新論 —《御制三角形推算法論》的成書及其背景." *Studies in the History of Natural Sciences* 自然科學史研究 25:1 (2016), pp. 1–9.

Harley, John B. & David Woodward (eds.), *The History of Cartography, Volume 2 Book 2: Cartography in the Traditional East and Southeast Asian Societies*. Chicago: Univ. of Chicago Press, 1994.

Hart, Roger, *Imagined Civilizations: China, the West, and their First Encounter*. Baltimore: The Johns Hopkins Univ. Press, 2013.

Ha-si-ba-gen 哈斯巴根, "Chuanjiaoshi yu Kangxi chao Menggu yutu de huizhi 傳教士於康熙朝蒙古輿圖的繪制." *Zhongyang minzu daxue xuebao* 中央民族大學學報 (2010:3), pp. 92–96.

Ha-si-ba-gen 哈斯巴根 & Cheng Chongde 成崇德, "Guanyu 'Kouwai zhuwang tu' 關於《口外諸王圖》." *Qingshi yanjiu* 清史研究 (2010:1), pp. 90–101.

Heawood, Edward, *A History of Geographic Discovery in the Seventeenth and Eighteenth Centuries*. Cambridge (UK): Cambridge Univ. Press, 1912. Reprint 2011.

Hébert, Elisabeth (ed.), *Instruments scientifiques à travers l'histoire*. Paris: Ellipses, 2004.

Heirmann, Ann, Paola de Troia & Jan Parmentier, "Francesco Sambiasi, a Missing Link in European Map Making in China?" *Imago Mundi* 61:1 (2009), pp. 29–46.

Hostetler, Laura, *Qing Colonial Enterprise: Ethnography and Cartography in Early Modern China*. Chicago: Univ. of Chicago Press, 2001.

Hostetler, Laura, "Global or Local? Exploring Connections between Chinese and European Geographic Knowledge during the Early Modern Period." *East Asian Science, Technology, and Medicine* 26 (2007), pp. 117–135.

Hostetler, Laura, "Contending Cartographic Claims? The Qing Empire in Manchu, Chinese and European Maps." In James R. Akerman (ed.), *The Imperial Map: Cartography and the Mastery of Empire*, Chicago: Univ. of Chicago Press, 2009, pp. 93–133.

Hsia, Florence C., *Sojourners in a Strange Land: Jesuits and Their Scientific Missions in Late Imperial China*. Chicago: Univ. of Chicago Press, 2009.

Hsü Chieh, "Xaver Ehrenbert Fridelli: Ein Pater aus Linz und sein Beitrag zur Kartographischen Erfassung Chinas." *Historisches Jahrbuch der Stadt Linz* (1986), pp. 113–120.

Huang Runhua 黄潤華, *Guojia tushuguan cang manwen wenxian tulu* 國家圖書館藏滿文文獻圖錄. Beijing: Guojia tushuguan chubanshe, 2010.

Huang Shijian 黄時鑒 & Gong Yingyan 龔纓晏, *Li Madou shijie ditu yanjiu* 利瑪竇世界地圖研究. Shanghai: Shanghai guji chubanshe, 2004.

Huang Song-Kang, "Kangxi's Attitude in the Rites Controversy." *Heythrop Journal*, 28:1 (1997), pp. 57–67.

Huard, Pierre & Ming Wong, "Les enquêtes françaises sur la science et la technologie chinoises au XVIIIe siècle." *Bulletin de l'Ecole française d'Extrême-Orient* 53 (1966), pp. 137–226.

Hucker, Charles O., *A Dictionary of Official Titles in Imperial China*. Stanford: Stanford Univ. Press, 1985.

Hummel, Arthur W., *Eminent Chinese of the Ch'ing Period (1644–1912)*. 2 Vols., Washington: US Govt. Printing Office, 1943.

Isnard, Albert, *Joseph-Nicolas Delisle: sa biographie et sa collection de cartes géographiques à la Bibiliothèque nationale*. Paris: Imprimerie nationale, 1917.

Jami, Catherine, "Western Learning and Imperial Control: The Kangxi Emperor's (r. 1662–1722) Performance." *Late Imperial China* 23:1 (2002), pp. 28–49.

Jami, Catherine, "For Whose Greater Glory? Jesuit Strategies and Science during the Kangxi Reign." In Wu Xiaoxin (ed.), *Encounters and Dialogues: Changing Perspectives on Chinese-Western Exchanges from the Sixteenth to the Eighteenth Centuries*, Nettetal: Steyler Verlag, 2005, pp. 211–226.

Jami, Catherine, "Western Learning and Imperial Scholarship: The Kangxi Emperor's Study." *East Asian Scrience, Technology and Medicine* 27 (2007), pp. 146–172.

Jami, Catherine, "Imperial Science Written in Manchu in Early Qing China: Does It Matter?" In Florence Bretelle-Establet (ed.), *Looking at it from Asia: the Processes that Shaped the Sources of History of Science*, Boston Studies in the Philosophy and History of Science 265, New York: Springer, 2010, pp. 371–391.

Jami, Catherine, *The Emperor's New Mathematics*. Oxford: Oxford Univ. Press, 2012.

Jami, Catherine, "The Jesuits' Negotiation of Science between France and China (1685–1722): Knowledge and Modes of Imperial Expansion." In László Kontler et al. (eds.), *Negotiating Knowledge in Early Modern Empires: A Decentered View*, New York: Palgrave Macmillan, 2014, pp. 53–77.

Kadushin, Charles, *Understanding Social Networks: Theories, Concepts, and Findings*. Oxford: Oxford Univ. Press, 2012.

Kangxi chao manwen zhupi zouzhe quanyi 康熙朝滿文朱批奏摺全譯. Beijing: Zhongguo shehui kexue chubanshe, 1996.

Kangxi chao hanwen zhupi zouzhe huibian 康熙朝漢文朱批奏摺彙編. Beijing: Dang'an chubanshe, 1984.

Kniajetskaia, E.A. & V. Chenakal, "Pierre le Grand et les fabricants français d'instruments scientifiques." *Revue d'histoire des sciences* 28:3 (1975), pp. 243–258.

Knüppel, Michael, "Eine unbekannt gebliebene Schrift Philipp Johann von Strahlenbergs." *Acta Orientalia* (2008), pp. 187–209.

Landry-Deron, Isabelle, "Les Mathématiciens envoyés en Chine par Louis XIV en 1685," *Archive for History of Exact Sciences* 55 (2001), pp. 423–463.

Landry-Deron, Isabelle, *La preuve par la Chine*. Paris: Éditions EHESS, 2002.

Landry-Deron, Isabelle, "'Pour la perfection des sciences et des arts': La mission jésuite française en Chine sous le patronage de l'Académie royale." In Pierre-Sylvain Filliozat et al. (eds.), *L'œuvre scientifique des missionnaires en Asie*, Paris: AIBL/De Boccard, 2011, pp. 76–96.

Latour, Bruno, *Reassembling the Social: An Introduction to Actor-Network-Theory*, Oxford/NY: Oxford Univ. Press, 2005.

Launay, Adrien, *Histoire des missions de Chine*. 7 Vols., Paris: Missions Etrangères de Paris, 2002.

Lecomte, Louis, *Un Jesuite à Pekin: nouveaux mémoires sur l'état présent de la Chine 1687–1692*. Reprint Paris: Phébus, 1990.

Ledyard, Gary, "Cartography in Korea." In John B. Harley & David Woodward (eds.), *The History of Cartography, Volume 2 Book 2: Cartography in the Traditional East and Southeast Asian Societies*, Chicago: Univ. of Chicago Press, 1994, 235–345.

Levallois, Jean-Jacques, "l'Académie royale des sciences et la figure de la terre." In Henri Lacombe & Pierre Costabel, *La figure de la terre du XVIIIe siècle à l'ère spatiale*, Paris: Gauthier-Villars, 1988, pp. 41–75.

Li Huazi 李花子, *Qingchao yu Chaoxian guanxishi yanjiu: yi yuebian jiaoshe wei zhongxin* 清朝與朝鮮關係史研究：以越邊交涉為中心. Yanji: Yanbian daxue chubanshe, 2006.

Li Xiaocong 李孝聰, *Ouzhou shoucang bufen zhongwen gu ditu xulu* 歐洲收藏部分中文古地圖敍錄. Beijing: Guoji wenhua chubanshe, 1996.

Li Xiaocong 李孝聰, "Ma Guoxian yu tongban Kangxi 'Huangyu quanlan tu' de yinshua jianlun zaoqi zhongwen ditu zai Ouzhou de chuanbu yu yingxiang 馬國賢與銅版康熙《皇輿全覽圖》的印制兼論早期中文地圖在歐洲的傳布與影響." In Michele Fatica (ed.), *La missione cattolica in Cina tra I secoli XVIII–XIX: Matteo Ripa e il Collegio dei Cinesi*, Naples: Instituto Univ. Orientale, 1997, pp. 123–134.

Li Xiaocong 李孝聰, *Meiguo guohui tushuguan cang zhongwen gu ditu xulu* 美國國會圖書館藏中文古地圖敍錄. Beijing: Wenwu chubanshe, 2004.

Li Xiaocong 李孝聰, "Ji Kangxi 'Huangyu quanlan tu' de cehui ji qi banben 記康熙《皇輿全覽圖》的測繪及其版本." *The National Palace Museum Research Quarterly* 30.1 (2012), pp. 55–86.

Li Xiaodan 李曉丹 & Wang Qiheng 王其亨, "Qing Kangxi nianjian Yidali chuanjiaoshi Ma Guoxian ji 'Bishu shanzhuang' tongban hua 清康熙年間意大利傳教士馬國賢及《避暑山莊》銅版畫." *Gugong bowuyuan yuankan* 故宮博物院院刊 (2006:3), pp. 44–51.

Lidai zhiguan biao 歷代職官表. Shanghai: Shanghai guji chubanshe, 1993.

Lim Yongtae, "'Western Astronomy vs. Korean Geography': Intellectual Exchanges between a Korean and the Jesuits as Seen from Yi Kiji's *1720 Beijing Travelogue*." In Luís Saraiva (ed.), *History of Mathematical Sciences: Portugal and East Asia IV*, Singapore: World Scientific Publishing, 2012, pp. 295–309.

Lim Yongtae, "Journeys of the Modest Astronomers: Korean Astronomers' Missions to Beijing in the Seventeenth and Eighteenth Centuries." In Catherine Jami (ed.), *Mobilité humaine et circulation des saviors techniques (XVIIe–XIXe siècles)*, *Extrême-Orient—Extrême-Occident* 36, Paris: Presses Univ. de Vincennes, 2014, pp. 81–108.

Lin Dongyang 林東陽, "Ferdinand Verbiest's Contribution to Chinese Geography and Cartography." In Witek, John (ed.), *Ferdinand Verbiest (1623–1688): Jesuit Missionary, Scientist, Engineer and Diplomat*, Monumenta Serica Monograph Series XXX, Nettetal: Steyler Verlag, 1994, pp. 135–164.

Lin Tianren 林天人, *Huangyu soulan: Meiguo guohui tushuguan suo cang Ming Qing yutu* 皇輿搜覧:美國國會圖書館所藏明清輿圖. Taipei: Zhongyanyuan shuwei wenhua zhongxin, 2013.

Liu, Adam Yuen-Chung, *Ch'ing Institutions and Society 1644–1795*. Hong Kong: Univ. of Hong Kong Centre of Asian Studies, 1990.

Liu Dun, "Karel Slavícek and His Scientific Works in China." In Louís Saraiva (ed.), *Europe and China: Science and the Arts in the Seventeenth and Eighteenth Centuries*, Singapore: World Scientific Publishing, 2002, pp. 227–240.

Liu Ruofang 劉若芳, "Qingting san da shice ditu de huizhi yu guanli 清廷三大實測地圖的繪制與管理." In *Zhongguo guojia tianwen* 中國國家天文 (Extra edition), 2012, pp. 46–61.

Livingstone, David, *Putting Science in Its Place: Geographies of Scientific Knowledge*. Chicago: Univ. of Chicago Press, 2003.

Lo Sardo, Eugenio (ed.), *Atlante della Cina di Michele Ruggieri, S.I.* Rome: Istituto Poligrafico e Zecca dello Stato, 1993.

Lo Sardo, Eugenio, "Antoine Thomas's and George David's Maps of Asia." In Willy F. Vande Walle, *The History of the Relations between the Low Countries and China in the Qing Era (1644–1911)*, Leuven Chinese Studies XIV, Leuven: Leuven Univ. Press, 2003, pp. 89–103.

Lockman, John (ed.), *Travels of the Jesuits*. 2 Vols., London, 1743. Reprinted New Delhi: Asian Educational Services, 1995.

Ma Menglong 馬孟龍, "Mu-ke-deng chabian yu 'Huangyu quanlan tu' bianhui: jian dui Mu-ke-deng 'shenshibei' chuli weizhi de kaobian 穆克登查邊與《皇輿全覽圖》編繪—兼對穆克登'審視碑'初立位置的考辨." *Zhongguo bianjiang shidi yanjiu* 中國邊疆史地研究 19:3 (2009), pp. 85–99.

Malatesta, Edward & Gao Zhiyu (eds.), *Departed, Yet Present: Zhalan, the Oldest Christian Cemetery in Beijing*. Macao: Insituto Cultural de Macao, 1995.

Man-Han hebi Qing neifu yitong yudi mitu 满漢合璧清内府一統與地秘圖. Shenyang: Gugong, 1929.

Martini, Martino, *Novus Atlas Sinensis: Tavole*. Trento: Trento Univ., 2003.

McCune, Shannon, "Geographical Observations on Korea: Those of Father Régis Published in 1735." *Journal of Social Sciences and Humanities* 44 (1976), pp. 1–19.

McCune, Shannon, "The Korean Cartographic Tradition: Its Cross-Cultural Relations." In *Papers of the 1st International Conference on Korean Studies*, Seoul: The Academy of Korean Studies, 1980, pp. 724–770.

McCune, Shannon, "Jean-Baptiste Régis, S.J., An Extraordinary Cartographer." In *Chine et Europe: évolution et particularités des rapports est-ouest du XVIe au XXe siècle*, Actes du IVe colloque internationale de sinologie, Centre de recherches interdisciplinaire de Chantilly, 1983, pp. 237–248.

Mémoires de la congrégation de la mission. Vols. 4–6, Paris: À la maison principale de la congrégation de la mission, 1865.

Menegon, Eugenio, "Newly Available Manchu Documents Pertaining to Sino-Western Relations in the Kangxi Period." *Sino-Western Cultural Relations Journal* XIX (1997), pp. 22–46.

Meskens, Ad, "De wereld op het platte vlak: Wiskundige cartografie ten tijde van Abraham Ortelius." In *Abraham Ortelius*, Turnhout: Brepoels, 1998, pp. 71–82.

Miller, Daniel (ed.), *Materiality*. Durham: Duke Univ. Press, 2005.

Millward, James A., "'Coming onto the Map': 'Western Regions' Geography and Cartographic Nomenclature in the Making of Chinese Empire in Xinjiang." *Late Imperial China* 20.2 (1999), pp. 61–98.

Ming Qing shiliao ding bian 明清史料丁編. Vol. 2, Reprint Taipei: Zhongyang yanjiuyuan lishi yuyan yanjiusuo, 1972.

Mosca, Matthew W., *From Frontier Policy to Foreign Policy: The Question of India and the Transformation of Geopolitics in Qing China*. Stanford: Stanford Univ. Press, 2013.

Mungello, David E. (ed.), *The Chinese Rites Controversy: Its History and Meaning*. Monumenta Serica Monograph Series XXXIII, Nettetal: Steyler Verlag, 1994.

Mungello, David E., *The Great Encounter of China and the West 1500–1800*. Totowa: Rowman & Littlefield, 2005.

Nappi, Carla, "Surface Tension: Objectifying Ginseng in Chinese Early Modernity." In Paula Findlen (ed.), *Early Modern Things: Objects and their Histories, 1500–1800*. London: Routledge, 2013, pp. 31–52.

Naquin, Susan & Evelyn S. Rawski, *Chinese Society in the Eighteenth Century*. New Haven: Yale Univ. Press, 1987.

Needham, Joseph, *Science and Civilization in China*. Vol. 3, Cambridge (UK): Cambridge Univ. Press, 1959. Reprint 1979.

Nicéron, Jean-Pierre, *Mémoires pour servir à l'histoire des hommes illustres de la république des lettres avec le catalogue raisonné de leurs ouvrages*. 39 Vols., Paris: Briasson, 1728–38.

Norman, Jerry, *A Comprehensive Manchu-English Dictionary*. Harvard-Yenching Institute Monograph Series 85, Cambridge (MA): Harvard Univ. Asia Center, 2013.

Nouvel atlas de la Chine, de la Tartarie chinoise et du Thibet. The Hague: Henri Scheurleer, 1737.

Nouvel atlas de la Chine, de la Tartarie chinoise et du Thibet. Amsterdam: Barthelemy Vlam, 1785.

Pagani, Catherine, "Clockmaking in China under the Kangxi and Qianlong Emperors." *Arts Asiatiques* 50 (1995), pp. 76–84.

Pang, Tatjana A., *Descriptive Catalogue of Manchu Manuscripts and Blockprints in the St. Petersburg Branch of the Institute of Oriental Studies Russian Academy of Sciences: Issue 2*. Wiesbaden: Harrassowitz Verlag, 2001.

Pang, Tatjana A., "An Evaluation of F. Verbiest's Account of his Journey to Manchuria in 1682, Its Errors and Problems." In Willy F. Vande Walle (ed.), *The History of the Relations between the Low Countries and China in the Qing Era (1644–1911)*, Leuven Chinese Studies XIV, Leuven: Leuven Univ. Press, 2003, pp. 59–72.

Pelletier, Monique, *Les Cartes des Cassini: La science au service de l'État et des provinces*. Paris: Éditions du CTHS, 2013.

Pelletier, Philippe, *L'Extrême-Orient: l'Invention d'une histoire et d'une géographie*. Paris: Gallimard, 2011.

Peng, Rita H., "The K'ang-hi Emperor's Absorption in Western Mathematics and Astronomy and His Extensive Applications of Scientific Knowledge." *Lishi xuebao* 歷史學報 3 (1975), pp. 422–349.

Perdue, Peter C., "Boundaries, Maps and Movement: Chinese, Russian, and Mongolian Empires in Early Modern Central Eurasia." *International History Review* 20:2 (1998), pp. 263–286.

Perdue, Peter C., *China Marches West: The Qing Conquest of Central Eurasia*. Cambridge (MA): Harvard Univ. Press, 2005.

Perdue, Peter C., "Boundaries and Trade in the Early Modern World: Negotiations at Nerchinsk and Beijing." *Eighteenth-Century Studies* 43:3 (2010), pp. 341–356.

Petech, Luciano, "Una carta Cinese del seculo XVIII." In *Instituto Universitario Orientale "Annali"* 5 (1953), pp. 185–187.

Petech, Luciano, *China and Tibet in the Early XVIIIth Century: History of the Establishment of Chinese Protectorate in Tibet. Monographies du T'oung Pao* 1, Leiden: Brill, 1972.

Petto, Christine M., *When France was King of Cartography: The Patronage and Production of Maps in Early Modern France*. Lanham: Lexington Books, 2007.

Pfister, Louis, *Notice biographiques et bibliographiques sur les jésuites de l'ancienne mission de Chine (1552–1773)*. 2 Vols., Shanghai: Imprimerie de la mission catholique, 1932.

Picard, Jean, *Mesure de la terre*. Paris: s.l., 1671.

Pinot, Virgile, *La Chine et la formation de l'esprit philosophique en France (1640–1740)*. Paris: Geuthner, 1932.

Pinot, Virgile, *Documents inédits relatifs à la connaissance de la Chine en France de 1685 à 1740*. Paris: Geuthner, 1932.

Prandi, Fortunato (trans.), *Memoirs of Father Ripa, During Thirteen Years' Residence at the Court of Peking in the Service of the Emperor of China*. London: John Murray, 1855.

Puente-Ballesteros, Beatriz, "Jesuit Medicine in the Kangxi Court (1662–1722): Imperial Networks and Patronage." *East Asian Science, Technology and Medicine* 34 (2011), pp. 86–162.

Puente-Ballesteros, Beatriz, "Antoine Thomas, SJ as a 'Patient' of the Kangxi Emperor (r. 1662–1722): A Case Study on the Appropriation of Theriac at the Imperial Court." *Revista de historia de la medicina y de la ciencia* 64:1 (2012), pp. 213–250.

Qin Guojing 秦國經, "18 shiji Xiyangren zai cehui Qingchao yutu zhong de huodong yu gongxian 18 世紀西洋人在測繪清朝輿圖中的活動與貢獻." *Qingshi yanjiu* 清史研究 (1997:1), pp. 37–44.

Qin Guojing 秦國經 & Liu Ruofang 劉若芳, "Qingchao yutu de huizhi yu guanli 清朝輿圖的繪制與管理." In Cao Wanru 曹婉如 et al. (ed.), *Zhongguo gudai dituji: Qingdai* 中國古代地圖集: 清代, Beijing: Wenwu chubanshe, 1997, pp. 71–78.

Qing shilu: Shengzu ren huangdi shilu 清實錄: 聖祖仁皇帝實錄. Vols. 3–6, Beijing: Zhonghua shuju chuban, 1985.

Qingdai qijuzhu ce: Kangxi chao 清代起居註冊: 康熙朝. 54 Vols., Beijing/Taipei: Zhonghua shuju/Lianjing chuban, 2009.

Qinggong neiwufu zaobanchu dang'an zonghui 清宮內務府造辦處檔案總匯. 55 Vols., Beijing: Renmin chubanshe, 2005.

Qinggong Xiyang yiqi 清宮西洋儀器, Gugong bowuyuan cang wenwu zhenpin daxi 故宮博物院藏文物珍品大系. Shanghai: Kexue jishu chubanshe, 2011.

Qingzhong qianqi Xiyang tianzhujiao zaihua huodong dang'an shiliao 清中前期西洋天主教在華活動檔案史料. 4 Vols., Beijing: Zhonghua shuju, 2003.

Ren Jiyu 任繼愈 (ed.), *Zhongguo guojia tushuguan guji zhenpin tulu* 中國國家圖書館古籍珍品圖錄. Beijing: Beijing tushuguan chubanshe, 1999.

Ribeiro, Roberto M. & John W. O'Malley, *Jesuit Mapmaking in China: D'Anville's Nouvelle* [sic] *Atlas de la Chine* (1737). Early Modern Catholicism and the Visual Arts Series 11, Philadelphia: St Joseph's Univ. Press, 2014.

Ripa, Matteo, *Storia della fondazione della Congregazione e del Collegio de' Cinesi sotto il titolo della sagra famiglia di G.C. scritta dallo stesso fondatore e de' viaggi da lui fatti.* Napoli: Manfredi, 1832.

Ripa, Matteo, *Giornale* (1705–1724). 2 Vols., Napoli: Instituto univ. orientale, 1991–96.

Roque di Oliveira, Francisco, "Cartografia Antiga da Cidade de Macao, c. 1600–1700: Confronto entre Modelos de Representação Europeus e Chineses." *Revista Electrónica de Geografía y Ciencias Sociales Vol. X* 218:53 (2006).

Roque di Oliveira, Francisco, "Seventeenth-Century Jesuit Surveys for a Secure Overland Route from Europe to China." In Arthur K. Wardega & António Vasconcelos de Saldanha, *In the Light and Shadow of an Emperor*, Newcastle upon Tyne: Cambridge Scholars Publishing, 2012, pp. 468–501.

Rule, Paul A., *K'ung-tzu or Confucius? The Jesuit Interpretation of Confucianism*. Sydney: Allen & Unwin, 1986.

Rule, Paul A., "The Correspondence of Antoine Thomas: A Major Source for the History of the Jesuits in China and for the Reign of Kangxi." In Gu Wei-ying 古偉瀛 & Zhao Xiaoyang 趙曉陽 (eds.), *Jidu zongjiao yu jindai Zhongguo* 基督宗教與近代中國 (Multi-aspect Studies on Christianity in Modern China), Beijing: Shehui kexue wenxian chubanshe, 2011, pp. 140–158.

Saldanha, António Vasconcelos de, "The Last Imperial Honours from Tomás Pereira to the *Eulogium Europeorum Doctorum* in 1711." In Arthur K. Wardega & António Vasconcelos de Saldanha, *In the Light and Shadow of an Emperor*, Newcastle upon Tyne: Cambridge Scholars Publishing, 2012, pp. 144–226.

Schechner, Sara, "The Material Culture of Astronomy in Daily Life: Sundials, Science, and Social Change." *Journal for the History of Astronomy* 32 (2001), pp. 189–222.

Schmid, Andre, "Tributary Relations and the Qing-Chosŏn Frontier on Mount Paektu." In Diana Lary (ed.), *The Chinese State at the Borders*, Vancouver: UBC Press, 2007, pp. 126–150.

Sebes, Joseph S., *The Jesuits and the Sino-Russian Treaty of Nerchinsk (1689): The Diary of Thomas Pereira, S.J.* Bibliotheca Instituti Historici Societatis Iesu 18, Roma: Institutum Historicum S.I., 1962.

Semans, Cheryl Ann, *Mapping the Unknown: Jesuit cartography in China, 1583–1773*. Ph.D. dissertation Univ. of California, 1987.

Simon, Walter & Howard G.H. Nelson, *Manchu Books in London: A Union Catalogue*. London: British Library, 1977.

Sionita, Gabriele & Joannes Hesronita, *Geographia Nubiensis: id est accuratissima totius orbis in septem climata divisi descriptio, continens praesertim exactam vniuersae Asiae, & Africae, rerumq[ue] in ijs hactenus incognitarum explicationem*. Paris: Hieronymi Blagaert, 1619.

Smith, Richard J., "Mapping China's World: Cultural Cartography in Late Imperial Times." In Wen-hsin Yeh (ed.), *Landscape, Culture, and Power in Chinese Society, China Research Monographs* 49, Berkeley: Center for Chinese studies, 1998, pp. 52–105.

Smith, Richard J., *Mapping China and Managing the World: Culture, Cartography and Cosmology in Late Imperial Times*. London: Routledge, 2013.

Souciet, Étienne, *Observations mathématiques, astronomiques, géographiques, chronologiques, et physiques; Tirées des Anciens Livres Chinois, ou faites nouvellement aux Indes, à la Chine & ailleurs, par les Pères de la Compagnie de Jesus. Rédigées & publiées par le P. Étienne Souciet*. 3 Vols., Paris: Rollin, 1729–32.

Spence, Jonathan D., *Emperor of China: Self-Portrait of K'ang-his*. New York: Vintage Books, 1988.

Standaert, Nicolas (ed.), *Handbook of Christianity in China: Volume One: 635–1800*. Leiden: Brill, 2001.

Standaert, Nicolas, *Methodology in View of Contact between Cultures: The China Case in the seventeenth Century*. CSRCS Occasional Paper No. 11, Hong Kong: The Chinese Univ. of Hong Kong, 2002.

Standaert, Nicolas, *The Interweaving of Rituals: Funerals in the Cultural Exchange between China and Europe*. Seattle: Univ. of Washington Press, 2008.

Standaert, Nicolas, "Jesuit Accounts of Chinese History and Chronology and their Chinese Sources." *East Asian Science, Technology and Medicine* 35 (2012), pp. 11–87.

Standaert, Nicolas, *The Intercultural Weaving of Historical Texts: Chinese and European Stories about Emperor Ku and His Concubines*. Leiden: Brill, 2016.

Stary, Giovanni, "Manchu Toponymy in the Atlas of Matteo Ripa." In Michele Fatica (ed.), *La missione cattolica in Cina tra I secoli XVIII–XIX: Matteo Ripa e il Collegio dei Cinesi*, Naples: Instituto Univ. Orientale, 1997, pp. 185–193.

Stary, Giovanni, "Nota sul ritrovamento a San Pietroburgo di un 'Atlante' di Matteo Ripa." In *Scritture di storia, Universita degli studi di Napoli "l'Orientale". Dipartimento di scienze umane e sociali. Quaderno* 2, Napoli: Edizioni Scientifiche Italiane, 2001, pp. 123–126.

Stary, Giovanni, "Sulla toponomastica mancese dell'atlante di Matteo Ripa." In Michele Fatica (ed.), *Matteo Ripa e il Collegio die Cinesi di Napoli (1682–1869): Percorso documentario e iconografico*, Napoli: Univ. degli studi di napoli l'orientale, 2006, pp. 71–86.

Strassberg, Richard E. (trans.), *Thirty-Six Views: The Kangxi Emperor's Mountain Estate in Poetry and Prints*. Cambridge (MA): Harvard Univ. Press, 2016.

Sun Zhe 孫喆, "Qianxi yingxiang Kangxi 'Huangyu quanlan tu' huizhi de jige yinsuo 淺析影響康熙《皇輿全覽圖》繪制的幾個因所." *Lishi dang'an* 歷史檔案 2012:1, pp. 87–93.

Swiderski, Richard M., "Bouvet and Leibniz: A Scholarly Correspondence." *Eighteenth Century Studies* 14:2 (1980–1), pp. 135–150.

Szczesniak, Boleslaw, "The Description and Map of Kansu by Giovanni Battista Maoletti de Serravalle." *Monumenta Serica* 18 (1959), pp. 294–313.

Tachard, Guy, *Voyage de Siam, des peres jesuites, envoyez par le roy aux Indes & à la Chine. Avec leurs observations astronomiques, et leurs remarques de physique, de géographie, d'hydrographie, & d'histoire*. Paris: Chez Arnould Seneuze & Daniel Horthemels, 1686.

Tan Qixiang 譚其驤 (ed.), *Zhongguo lishi dituji* 中國歷史地圖集. 8 Vols., Shanghai: Ditu chubanshe, 1987.

Teng, Emma Jinhua, *Taiwan's Imagined Geography: Chinese Colonial Travel Writing and pictures, 1683–1895*. Cambridge (MA): Harvard Univ. Asia Center, 2004.

Terrall, Mary, *The Man Who Flattened the Earth: Maupertius and the Sciences in the Enlightenment*. Chicago: Univ. of Chicago Press, 2006.

Torbert, Preston M., *The Ch'ing Imperial Household Department: A Study of its Organization and Principal Functions, 1662–1796*. Cambridge (MA): Harvard Univ. Press, 1977.

Turner, A.J., "The Observatory and the Quadrant in Eighteenth-Century Europe." *Journal for the History of Astronomy* 33 (2002), pp. 373–385.

Unno, Kazutaka, *The Kuang-yü-t'u in the Cultural History of Chinese Maps*. Tokyo: Toyo bunko, 2010.

von Collani, Claudia, "The Report on Kilian Stumpf About the Case of Father Joachim Bouvet." *Zeitschrift für Missionswissenschaft und Religionswissenschaft* 83:3 (1999), pp. 231–251.

von Collani, Claudia, *Joachim Bouvet, S.J.: Journal des voyages*. Variétés Sinologiques New Series 95, Taipei: Ricci Institute, 2005.

von Collani, Claudia, "Per pedes apostolorum: Bishop Johannes Müllener CM Mu Tianchi in Sichuan." In Lu Yan R. & Philip Vanhaelmeersch (eds.), *Silent Force: Native Converts in the Catholic China Mission*, Leuven: Ferdinand Verbiest Institute, 2009, pp. 95–130.

von Strahlenberg, Philipp J., *Das Nord- und Ostliche Theil von Europa und Asia*. Stockholm: s.l., 1730.

Vu Thanh, Hélène, "Principles of Missionary Geography in Jesuit Spirituality and Their Implementation in Japan." *Bulletin of Portuguese/Japanese Studies* 18/19 (2009), pp. 175–191.

Waldron, Arthur, *The Great Wall of China: From History to Myth*. Cambridge (UK): Cambridge Univ. Press, 1994.

Waley-Cohen, Joanna, *The Sextants of Beijing: Global Currents in Chinese History*. New York: Norton, 1999.

Waley-Cohen, Joanna, "The New Qing History." *Radical History Review* 88 (2004), pp. 193–206.

Wallis, Helen M., "Chinese maps and Globes in the British Library and the Phillips Collection." In Francis Wood (ed.), *Chinese Studies: Papers Presented at a Colloqium at the SOAS, University of London, 24–26 August, 1987*, London: British Library, 1988, pp. 88–96.

Wang Qianjin 汪前進, *"Huangyu quanlan tu" cehui yanjiu* 《皇輿全覽圖》測繪研究. Ph.D. dissertation Chinese Academy of Sciences, 1990.

Wang Qianjin 汪前進, "Kangxi tongban 'Huangyu quanlan tu' touying zhonglei xintan 康熙銅版《皇輿全覽圖》投影種類新探." *Ziran kexue shi yanjiu* 自然科學史研究 10:2 (1991), pp. 186–194.

Wang Qianjin 汪前進, "'Huangyu quanlan tu' cehui 《皇輿全覽圖》測繪." In Dong Guangbi 董光璧 (ed.), *Zhongguo jinxiandai kexue jishu shi* 中國近現代科學技術史, Changsha: Hunan jiaoyu chubanshe, 1995, pp 131–174.

Wang Qianjin 汪前進 & Liu Ruofang 劉若芳, *Qingting sanda shice quantuji* 清廷三大實測全圖集. 3 Vols., Beijing: Waiwen chubanshe, 2007.

Wang Qianjin 汪前進, "Luo Mingjian bianhui 'Zhongguo ditu ji' suo yiju zhongwen yuanshi ziliao xintan 羅明堅編繪《中國地圖集》所依據中文原始資料新探." *Beijing xingzheng xueyuan xuebao* 北京行政學院學報 (2013:3), pp. 120–128.

Whiteman, Stephen, "Kangxi's Auspicious Empire: Rhetorics of Geographic Integretion in the Early Qing." In Du Yongtao and Jeefrey Kyong-McClain (eds.), *Chinese History in Geographic Perspective*, Lanham (MD): Lexington Books, 2013, pp. 33–54.

Widmaier, Rita (ed.), *Leibniz korrespondiert mit China: Der Briefwechsel mit den Jesuitenmissionaren (1689–1714)*. Frankfurt: Vittorio Klostermann, 1990.

Wilkinson, Endymion, *Chinese History: A Manual*. Harvard-Yenching Institute Monograph Series 52, Cambridge (MA) & London: Harvard Univ. Asia Center, 2000.

Witek, John W., *An Eighteenth Century Frenchman at the Court of the K'ang-hsi Emperor: A Study of the Early Life of Jean-François Foucquet*. Ph.D. disseration Georgetown Univ., 1973.

Witek, John W., "The Role of Antoine Thomas, SJ, (1644–1709) in Determining the Terrestrial Meridian Line in Eighteenth-Century China." In Willy F. Vande Walle (ed.), *The History of the Relations between the Low Countries and China in the Qing Era (1644–1911)*, Leuven Chinese Studies XIV, Leuven: Leuven Univ. Press, 2003, pp. 89–103.

Wu, Silas Hsiu-liang, "The Memorial System of The Ch'ing Dynasty (1644–1911)." *Harvard Journal of Asiatic Studies* 27 (1967), pp. 7–75.

Xinzuan Yunnan tongzhi 新纂雲南通志. Kunming: Yunnan renmin chubanshe, 1949.

Xu Pingfang 徐蘋芳 (ed.), *Ming-Qing Beijing cheng tu* 明清北京成圖. Shanghai: Shanghai guji chubanshe, 2012.

Yan Ping (et al.), *China in Ancient and Modern Maps*. London: Sotheby's, 1998.

Yao Boyue 姚伯嶽 (ed.), *Huangyu xialan: Beijing daxue tushuguan cang qingdai caihui ditu* 皇輿遐覽: 北京大學圖書館藏清代彩繪地圖. Beijing: Zhongguo renmin daxue chubanshe, 2008.

Yao Jingming 姚京明 & Hao Yufan 郝雨凡, *Luo Mingjian "Zhongguo dituji" xueshu yantaohui: Lunwenji* 羅明堅《中國地圖集》學術研討會: 論文集. Macao: Instituto Cultural, 2014.

Yee, Cordell D.K., "Taking the World's Measure: Chinese Maps between Observation and Text." In John B. Harley & David Woodward (eds.), *The History of Cartography, Volume 2 Book 2: Cartography in the Traditional East and Southeast Asian Societies*, Chicago: Univ. of Chicago Press, 1994, pp. 96–127.

Yee, Cordell D.K., "Traditional Chinese Cartography and the Myth of Westernization." In John B. Harley & David Woodward (eds.), *The History of Cartography, Volume 2 Book 2: Cartography in the Traditional East and Southeast Asian Societies*, Chicago: Univ. of Chicago Press, 1994, pp. 170–202.

Yu Fushun 于福順, "Qing Yongzheng shipai 'Huangyutu' de chubu yanjiu 清雍正十排《皇輿圖》的初步研究." *Wenwu* 文物 12 (1983), pp. 71–75 & 83.

Yutu yaolu: Beijing tushuguan cang 6827 zhong zhongwaiwen gujiu ditu mulu 輿圖要錄: 北京圖書館藏 6827 種中外文古舊地圖目錄. Beijing: Beijing tushuguan chubanshe, 1997.

Zerlik, Alfred, "P. Xaver Ernbert Fridelli: Chinamissionar und Cartograph aus Linz." *Oberösterreichische Heimatblätter* 16:3–4 (1962), pp. 1–68.

Zhang Baichun, "The Introduction of European Astronomical Instruments and the Related Technology into China during the Seventeenth Century." *East Asian Science, Technology and Medecine* 20 (2003), pp. 99–131.

Zhang Naiwei 章乃煒, *Qinggong shuwen* 清宮述聞. 2 Vols., Beijing: Zijincheng chubanshe, 2009.

Zhao Erxun 趙爾巽 (ed.), *Qingshi gao* 清史稿. 28 Vols., Beijing: Zhonghua shuju, 1977.

Zhongguo guojia tianwen 中國國家天文 (2012: Extra edition).

Zhu Chengru 朱誠如, *Qingshi tudian: Kangxi chao* (*xia*) 清史圖典: 康熙朝(下), Qingchao tongshi tulu 清朝通史圖錄 4. Beijing: Zijincheng chubanshe, 2002.

Index

www.ingramcontent.com/pod-product-compliance
Lightning Source LLC
LaVergne TN
LVHW020634300926
849161LV00004B/701
* 9 7 8 9 0 0 4 3 4 5 3 5 5 *